Springer Series in Statistics

Springer Series in Statistics

Andrews/Herzberg: Data: A Collection of Problems from Many Fields for the Student and Research Worker.

Anscombe: Computing in Statistical Science through APL.

Berger: Statistical Decision Theory and Bayesian Analysis, 2nd edition.

Brémaud: Point Processes and Queues: Martingale Dynamics.

Brockwell/Davis: Time Series: Theory and Methods.

Dzhaparidze: Parameter Estimation and Hypothesis Testing in Spectral Analysis of Stationary Time Series.

Farrell: Multivariate Calculation.

Goodman/Kruskal: Measures of Association for Cross Classifications.

Hartigan: Bayes Theory.

Heyer: Theory of Statistical Experiments.

Jolliffe: Principal Component Analysis.

Kres: Statistical Tables for Multivariate Analysis.

Leadbetter/Lindgren/Rootzén: Extremes and Related Properties of Random Sequences and Processes.

LeCam: Asymptotic Methods in Statistical Decision Theory.

Manoukian: Modern Concepts and Theorems of Mathematical Statistics.

Miller, Jr.: Simultaneous Statistical Inference, 2nd edition.

Mosteller/Wallace: Applied Bayesian and Classical Inference: The case of *The Federalist Papers*.

Pollard: Convergence of Stochastic Processes.

Pratt/Gibbons: Concepts of Nonparametric Theory.

Sachs: Applied Statistics: A Handbook of Techniques, 2nd edition.

Seneta: Non-Negative Matrices and Markov Chains.

Siegmund: Sequential Analysis: Tests and Confidence Intervals.

Vapnik: Estimation of Dependences Based on Empirical Data.

Wolter: Introduction to Variance Estimation.

Yaglom: Correlation Theory of Stationary and Related Random Functions I: Basic Results.

Yaglom: Correlation Theory of Stationary and Related Random Functions II: Supplementary Notes and References.

A.M. Yaglom

Correlation Theory of Stationary and Related Random Functions

Volume II: Supplementary Notes and References

Springer-Verlag
New York Berlin Heidelberg
London Paris Tokyo

A.M. Yaglom
Institute of Atmospheric Physics
Academy of Sciences of the U.S.S.R.
109017 Moscow
U.S.S.R.

AMS Classification: 62 jxx

Library of Congress Cataloging-in-Publication Data
(Revised for Volume 2)
I͡Aglom, A. M.
  Correlation theory of stationary and related random functions.
  (Springer series in statistics)
  Bibliography: v. 1, p.
  Includes indexes.
  Contents: v. [1] [without special title]—
v. 2. Supplementary notes and references
  1. Stochastic analysis. 2. Correlation
(Statistics) 3. Stationary processes.   I. Title.
QA274.2.I24 1986   519.5   86-10167

9 8 7 6 5 4 3 2 1
ISBN-13: 978-1-4612-9090-2   e-ISBN-13: 978-1-4612-4628-2
DOI: 10.1007/978-1-4612-4628-2

Preface to Volume II

The material of this volume supplements that of Volume I. The central place is occupied here by about 400 notes to some statements in Volume I, which were marked by appropriate superscripts in that volume. It has already been noted in the Preface to Volume I that these notes are very diverse in their content. Some of them give a more detailed mathematical derivation of results whose proofs are only briefly outlined, or altogether omitted, in Volume I; e.g., two different complete proofs of the fundamental spectral representation theorem for stationary processes are given in Note 17 to Chapter 2, together with a sketch of the proof of the theorem on generalized spectral representations of random functions, which was widely used in Chapter 4 of Volume I. Other notes contain generalizations of some results from Volume I or references to sources where one can find complete proofs of propositions stated in that volume without proof. Some notes deal with the history of questions at hand, or contain additional examples of new interesting applications. For instance, Note 9 to Chapter 2 includes the description of a specific radio engineering application of some formulae from Volume I related to pulse processes, and also an analysis of a new general class of random pulse processes of radio engineering importance. There are also many notes that contain more specialized additional material enriching and extending the material of Volume I, and also numerous references to relevant supplementary literature.

In some places, the refinements of the material of Volume I considered in this volume contain rather sophisticated mathematical results and statements which are very rarely mentioned in the literature which is elementary enough to be suitable for people working in the applied fields. Note that there are very different opinions about the advisability of familiarizing workers in such fields with sophisticated mathematical results and approaches. In particular, in the well-known paper by M.C. Wang and G.E. Uhlenbeck (1945) intended for physicists (of which paper I happen to be a great admirer), only the simplest definition of the notion of a random process is given (even more restricted than the simple definition given on pp. 39–41 of Vol. I). For this attitude, see footnote 9 on page 324 of Wang and Uhlenbeck's paper, ending in these words:

The authors are aware of the fact that in the mathematical literature (especially in papers by N. Wiener, J.L. Doob, and others) the notion of a random (or stochastic) process has been defined in a much more refined way. This allows [one], for instance, to determine in certain cases the probability that the random function $y(t)$ is of bounded variation, or continuous, or differentiable, etc. However, it seems to us that these investigations have not helped in the solution of problems of direct physical interest, and we will therefore not try to give an account of them.

On the other hand, in the Preface to the very interesting book by H. Cramér and M.R. Leadbetter (1967) the authors write:

Starting in 1962, our joint work was originally concerned with certain reliability problems, which were found to be intimately connected with the properties of the trajectories (or sample functions) of stationary stochastic processes . . . . When trying to work out a possible programme for a book, we soon came to the conclusion that the most desirable plan would be to include an account of the general theory of stationary processes, with special emphasis on the properties of their sample functions. It is well known that some of these properties are important also in other fields of application, such as communication engineering.

We remark that the investigations of the sample function properties mentioned by Cramér and Leadbetter are just the typical examples of those more sophisticated investigations that Wang and Uhlenbeck considered useless in solving problems of direct physical interest. The striking discrepancy between the viewpoints of famous scientists Uhlenbeck and Cramér is naturally due to the fact that quite different applications of the mathematical theory of random functions were actually in the focus of their scientific interests. Since the author wishes that this book would be useful for readers dealing with very diverse applications of the theory of random functions, and because he understands that any reader can skip notes that seem uninteresting to him, he has decided to discuss in the present volume a number of results related to the so-called almost sure properties of sample functions (or realizations) of random processes. These properties include those mentioned by Cramér and Leadbetter in their Preface, and also the various forms of the strong law of large numbers for random functions.

The wish to make the book useful to as wide as possible a circle of readers also stimulated the author's decision to be liberal with references to available works. It was already mentioned in the Preface to Volume I that the literature on stationary and related random functions is enormous, extremely diverse, and scattered throughout a multitude of books of very different style and content, and in a vast number of papers published in mathematical, statistical, engineering, physical, geophysical, economic, and other journals often having nothing in common with each other. At the same time, most of the books (both mathematical and applied) on the theory of random functions refer to only a small part of the available literature, consisting of books and papers close in degree of sophistication and selection of the considered topics to the content of the given book. In contrast, the references in this volume (listed in the extensive Bibliography on pp. 169–228) include books and papers of varied degrees of complexity, meant for different classes of readers, pursuing different aims, and often written in different languages; in particular, the Bibliography includes many Russian books and papers that are often insufficiently known in the West. These references are intended to give readers a reasonably complete picture of the present state of investigations in the area considered, to aid them in finding the necessary information (and many more additional references) related to matters of special interest, and help them find their way in the tremendous amount of books and papers dealing with stationary and related random functions. Therefore, both volumes of the present book taken together can very well be used as a sufficiently complete reference book by anyone interested in the theory and/or the applications of stationary and related random functions. The detailed indices to both volumes at the end of the present volume are meant to facilitate this kind of use of the book.

The references to the works listed in the Bibliography are given by the indications of the author's name and the date of publication of the work. In cases where several works of a given author all appear in the same year, these are denoted further by letters of the alphabet. The initials of the authors are given only in some cases where one must distinguish between two authors with the same surname.

All equation numbers in the present volume are supplied with primes; therefore, all the references to formula numbers with primes relate to Volume II. Formula numbers without primes, unless otherwise specified, refer to equations in Volume I.

The author wishes to express his thanks to B.L. Dribinskaya for her help in the preparation of the Bibliography and Indices.

Contents

INTRODUCTION

[1]As is well known, Dirac's δ-function is the "improper function" $\delta(x)$, which satisfies the relations:

$$\delta(x) = 0 \quad \text{for all } x \neq 0, \quad \int_{-\infty}^{\infty}\delta(x)dx = 1$$

(so that $\delta(0)$ cannot be finite). The function $\delta(x)$ can be interpreted as the limit (nonexistent, in fact) of the sequence of ordinary functions $\delta_n(x)$ such that

$$\lim_{n\to\infty}\delta_n = 0 \text{ for all } x \neq 0, \quad \int_{-\infty}^{\infty}\delta_n(x)dx = 1 \text{ for all } n.$$

Such interpretation means that the ordinary function $\delta_n(x)$ has to be first substituted into all equations containing $\delta(x)$, and then the limit as $n \to \infty$ must be evaluated.

However, at present the δ-function is best interpreted as a special case of the so-called *generalized function* (or *distribution*). A generalized function is a "function" d, for which there may exist no "value $d(x)$ at a given point x", but nevertheless it is possible to determine, for any sufficiently "good" (i.e. smooth) weighting function $\varphi(x)$, a finite "smoothed value" $d(\varphi) = \int_{-\infty}^{\infty}d(x)\varphi(x)dx$. The value $d(\varphi)$ must possess the usual properties of an integral and, hence, the following condition must hold:

$$(0.1') \qquad d(a_1\varphi_1 + a_2\varphi_2) = a_1 d(\varphi_1) + a_2 d(\varphi_2)$$

for any real numbers a_1, a_2 and any (smooth) functions φ_1 and φ_2. To state it differently, the generalized function d is a *linear functional* which assigns the number $d(\varphi)$ (often also symbolically denoted by $\int_{-\infty}^{\infty}d(x)\varphi(x)dx$) to each smooth function $\varphi(x)$. In particular, for the particular, for the δ-function $\delta(\varphi) \equiv \int_{-\infty}^{\infty} \delta(x)\varphi(x)dx = \varphi(0)$ (so that $\int_{-\infty}^{\infty} e^{i\omega x}\delta(x)dx = 1$ for all ω and, hence, symbolically $\delta(x) = (2\pi)^{-1} \int_{-\infty}^{\infty} e^{i\omega x}dx$). Similarly, the equations $\int_{-\infty}^{\infty} \delta(x - x_0)\varphi(x)dx = \varphi(x_0)$ and $\int_{-\infty}^{\infty} \delta^{(n)}(x - x_0)\varphi(x)dx = (-1)^n\varphi^{(n)}(x_0)$, where $\varphi^{(n)}$ is the nth derivative of φ, define the generalized functions $\delta(x - x_0)$ and $\delta^{(n)}(x-x)$. For the theory of generalized functions see, e.g., Schwartz, 1950–51, 1962, Gel'fand and Shilov, 1964, or Lighthill, 1968; cf. also Sec. 24 of this book devoted to generalized random functions.

[2]It can be shown that any distribution function can be represented as a sum of three nondecreasing functions

$$F(x) = F_{I}(x) + F_{II}(x) + F_{III}(x)$$

where $F_{I}(x)$ is an absolutely continuous nondecreasing function (i.e.

$$F_{I}(x) = \int_{-\infty}^{x} p(x')dx', \quad p(x') \geqslant 0),$$ $F_{II}(x)$ is a nondecreasing

step-function, and $F_{III}(x)$ is a "nondecreasing singular function" (i.e. a continuous, but not absolutely continuous nondecreasing function whose derivative vanishes at almost all points); see, e.g., Cramér (1962) or Loève (1963). The examples of singular distribution function $F(x) = F_{III}(x)$ can be found, e.g., in the books by Gnedenko (1962), Sec. 22, and Shiryaev (1980), Sec. II.3. However, singular components $F_{III}(x)$ have never appeared in applied problems and in practice their possible existence can be usually ignored.

[3]Since some facts from the theory of characteristic functions will be needed further on in another connection, it is worthwhile discussing them here at some length.

The *characteristic function* of a random variable X with distribution function $F(x)$ is a complex-valued function of the auxiliary variable t defined by the formula

$$(0.2') \qquad \psi(t) = \langle e^{itX} \rangle = \int_{-\infty}^{\infty} e^{itx}dF(x).$$

Here, the angular brackets symbolize the probabilistic averaging (see (0.6)), and the integral on the right-hand side is the improper Stieltjes integral defined in Vol. I after equation (0.6). (It is easily seen that this integral converges for all real t.) If X has probability density $p(x)$, then (0.2') takes the form

$$(0.3') \qquad \psi(t) = \int_{-\infty}^{\infty} e^{itx}p(x)dx.$$

Hence, $\psi(t)$ is a Fourier transform of the function $p(x)$ and the inversion formula for Fourier integrals gives the result

$$(0.4')\qquad p(x) = (2\pi)^{-1} \int_{-\infty}^{\infty} e^{-itx}\psi(t)dt.$$

Integrating both sides of (0.4') with respect to x from x_1 to x_2 and taking into account that $F(x)$ is an integral of $p(x)$, we obtain

$$(0.5')\qquad F(x_2) - F(x_1) = \frac{1}{2\pi}\int_{-\infty}^{\infty} \frac{e^{-itx_2} - e^{-itx_1}}{-it}\,\psi(t)dt.$$

Equation (0.5') shows that all the values of $F(x_2) - F(x_1)$ (and of $F(x_2) = \lim_{x_1 = -\infty} [F(x_2) - F(x_1)]$) can be uniquely determined if $\psi(t)$ is known.

In the general case of an arbitrary random variable X the density $p(x)$ may be nonexistent, while the integrand on the right-hand side of (0.5') may decrease too slowly as $|t| \to \infty$ to be integrable. It can be shown, however, that for any characteristic function $\psi(t)$ the principal value of the integral on the right-hand side of (0.5') (i.e. the limit of the integral from $-T$ to T as $T \to \infty$) necessarily exists and we always have

$$(0.6')\qquad F(x_2) - F(x_1) = \frac{1}{2\pi}\lim_{T\to\infty}\int_{-T}^{T} \frac{e^{-itx_2} - e^{-itx_1}}{-it}\,\psi(t)dt$$

for any continuity points x_2 and x_1 of the function $F(x)$ [see e.g., Gnedenko (1962), Sec. 36, Shiryaev (1980), Sec. II.12, Cramér (1946), Sec. 10.3, or Cramér (1962), Chap IV]*. Assuming that $x_2 = x$, and $x_1 \to -\infty$, we obtain the values of $F(x)$; hence, *the characteristic function $\psi(t)$ uniquely determines the distribution function F(x).*

It is clear that not every complex function $\psi(t)$ can be a characteristic function. From the definition (0.2') it follows that $\psi(t)$ must satisfy the conditions

$$(0.7')\qquad \psi(0) = 1, \quad \psi(-t) = \overline{\psi(t)}, \quad |\psi(t)| \leqslant \psi(0) = 1,$$

where the bar indicates complex conjugation. From (0.2') it readily follows also that the function $\psi(t)$ is continuous for all t. It is not much harder to prove that if $\psi(t)$ is a twice differentiable characteristic function and $\psi'(0) = \psi''(0) = 0$, then $\psi(t) = 1$ for all t (see Note 6 below). Consequently, if $\psi'(0) = \psi''(0) = 0$, but $\psi(t) \neq 1$ at least for one t (for instance, if $\psi(t) = \exp(-\alpha|t|^m)$, where $\alpha > 0$, $m > 2$), then $\psi(t)$ cannot be a characteristic function.

There are also many other conditions restricting the class of characteristic functions $\psi(t)$. We shall now formulate one, very general (but not very easily verifiable) condition. *Let n be a positive integer, t_1, ..., t_n real numbers, and c_1, ..., c_n complex numbers. Then, for any characteristic function $\psi(t)$,*

---

*Any one of the available proofs of (0.6') also shows that if the value of the function $F(x)$ at the points of its jump is defined as the half-sum $[F(x - 0) + F(x + 0)]/2$ of the "pre-jump" and "post-jump" values, then (0.6') will hold true for all the values of x_1 and x_2.

$$(0.8') \qquad \sum_{j,k=1}^{n} \psi(t_j - t_k) c_j \bar{c}_k \geqslant 0.$$

Indeed, by virtue of (0.2') the left-hand side of (0.8') is equal to

$$(0.9') \qquad \int_{-\infty}^{\infty} \sum_{j,k=1}^{n} e^{i(t_j - t_k)x} c_j \bar{c}_k dF(x) = \int_{-\infty}^{\infty} \Big| \sum_{j=1}^{n} e^{it_j x} c_j \Big|^2 dF(x),$$

where all the increments $dF(x)$ are nonnegative.

The complex function $\psi(t)$, for which the condition (0.8') is fulfilled for any n, t_1, ..., t_n, and c_1, ..., c_n, is said to be *positive definite*. Thus, *the characteristic function is a continuous positive definite function satisfying the condition $\psi(0) = 1$*. It is easy to check that condition (0.8') with $n = 2$ and the fact that $\psi(0) = 1$ imply the necessity of all the conditions in (0.7'). It can also be shown that the continuity and positive definiteness of $\psi(t)$ supplemented by the condition $\psi(0) = 1$ actually ensure the fulfillment of all the conditions which must be satisfied by a characteristic function. In other words, *the class of characteristic functions of probability distributions coincides with the class of continuous positive definite functions taking the value 1 at* t = 0. The above statement actually constitutes the content of the following important theorem proved almost simultaneously and independently by S. Bochner and A. Ya. Khinchin, but first published by Bochner in 1932 and hence usually called *Bochner's theorem:*

Bochner's theorem. *A continuous complex-valued function ψ(t) of a real variable* t *is positive definite if, and only if, it can be represented as the Fourier-Stieltjes integral of the form (0.2'), where* F(x) *is a monotone nondecreasing function of the variable* x.

Indeed if we add to the conditions of this theorem the condition $\psi(0) = 1$ (which, by virtue of (0.2') is equivalent to the condition $F(\infty) - F(-\infty) = 1$), then $F(x)$ will be a distribution function, and $\psi(t)$ will be a characteristic function corresponding to this distribution.

A rigorous proof of Bochner's theorem can be found, for instance, in Bochner's classical book (1959), first published in German in 1932, in the books by Goldberg (1961), Gnedenko (1962), Feller (1966), Lévy (1965), Loève (1963), or Cramér and Leadbetter (1967), and in the special monographs on characteristic functions by Ramachandran (1967) and Lukacs (1970). An original proof of this theorem, based on certain facts from the theory of stationary random functions, is outlined at the end of Sec. 11.

[4]There is no universally accepted symbol to denote probabilistic averaging. In the mathematical literature the letter E (or E, or $\mathcal{E}$) is most often used for this purpose, i.e. the mean value of X is denoted by EX (or EX, or $\mathcal{E}X$). In the applied literature the mean value of X is usually denoted either by $\bar{X}$ (complex conjugation is then indicated by an asterisk) or (as in our book) by $\langle X \rangle$.

[5]As a measure of spread in X one can use, for example, the half-difference $(x_2^{(\epsilon)} - x_1^{(\epsilon)})/2$, where $P\{X < x_1^{(\epsilon)}\} = \epsilon$, $P\{X > x_2^{(\epsilon)}\} = 1 - \epsilon$, and ϵ is a fixed number (at $\epsilon = 0.25$ the values $x_1^{(0.25)}$ and $x_2^{(0.25)}$ are called distribution *quartiles*, and $(x_2^{(0.25)} - x_1^{(0.25)})/2$ is the *semi-interquartile range*, or the *probable deviation* of X). One can also use the *mean absolute deviation of a given order* m, equal to $\langle |X - \langle X \rangle |^m \rangle^{1/m}$. None of the above measures has fundamental advantages over the others; therefore, the main criterion for selecting some one of them is the convenience of its use.

[6]The use of the characteristic function $\psi(t)$ may often facilitate the evaluation of the moments of a random variable X. Indeed, from (0.2') it readily follows that *if* $\langle |X|^n \rangle < \infty$, *then the function* $\psi(t)$ *is differentiable at least* n *times and*

$$(0.10') \quad \langle X^n \rangle = \mu^{(n)} = (-i)^n \psi^{(n)}(0).$$

If $\psi'(0) = \psi''(0) = 0$, then it follows from (0.10') that $\langle X \rangle = 0$ and $\langle X^2 \rangle = \int_{-\infty}^{\infty} x^2 dF(x) = 0$. This is obviously possible only if $X = 0$ with probability one and, hence, $\psi(t) = \langle e^{itX} \rangle \equiv 1$ (the last circumstance has already been indicated in Note 3).

With reference to the distributions in examples (b), (c), and (d) simple calculations lead to the following results:

in the case of the Poisson distribution

$$(0.11a') \quad \psi(t) = \sum_{k=0}^{\infty} e^{itk} e^{-\lambda} \frac{\lambda^k}{k!} = e^{\lambda(e^{it}-1)};$$

in the case of the exponential distribution

$$(0.11b') \quad \psi(t) = \lambda \int_0^{\infty} e^{(it-\lambda)x} dx = \frac{\lambda}{\lambda - it};$$

in the case of the normal distribution

$$(0.11c') \quad \psi(t) = \frac{1}{\sqrt{2\pi}\,\sigma} \int_{-\infty}^{\infty} e^{itx - (x-m)^2/2\sigma^2} dx = e^{imt - \sigma^2 t^2/2}.$$

By using (0.10') it is now easy to evaluate the corresponding moments $m_X = \mu^{(1)}$ and $\sigma_X^2 = \mu^{(2)} - (\mu^{(1)})^2$ (the calculation of σ_X^2 can be further simplified by noting that $\sigma_X^2 = -d^2\ln\psi(t)/dt^2|_{t=0}$).

[7]Let, e.g., X be an arbitrary random variable with $m_X = \langle X \rangle = 0$ and an even probability density $p(x)$ (i.e. $p(-x) = p(x)$), and $Y = X^2$. Then X and Y are related by a functional dependence and the value of Y is uniquely determined by the value of X. However, in this case

$$\mu^{(1,1)} = \langle XY \rangle = \langle X^3 \rangle = \int_{-\infty}^{\infty} x^3 p(x) dx = 0.$$

Hence $b_{XY} = \mu^{(1,1)} - m_X m_Y = 0$, and $r = 0$, i.e. the variables X and Y are mutually uncorrelated.

[8]In considering multidimensional normal distributions it is very convenient to use *multidimensional characteristic functions*. These functions simplify many proofs and eliminate the need to distinguish between degenerate (improper) and nondegenerate (proper) distributions.

The characteristic function of the n-dimensional random variable $\mathbf{X} = (X_1,..., X_n)$ is the function

$$(0.12') \quad \psi(t_1,...,t_n) = \left\langle e^{i(t_1 X_1 + ... + t_n X_n)} \right\rangle$$

$$= \int_{-\infty}^{\infty} ... \int_{-\infty}^{\infty} e^{i(t_1 x_1 + ... + t_n x_n)} d^n F(x_1,...,x_n)$$

depending on the point $\mathbf{t} = (t_1,..., t_n)$ of n-dimensional space. In the particular case of the continuous variable $\mathbf{X}$ having a probability density $p(x_1, ..., x_n) = p(\mathbf{x})$ the function $\psi(\mathbf{t})$ is an n-dimensional Fourier transform of this density, and, therefore, the probability density $p(\mathbf{x})$ can be expressed in terms of $\psi(\mathbf{t})$ with the aid of the well-known inversion formula for n-dimensional Fourier integrals (see, e.g., Cramér, 1946, eq. (10.6.3)). By integrating the obtained equation for $p(\mathbf{x})$ over an n-dimensional parallelepiped we arrive at the n-dimensional generalization of the one-dimensional equation $(0.6')$. This new equation (which can be found, e.g., in Gnedenko, 1962, Sec. 40, in Cramér, 1962, Chap IX, or in Cramér, 1946, Sec. 10.6) also holds in the general case of an arbitrary n-dimensional probability distribution (with the same reservations as were made for $(0.6')$). Hence, in the multidimensional case the characteristic function also uniquely determines the corresponding probability distribution.

Multidimensional characteristic functions possess many properties similar to those of one-dimensional characteristic functions. In particular, $\psi(\mathbf{t})$ *is always a continuous positive definite function of a point* $\mathbf{t} = (t_1, ..., t_n)$, i.e. it is continuous with respect to all its arguments t_j, $j = 1$, ..., n, and the following inequality holds for any positive integer m, points of n-dimensional space $\mathbf{t}_1, ..., \mathbf{t}_m$, and complex numbers $c_1, ..., c_m$:

$$(0.13') \quad \sum_{j,k=1}^{m} \psi(\mathbf{t}_j - \mathbf{t}_k) c_j \bar{c}_k \geqslant 0.$$

The proof of this fact does not differ from the proof of the inequality $(0.8')$. It is also true that *the class of n-dimensional characteristic functions* $\psi(\mathbf{t})$ *coincides with that of continuous positive definite functions of a point of the n-dimensional space which takes the value one at* $\mathbf{t} = \mathbf{0}$, where $\mathbf{0} = (0, ..., 0)$. This last statement is implied by the following important theorem due to Bochner (see Supplement to Bochner's book, 1959, first published in German in 1932):

Multidimensional Bochner's theorem. *A continuous complex function* $\psi(\mathbf{t})$ *of a point* $\mathbf{t}$ *of n-dimensional space is positive definite if, and only* *if, it can be represented in the form*

$$(0.14') \quad \psi(t) = \int e^{itx} dF(x),$$

where F is a nonnegative measure in the n-*dimensional space of points* **x.**

By the nonnegative measure is meant the function of the region Δ of n-dimensional space (defined, in particular, for all n-dimensional parallelepipeds) which takes only nonnegative values and is additive, i.e. such that

$$(0.15') \quad F(\Delta_1 + \Delta_2) = F(\Delta_1) + F(\Delta_2)$$

for disjoint regions Δ_1 and Δ_2, where $\Delta_1 + \Delta_2$ indicated the union of the regions Δ_1 and Δ_2. The nonnegative measure F, normalized by the condition $F(\mathbb{R}^n) = 1$, where $\mathbb{R}^n$ is the whole n-dimensional space of points **x**, is evidently the *probability measure* which assigns some n-dimensional probability distribution. Note also that the condition $\psi(0) = 1$ is precisely equivalent to the normalization condition $F(\mathbb{R}^n) = 1$.

Moments of an n-dimensional probability distribution are easily expressed in terms of the corresponding characteristic function $\psi(t)$. In fact, it is easy to see that

$$(0.16') \quad \mu^{(j_1,\ldots,j_n)} = (-1)^{j_1+\ldots+j_n} \left. \frac{\partial^{j_1+\ldots+j_n} \psi(t_1,\ldots,t_n)}{\partial t_1^{j_1} \ldots \partial t_n^{j_n}} \right|_{t=0},$$

where $\mu^{(j_1+\ldots+j_n)} = \langle X_1^{j_1} \ldots X_n^{j_n} \rangle$ and the partial derivative on the right-hand side exists if the moment $\mu^{(j_1,\ldots,j_n)}$ is finite. Note that by virtue of (0.12') the characteristic function $\psi_{Y_1,\ldots,Y_m}(t_1, \ldots, t_m)$ of the variables $Y_1, \ldots, Y_m$, where $Y_1 = \sum_{j=1}^n a_{1j} X_j + b_1, \ldots, Y_m = \sum_{j=1}^n a_{mj} X_j + b_m$, is equal to

$$(0.17') \quad \psi_{Y_1,\ldots,Y_m}(t_1,\ldots,t_m)$$
$$= e^{i(b_1 t_1 + \ldots + b_m t_m)} \psi(a_{11} t_1 + \ldots + a_{m1} t_m, \ldots, a_{1n} t_1 + a_{mn} t_m).$$

In particular, the characteristic function $\psi_Y(t)$ of the linear combination $Y = a_1 X_1 + \ldots + a_n X_n + b$ is given by the formula

$$(0.18') \quad \psi_Y(t) = e^{ibt} \psi(a_1 t, \ldots, a_n t).$$

Moreover, the characteristic function $\psi_{i_1,\ldots,i_m}(t_{i_1},\ldots,t_{i_m})$ of the arbitrary group $X_{i_1}, \ldots, X_{i_m}$, $i_1 < \ldots < i_m$, formed by $m < n$ random variables

X_i has an especially simple form.   In fact, according to (0.12'), in order to obtain $\psi_{i_1,\ldots,i_m}(t_{i_1}, \ldots, t_{i_m})$ one only has to equate to zero all the variables in $\psi(t_1, \ldots, t_n)$ except $t_{i_1}, \ldots, t_{i_m}$.   For instance, the characteristic function $\psi_{\mathbf{X}'}(t_1, \ldots, t_m)$ of the m-dimensional random variable $\mathbf{X}' = (X_1, \ldots, X_m)$, where $m < n$, is equal to

$$(0.19') \quad \psi_{\mathbf{X}'}(t_1, \ldots, t_m) = \psi(t_1, \ldots, t_m, 0, \ldots, 0).$$

Let us now consider n-dimensional normal distributions.   Such distributions can be defined as distributions corresponding to characteristic functions of the form

$$(0.20') \quad \psi(t_1, \ldots, t_n) = \exp\left\{ i \sum_{j=1}^{n} m_j t_j - \frac{1}{2} \sum_{j,k=1}^{n} b_{jk} t_j t_k \right\},$$

where $\mathbf{m} = (m_1, \ldots, m_n)$ is an arbitrary n-dimensional vector, and $\overset{\circ}{B} = \|b_{jk}\|$ is an arbitrary symmetric, positive definite (but not necessarily strictly positive definite) matrix. To prove that the function (0.20') is actually a characteristic function, one must check the validity of the condition (0.13') for it or the possibility of representing this function in the form (0.14'). If $\overset{\circ}{B}$ is a strictly positive definite matrix, the function (0.20') will have a Fourier transform; in this case one can check directly that this function is a characteristic function of an n-dimensional probability distribution with a density described by (0.37) to (0.39), where $G = \|g_{jk}\|$ is a matrix reciprocal to the matrix $\overset{\circ}{B}$ If, however, $\overset{\circ}{B}$ is a positive definite, but not a strictly positive definite matrix, the function (0.20') will no longer have any n-dimensional Fourier transform and, hence, will certainly not be a characteristic function of a probability distribution with a continuous probability density. But then it is easy to show that the function (0.20') will be the limit, as $N \to \infty$, of the sequence of functions $\psi_N(t_1, \ldots, t_n)$ of the same form corresponding to a sequence of strictly positive definite matrices $\overset{\circ}{B}_N$ (where $\overset{\circ}{B}_N \to \overset{\circ}{B}$ as $N \to \infty$). It readily follows from this that condition (0.13') will also hold for the function (0.20') with a positive definite (but not a strictly positive definite) matrix $\overset{\circ}{B}$. Hence any function (0.20'), where $\overset{\circ}{B}$ is a positive definite matrix, is a characteristic function of some n-dimensional probability distribution (for more details see, e.g., Cramér, 1962, Chap. X, or 1946, Chap. 24).

Equation (0.20') for the characteristic function of the n-dimensional normal distribution implies all the properties of this distribution mentioned in the Introduction.   In the first place, (0.16') shows that the vector $\mathbf{m} = (m_1, \ldots, m_n)$ coincides with that of the mean value of the variable $\mathbf{X} = (X_1, \ldots, X_n)$, and the matrix $\overset{\circ}{B} = \|b_{jk}\|$ with its centered correlation matrix.   Further, from (0.19') and similar equations related to arbitrary groups of $m < n$ random variables from among $X_1, \ldots, X_n$, it follows at once that any group of $m < n$ variables X_i has an m-dimensional normal distribution.

Finally, (0.17') and (0.18') show that any m linear combinations of variables X_1, ..., X_n, where m is an arbitrary integer, has an m-dimensional (nondegenerate or degenerate) normal distribution. Hence, in particular, any linear combination of X_1, ..., X_n is a normally distributed random variable.

[9]Here, as in some other questions discussed in this book, there is no unified, universally accepted terminology. In particular, the word "random" is often replaced by the synonymous Greek word "stochastic", i.e. the term "random process", or "random function" is replaced by "stochastic process", or "stochastic function". Moreover, the term "random (or stochastic) process" is often used to denote an arbitrary random function of real argument t (not necessarily time). If one wishes to emphasize that one is dealing with a function of discrete argument t ("random sequence") or, alternatively, of continuous argument, one speaks then of a "discrete time (or discrete parameter) process" or, accordingly, a "continuous time (or continuous parameter) process". Sometimes, however, by a "random process" is meant an arbitrary random function of any argument and the term "random field" is replaced by "multidimensional time (or multidimensional parameter) random process". Finally, the term "random function of time" (especially in the case of discrete t) is sometimes replaced by the term "time series" (which we shall also use occasionally), though even more often by a "time series" a realization of a random sequence is meant.

[10]The term "Brownian motion" derives from the name of the distinguished British botanist Robert Brown, who described the motion in detail (see R. Brown, 1828, 1829). However this motion was not discovered by Brown, but had been known before him (cf. Nelson, 1967, Sec. 2).

The quantitative theory of Brownian motion explains the motion by the effect of molecular impacts on small particles produced by the thermal motion of the molecules of the surrounding fluid. This theory was developed almost simultaneously (about 1905) and independently by A. Einstein and M. von Smoluchowski (see Einstein, 1956; Smoluchowski, 1923; or Chandrasekhar, 1943). Curiously, at the time of writing his first paper on this subject, Einstein was apparently unaware of the existence of Brownian motion (which was widely known by then). He actually predicted it on theoretical grounds and described all its basic properties correctly. The Einstein–Smoluchowski theory (some important results of which had already been contained in the paper by Bachelier published in 1900, but which attracted no attention at the time) dealt exclusively with the probability distributions (and statistical moments) of the coordinates $X(t)$ of a Brownian particle at one or two time moments. A mathematical description of the random function $X(t)$ of the Einstein–Smoluchowski theory, which included a rigorous proof of a number of subtle mathematical results concerning the properties of individual realizations $x(t)$, was

iven between 1920 and 1923 in a series of papers by N. Wiener
see, e.g., Chap. IX of Paley and Wiener's book, 1934, or Doob, 1966).
A substantial refinement of the Einstein–Smoluchowski theory of
Brownian motion was developed by Uhlenbeck and Ornstein, 1930
see also Wang and Uhlenbeck, 1945; Chandrasekhar, 1943; Doob,
942; Nelson, 1967). The present-day state of the art in the
mathematical theory of the Brownian motion is elucidated, for
instance, in (rather complicated) books by Lévy (1965) and Itô and
McKean (1965).

[11]See, e.g., the books by Bendat (1958), Davenport and Root (1958),
Levin (1974, 1975, 1976), Livshits and Pugachev (1962), Laning and
Battin (1956), Max (1981), Middleton (1960), Pugachev (1965), Rytov
(1976), Rytov, Kravtsov, and Tatarskii (1978), Solodovnikov (1960),
Stratonovich (1967), Tikhonov (1982), and by a number of other
authors.
Electrical fluctuations and noises, which are mainly responsible for
the vigorous development of the applied theory of random functions,
are treated, e.g., in the books by MacDonald (1962), Malakhov (1968),
Robinson (1974), Van der Ziel (1959, 1970, 1976), Wolf (1978) and
Buckingham (1983), which also contain many additional references.

[12]There are many books discussing turbulence; here we shall
mention only the introductory text by Bradshaw (1971), the larger
book by Hinze (1975), the books by Panofsky and Dutton (1984) and
Monin and Ozmidov (1985) on atmospheric turbulence and oceanic
turbulence, and the huge two-volume monograph by Monin and
Yaglom (1971, 1975), which contains an extensive bibliography. The
paper by Haubrich (1965) is typical for microseism studies; see also
the Russian books by Rykunov (1967) and Vinnik (1968).
Geomagnetic variations are described, for instance, in the classic
book by Chapman and Bartels (1940); for a more modern treatment
see, e.g. the book by Yanovsky (1978). The statistical analysis of
EEG's and of some other biological signals (e.g., electrocardiograms
or myoelectric signals) is treated, for instance, in the collection
edited by Williams (1977). Random variations in the thickness of a
textile thread along its length are considered, in particular, by Foster
(1946). The Russian book by Khusu, Witenberg, and Pal'mov (1975)
and many dozens of papers (e.g., Peklenik, 1967-68) deal with
statistical analysis of surface roughness; sea waves are discussed, for
instance, in the book by Phillips (1977) and the review paper by
Barnett and Kenyon (1975).

[13]The Wolfer number is usually defined as the sum of the number
of individual sunspots observed at a given moment and the tenfold
number of closely spaced clusters of such spots. However, at present
one usually includes in the obtained value an additional factor
depending on the telescope parameters and the meteorological
conditions during the observation; this factor makes the very diverse

available data much more uniform (see, e.g., Waldmeir, 1961). Some parts of the series of the mean annual Wolfer numbers are presented, in particular, in the books by Box and Jenkins (1970) and Anderson (1971), which also contain several deductions from its statistical analysis. There are also a lot of statistical papers treating this series (e.g., Yule, 1927, Whittle, 1954a, and Brillinger and Rosenblatt, 1967).

The Beveridge wheat price series was introduced in Beveridge's articles (1921, 1922); see also Granger and Hughes (1971). The series is reproduced in the books by Kendall and Stuart (1968) and Anderson (1971) and is also discussed in the books by Wold (1954), and Yule and Kendall (1950). Other examples of randomly fluctuating economic time series can be found, for example, in the books by Yule and Kendall (1950), R.G. Brown (1963), Granger and Hatanaka (1964), Kendall and Stuart (1968), Fishman (1969), Malinvaud (1969), and Box and Jenkins (1970). Similar time series of geophysical origin are considered, in particular, in the book by Båth (1974) containing many additional references.

[14]The set of all possible time shifts of a given function $x(t)$ and averaging over this set plays an important role in the "generalized harmonic analysis" developed by Wiener (1930) and (1933), Chap. IV (see also the beginning of Sec. 8 of the present book). Wiener's theory was primarily meant to be used in studying disordered fluctuations of the type depicted in Figs. 5 and 6. Probabilistic considerations are not mentioned explicitly by Wiener; but he indicates in his works a realization of a simple random process as the only nontrivial example of the function $x(t)$ to which his theory is applicable. Later on, the problem of finding analytical formulae describing functions $x(t)$ accessible to Wiener's generalized harmonic analysis was thoroughly discussed by Bass (1962), who even proposed a special term "pseudorandom functions" for such $x(t)$. However, all the examples of Bass turn out to be quite artificial and therefore they can hardly be of interest in practical applications. The interpretation of the term "statistical stability of a set of time shifts of the $x(t)$ curve" will be discussed at the end of Note 1 to Chap. 1.

[15]Thus, instead of the values $x(t)$ themselves, one sometimes considers deviations of these values from the corresponding *trend* $x_0(t)$ (a smooth slowly varying function representing the systematic part of the time series $x(t)$). One may hope that the time shifts of the differences $\overset{\circ}{x}(t) = x(t) - x_0(t)$ will then possess the necessary statistical stability (see, e.g., Kendall and Stuart, 1968, Chap. 46). Sometimes it is also expedient to divide the values $\overset{\circ}{x}(t)$ by $|x_0(t)|$ to achieve a better uniformity of the obtained time series (cf. the description of "the Beveridge wheat price series" in the Introduction). Another method used to exclude the systematic part of the series $x(t)$ consists of studying, not the values $x(t)$, but their increments $x(t) - x(t - \tau)$ or their second increments $x(t) - 2x(t - \tau) + x(t - 2\tau)$,

or increments of a still higher order (see, in particular, Kendall and Stuart, 1968, Secs. 46.24 to 46.31). The mathematical theory of random processes $X(t)$ having statistically stable time shifts of increments of a given order were developed by Kolmogorov (1940a) (see also Doob, 1953, Sec. XI.11), Pinsker (1955), and Yaglom (1955b); see Secs. 23 and 24 of this book. The practical applications of some results related to such random function were studied by Box and Jenkins (1970).

Statistical analysis of the fluctuations in the Caspian Sea level is considered, e.g., in the papers by Privalsky (1973, 1976) and in his book published in 1985; cf. also Note 6 to Chap. 1. The time series shown in Fig. 8b is analyzed in the books by R.G. Brown (1963) and Box and Jenkins (1970).

[16]The Poisson process $N(t)$ determined by the Poisson point process $\{t_n\}$ is studied, e.g., in the books by Gnedenko (1962), Sec. 51, and Parzen (1962a), Chap. 4. A random telegraph signal was apparently introduced by Rice (1944-45). General point processes with adjoined random variables were considered by Grenander (1950); one special example of such a process (corresponding to a non-Poissonian point process $\{t_n\}$ was also given by Rice (1944-45). Point processes with adjoined random variables corresponding to the Poisson point process $\{t_n\}$ were also introduced by P. W. Anderson (1954) and Kubo (1954) as models useful in statistical optics; therefore such processes are sometimes called the *Kubo—Anderson processes* (see, e.g., Brissaud and Frisch, 1974). Brissaud and Frisch (1971, 1974) studied also (again in connection with some optical problems) a more general process where the probability distribution of the random variable $t_{n+1} - t_n$ depends on the value x_n of the process $X(t)$ for $t_n < t \leqslant t_{n+1}$; they called this process the *kangaroo process*. The shot noise process was studied by Rice (1944-45); see also Parzen (1962a), Davenport and Root (1958), and Papoulis (1965, 1984). The general Poisson pulse process (0.42) was considered, e.g., by Doob (1953), Sec. IX.3, and Rytov (1976), Chap. II.

[17]Random point processes are discussed, e.g., in Cox and Lewis' book (1966), in which (as well as in the books by Cox, 1962, and Parzen, 1962a) considerable attention is given to renewal processes. General counting processes are considered by Parzen (1962a). A number of works, of which we shall mention only Masry's paper (1972) and the book by Kedem (1980), are devoted to general binary processes taking only two values, $+a$ and $-a$ (i.e. having realizations similar to the one shown in Fig. 11). Such processes may arise in many nonlinear devices where the signal is inevitably "clipped" (limited in amplitude), i.e. all input signals with an amplitude exceeding some value a are perceived as having an amplitude a. It is obvious that processes of the type depicted in Fig. 11 can serve as a good model of a "strongly clipped signal", i.e. one limited by the value a, where a is much less than the typical amplitude of the real input signal. (So in the absence of a limiter the signal would overshoot the level $\pm a$ almost all the time.)

Amplitude- or phase-modulated pulse processes are studied, for example, in the books by Lawson and Uhlenbeck (1950), Middleton (1960), Franks (1969), Levin (1974), Rytov (1976), Tikhonov (1982), and Papoulis (1984); see also Stratonovich (1967), Sec. 6.

[18]The first attempt at a mathematical investigation of probabilistic models leading to the notion of a random function of a continuously varying argument t appeared in the above-cited interesting, but somewhat vague paper by Bachelier (1900). Later, the physicists Einstein, Smoluchowski, Fokker, and Planck developed the Brownian motion theory, which in fact can be formulated as a theory of one special class of random processes. In the early 1920s Wiener constructed a mathematical model of the Brownian motion of a free particle, which was the first example of a rigorous mathematical investigation of a particular random process (see Note 10 above). At about the same time, Slutsky began the publication of a series of papers on the general theory of random functions (see, e.g., Slutsky, 1928, 1929, 1937, etc.). The concept of a "probability distribution in infinite-dimensional space", which is equivalent to that of the random function, acquired a clear meaning within the framework of the axiomatic formulation of probability theory developed by Kolmogorov in 1933 (see Kolmogorov, 1956; cf. also Kolmogorov, 1935). The late 1920s and early 30s saw the publication of the works by de Finetti, Kolmogorov, and Lévy on the theory of random processes with independent increments (see, e.g., Doob, 1953, Chap. VIII, and Gihman and Skorohod, 1969, Chap. VI). Also in the 1930s Kolmogorov (1931) developed a general theory of Markov random processes and Khinchin (1934) laid the foundation for the theory of stationary random processes.

In subsequent years, the literature on the mathematical theory of random functions was expanding at an ever-increasing rate. In the late 1940s and early 50s several surveys on the theory of random processes appeared, which have much in common with the present book on their content (Loève, 1945-46; Kolmogorov, 1947; Doob, 1949; and Yaglom, 1962a, published in Russian in 1952). The earliest books on the theory of random processes were apparently the first edition of Wold's book (1954) published in 1938 and the first edition of Lévy's book (1965) published in 1948. At present, there are probably several hundred books on the mathematical theory of random processes and even more applied books actually dealing with random functions (see, e.g., Note 11 above).

Chapter 1

[1]The exact meaning of this statement is related to some refined mathematical considerations which are, in fact, closely associated with the way a random function arises, usually in an actual physical context. As already emphasized in the Introduction, in order to apply probabilistic methods, we must have an experiment which can be repeated many times under similar conditions and which can lead to different outcomes. The set Ω of all possible outcomes ω^* of such an experiment (the set of so-called *elementary events*) plays a basic role in Kolmogorov's axiomatic formulation of probability theory (see, e.g., Kolmogorov, 1956; Cramér, 1962; Shiryaev (1980); or any other modern advanced text on probability). A *random variable X* is a quantity which takes different numerical values for different outcomes of an experiment. In other words, a random variable is a numerical function of the point ω^* of the set Ω. Therefore, it would be more accurate to write $X(\omega^*)$ instead of X. (This was not done anywhere above, since in probability theory, the dependence of random variables on an elementary event ω^* is traditionally suppressed.) From this point of view, a *random function $X(t)$* on T should be defined as a function $X(t,\omega^*)$ of two variables t and ω^*.

If we fix the value of the argument t in the function $X(t,\omega^*)$, we arrive at a random variable $X(t) = X_t(\omega^*)$. If, however, we fix the value of the argument ω^*, we obtain a numerical function $x(t) = X_{\omega^*}(t)$ of the variable t, which depends on ω^*. Thus, to each outcome ω^* of our experiment, there corresponds a definite real function $x(t)$ of the variable t. This function is called a *realization* (or a *trajectory*, or an *observed function*, or a *sample function*) of the random function $X(t)$ (cf. Introduction). The random function $X(t)$ = $X(t,\omega^*)$ can therefore be regarded either as a family of random variables $X(t)$ corresponding to all the elements t of the set T (this is what we actually did in the beginning of Sec. 1), or as a family of realizations $x(t) = X_{\omega^*}(t)$ corresponding to all the possible elementary events ω^*. In the latter case, to specify $X(t)$ we must define the probability of occurrence of the various realizations (or, more precisely, various realization subsets, since the probability of a single specified realization $x(t)$ is usually equal to zero). This leads to the definition, on the function space of all realizations $x(t)$, of a particular measure P (the *probability measure*). The measure P takes nonnegative values (not exceeding unity) on various realization sets and assigns the value 1 to the entire space of all realizations. The specification of this probability measure is equivalent to the specification of the random function $X(t)$.

Such an approach to the concept of a random function, also originating in Kolmogorov's work (1956) on foundations of the probability theory, proved to be very fruitful. It can be shown that specifying a random function $X(t)$ by defining a probability measure P on the set of realizations $x(t)$ includes as special cases all the other ways of specifying a random function discussed above. The specification of $X(t)$ by defining all the finite-dimensional probability distributions (1.3) is also not exclusive in this respect. In fact, according to Kolmogorov's "fundamental theorem" (Kolmogorov, 1956, p. 29), *to each system of finite-dimensional probability distributions satisfying the conditions of symmetry and compatibility there corresponds a unique probability measure* P *in the space of all real functions given on* T (see also Cramér and Leadbetter, 1967, Sec. 3.3, or Wentzell, 1981, Sec. 5.1, or Shiryaev, 1980, Sec. II.3). We meant exactly this fact when we stated that any system of probability distributions (or densities) satisfying symmetry and compatibility conditions defines a random function.

Specification of the probability measure with the aid of a system of finite-dimensional probability distributions is quite sufficient for all the problems in which one has to deal only with probabilities of events depending on a finite number of variables $X(t)$. (Only such problems are in fact considered in this book.) However, such a specification is often insufficient for determining the probabilities of events depending on an infinite number of values of $X(t)$ (e.g., the probability that the realization $x(t)$ is a continuous function of t). The necessity of determining, in some cases, probabilities of this latter type gave rise to a number of detailed investigations of probability measures on function spaces, which use advanced techniques of modern set theory, function theory, and functional analysis. However, the use of such techniques is not essential for the content of this book and we won't pursue this matter here. Therefore, we only refer to the brief discussion of the related problems in the book by Cramér and Leadbetter (1967), Sec. 3.6, and to the more detailed (and complicated) books by Doob (1953), Sec. II.2, Loève (1963), Sec. 35, Gihman and Skorohod (1969), Chap. IV, and (1974), Chap. III, and Gel'fand and Vilenkin (1964), Chap. IV.

To conclude this Note we shall also remark that the specification of a random function $X(t)$ by a system of its finite-dimensional probability distributions makes it possible to impart an exact meaning to the concept of "statistical stability" of a set of all possible time shifts of the curve $x(t)$, mentioned in the Introduction and in Note 14 to it.   Let us assume that time, t, is continuous (the case where t takes integral values requires only the obvious replacement for the integrals by sums in the subsequent equations) and that $\chi_a[x] = 1$ for $x < a$ and $\chi_a[x] = 0$ at $x \geqslant 0$.   Let us try to determine the functions (1.3) for any n, t_1, ..., t_n and for values x_1, ..., x_n such that the right-hand side of (1.3) is a continuous function of its arguments, proceeding from the set of time shifts of the function $x(t)$.   It is clear that for this the following limits must exist:

$$\lim_{T_1 \to -\infty,\ T_2 \to \infty} \frac{1}{T_2 - T_1} \int_{T_1}^{T_2} \chi_{x_1}[x(t + t_1)] \dots \chi_{x_n}[x(t + t_n)]dt$$

(1.1')

$$= F_{t_1, \dots, t_n}(x_1, \dots, x_n).$$

If however, these limits do exist, then it is easy to see that the functions $F_{t_1, \dots, t_n}(x_1, \dots, x_n)$ will satisfy the symmetry and

compatibility conditions, i.e. will specify a random function $X(t)$. The function $x(t)$ can be considered then as a realization of $X(t)$, and the existence of the limits (1.1') for it is precisely the required "statistical stability" condition.   Note also that the random function $X(t)$ thus obtained is necessarily a *stationary* and *ergodic* one (the meaning of these two terms is explained in Secs. 3 and 16 of this book).   Conversely, *for a stationary and ergodic random function $X(t)$ the limits* (1.1') *always exist and can be used to determine the corresponding multidimensional probability distributions* (1.3).   If, however, we are interested only in the correlation theory of stationary random functions $X(t)$ (see Secs. 2 and 3), i.e. if we are studying only those properties of $X(t)$ which depend on its first and second moments, then the situation is simplified.   In this case, having at our disposal a single realization $x(t)$, we need not even require the existence of the general limits (1.1') and can restrict ourselves to the requirement that the following limits exist:

$$m = \lim_{T_1 \to -\infty,\ T_2 \to \infty} \frac{1}{T_2 - T_1} \int_{T_1}^{T_2} x(t)dt,$$

(1.2')

$$B(t_1, t_2) = B(t_1 - t_2) = \lim_{T_1 \to -\infty,\ T_2 \to \infty} \frac{1}{T_2 - T_1} \int_{T_1}^{T_2} x(t + t_1)x(t + t_2)dt$$

(see Wiener, 1930 or 1933, Chap. IV; cf. also Wold, 1948, and Brillinger, 1975, Sec. 2.11).

[2]In particular, by analogy with specification of a random variable by its characteristic function $\psi(t)$ or $\psi(\mathbf{t}) = \psi(t_1, \dots, t_n)$ (see Notes 3 and 8 to Introduction), the random function $X(t)$ can be specified by its *characteristic functional*

(1.3') $\Psi[\theta(t)] = \langle \exp\{i(X(t), \theta(t)\} \rangle.$

Here $\theta(t)$ is a real function of t, and

$$(1.4')\qquad (X(t),\theta(t)) = \begin{cases} \sum_{t\in T} X(t)\theta(t) & \text{for discrete } t, \\ \int_T X(t))\ \theta(t)dt & \text{for continuous } t \end{cases}$$

(the meaning of the integral on the right-hand side of (1.4') is explained in Sec. 4 of this book). In the case of an unbounded set T we can restrict ourselves to the consideration of a function $\theta(t)$ different from zero only at a finite number of points t (for discrete t) or on a bounded set of t-values (for continuous t) and also assume that the function $\theta(t)$ is bounded, continuous, and smooth enough in the case of continuous t.   Then the expressions on the right-hand side of (1.4') are clearly convergent.

The characteristic functional $\Psi[\theta(t)]$ is a complex-valued function of the real function $\theta(t)$ (the term "functional" means "a function of a function").   It has a number of properties related to those of the characteristic functions $\psi(t)$ or $\psi(t_1, ..., t)$.   It is easy to show that the function $\Psi[\theta(t)]$ uniquely determines all the finite-dimensional probability distributions (1.3).* Thus it provides a very compact method for a complete specification of the random function $X(t)$.

The main shortcoming of this specification method is that an explicit formula for $\Psi[\theta(t)]$ can be given only for a few particular classes of random functions $X(t)$. The most important of such classes is that of Gaussian functions $X(t)$, for which

$$(1.5')\qquad \Psi[\theta(t)] = \exp\left\{ i\int_T m(t)\theta(t)dt - \frac{1}{2}\iint_T b(t,s)\theta(t)\theta(s)dt\,ds \right\},$$

where $m(t)$ is a real function of t, and $b(t,s)$ is a positive definite kernel on T. (The meaning of the last term is explained in Sec. 2.) The time t is assumed here to be continuous; in the case of discrete t both the integrals on the right-hand side of (1.5') must be replaced by sums.

Other examples of explicit formulae for the characteristic

---

*Indeed, let us assume that the function $\theta(t)$ is different from zero only at a finite number of points $t = t_1, ..., t = t_n$, $\theta(t_j)$ being equal to θ_j (in the case of discrete t), or, if t is continuous, let us select a sequence of smooth functions $\theta_N(t)$ tending to an improper ("generalized") function $\theta_1 \delta(t - t_1) + ... + \theta_n \delta(t - t_n)$ as $N \to \infty$. Then $\Psi[\theta(t)] = \psi_{t_1,...,t_n}(\theta_1, ..., \theta_n)$ is the characteristic function of the variables $X(t_1), ...,$ $X(t_n)$ if t is discrete, while, if t is continuous, $\lim_{N\to\infty}\Psi[\theta_N(t)]$ also coincides with the same characteristic function $\psi_{t_1, ..., t_n}(\theta_1, ..., \theta_n)$.   This characteristic function uniquely determines the distribution function $F_{t_1, ..., t_n}(x_1, ..., x_n)$.

functional $\Psi[\theta(t)]$ can be found in the books by Bartlett (1978), Sec. 5.2, and Stratonovich (1967), Sec. 6, and also in Skorohod's paper (1965).

The concept of a characteristic functional was first introduced in Kolmogorov's paper (1935), which was far ahead of its time. Later it began to be used by some other authors (e.g., by LeCam, 1947). The reader can familiarize himself with the elements of the mathematical theory of characteristic functionals in the books by Gel'fand and Vilenkin (1961), Chap. IV, and Grenander (1963), Sec. 6.2, which contain many additional references.

[3]The probability density $p_{t_1, \ldots, t_n}(x_1, \ldots, x_n)$ exists for all $t_1, \ldots, t_n$ only if the kernel $b(t,s) = B(t,s) - m(t)m(s)$ is strictly positive definite (i.e. if the left-hand side of (1.18a) is positive, provided that at least one of the numbers $c_1, \ldots, c_n$ is different from zero). However, normally distributed variables $X(t_1), \ldots, X(t_n)$ are most conveniently specified not by a probability density but by the characteristic function

$$(1.6') \qquad \psi_{t_1,\ldots,t_n}(\theta_1,\ldots,\theta_n) = \exp\left\{ i \sum_{j=1}^{n} m(t_j)\theta_j - \frac{1}{2} \sum_{j,k=1}^{n} b(t_j,t_k)\theta_j\theta_k \right\}.$$

This permits one to treat jointly the proper and improper normal probability distributions. It is also possible to replace the set of characteristic functions (1.6'), by a characteristic functional (1.5'). It is easy to see that in (1.5') the function $m(t)$ is the mean value of the random function $X(t)$, and $b(t,s) = \langle X(t)X(s)\rangle - m(t)m(s)$. To prove this it is sufficient to substitute into (1.5') the function $\theta_1\delta(t - t_1) + \ldots + \theta_n\delta(t - t_n)$ in place of $\theta(t)$. Then $\Psi[\theta(t)]$ transforms to the characteristic function $\psi_{t_1,\ldots,t_n}(\theta_1, \ldots, \theta_n)$ of the variables $X(t_1), \ldots,$ $X(t_n)$, which has the form (1.6'). It is also possible to disregard the finite-dimensional characteristic function and to use instead the general equation which determines the moment functions of different orders of the random function $X(t)$ according to its characteristic functional $\Psi[\theta(t)]$. This general equation is similar to (0.16') of Note 8 to the Introduction, but it contains, instead of the partial derivative, the so-called *functional derivative* $\delta^n\Psi[\theta(x)]/\delta\theta(x_1)dx_1 \ldots \delta\theta(x_n)dx_n$ of the functional $\Psi[\theta(t)]$ (see, e.g., Monin and Yaglom, 1971, Sec. 4.4).

The proof of the fact that any positive definite kernel is a correlation function of some random function $X(t)$, based on consideration of the corresponding system of multidimensional normal distributions, is due to Kolmogorov; it was first published (for the particular case of stationary random processes considered in Sec. 3) in Khinchin's paper (1934). The general case of functions $X(t)$ defined on an arbitrary set T was considered by Loève (1945-46); see also Loève (1963).

[4]The mean value $m(t)$ and the variance $\sigma^2(t)$ of the Poisson process $X(t)$ are evidently equal to the mean value and the variance of the number of Poisson points t_k such that $0 < t_k < t$, i.e. to the mean value and the variance of the Poisson distribution with a parameter λt. Hence, both of them are equal to λt (cf. p. 9 of the Introduction and Note 6 to it). Note now that the number of Poisson points t_k occurring on the interval $[0,s]$ does not affect the occurrence of points after the moment s (see, e.g., Gnedenko, 1962, or Parzen, 1962a). Hence the increment $X(t) - X(s)$, where $t > s$ (equal to the number of points t_k on the interval $[s,t]$) is independent of the random variable $X(s)$. Therefore, if $t > s$, then

$$B(t,s) = \langle X(t)X(s)\rangle = \langle [X(s)]^2\rangle + \langle [X(t) - X(s)]X(s)\rangle$$

$$= \sigma^2(s) + m^2(s) + \langle X(t) - X(s)\rangle \langle X(s)\rangle$$

$$= \lambda s + \lambda^2 s^2 + \lambda(t - s)\lambda s = \lambda s + \lambda^2 t s.$$

If $X(t)$ is the random telegraph signal, then $X(t)X(s) = a^2$, when an even number of Poisson points t_k occur on the interval $[s,t]$, and $X(t)X(s) = -a^2$ when the number of these points is odd. Therefore

$$B(t,s) = \langle X(t)X(s)\rangle = a^2\exp(-\lambda|t - s|)\left[\sum_{k=0}^{\infty} \frac{(\lambda|t - s|)^{2k}}{(2k)!}\right.$$
$$\left. - \sum_{k=0}^{\infty} \frac{(\lambda|t - s|)^{2k+1}}{(2k + 1)!}\right] = a^2 \exp(-2\lambda|t - s|).$$

If $X(t)$ is the Poisson point process with adjoined random variables, then $\langle X(t)\rangle = m$, where m is the mean value of the adjoint variables X_k. Moreover, in this case $X(t)X(s) = X_k^2$ if no Poisson points occur on the interval $[s,t]$ (this has a probability $\exp(-\lambda'|t - s|)$ while $X(t)X(s) = X_k X_j$ in the opposite case, where X_k and X_j are independent random variables. Since $\langle X_k^2\rangle = \sigma^2 + m^2$, where σ^2 is the variance of X_k, while $\langle X_k X_j\rangle = m^2$, then $B(t,s) = \langle X(t)X(s)\rangle = \sigma^2\exp(-\lambda'|t - s|) + m^2$; when $m = 0$ we arrive at (1.22).

[5]Following Slutsky (1928) and Loève (1945-46), the correlation theory of random functions is also often called the *second-order theory* of random functions, or the theory of *second-order random functions*.

[6]An attempt to apply the theory of stationary random functions to the variations of the Caspian Sea level was undertaken, among others, by Privalsky (1973, 1976, 1985). An extensive Russian literature is dedicated to the discussion of whether or not these variations can be regarded as stationary; see, e.g., Drozdov and Pokrovskaya (1961), Antonov (1963), Kritsky and Menkel (1964). Related discussion of the applicability of the stationarity assumption to hydrologic time series is given by Kisiel (1969); cf. also Bras and Rodriguez-Iturbe (1985).

[7]Cf. Hurd's paper (1974a) dealing with random functions of the form $Y(t) = X(t + \Phi)$, where $X(t)$ is a random function and Φ (the "phase" of $Y(t)$) is a random variable which is independent of $X(t)$ and uniformly distributed over an interval of length T_0. Hurd proved that the process $Y(t)$ is a stationary one if and only if $X(t)$ is a *periodically distributed* ("strictly cyclostationary") random process of period T_0 (i.e. if all the finite-dimensional distributions of this process remain unaltered under time shifts $X(t) \to X(t + h)$, where h is a multiple of T_0). Hurd also showed that *if the stationarity of* Y(t) *is replaced in the previous statement by wide sense stationarity* (the meaning of this is explained at the end of Sec. 3), *then one has to replace the requirement of the periodicity of distributions for the process* X(t) *by the requirement of the correlation periodicity* ("wide sense cyclostationarity"), i.e. by the requirement $m(t) = m(t + kT_0)$, $B(t,s) = B(t + kT_0, s + kT_0)$, where $m(t) = \langle X(t) \rangle$, $B(t,s) = \langle X(t)X(s) \rangle$ and k is an arbitrary integer. (Section 26.5 of this book is devoted to the study of processes $X(t)$ satisfying the last requirement.) More particular random processes of the form $Y(t) = X(t + \Phi)$, where $X(t)$ is an ordinary (nonrandom) periodic function and Φ is a random "phase", had previously been considered by Beutler (1961a). More general random processes $X(t)$ having the property that the process $Y(t) = X(t + \Phi)$ is stationary for at least one random variable Φ independent of $X(t)$ were studied by Gardner (1978), who called such $X(t)$ *stationarizable random processes*.

[8]The following synonyms of the term *wide sense stationary* functions are also used in the literature: *second-order stationary* functions (cf. Slutsky's paper, 1928, where the corresponding functions were called *homogeneous to the second order*), or *weakly stationary,* or *stationary in Khinchin's sense* (since the importance of the relevant concept was emphasized by Khinchin, 1934). Note also that Doob (1953) proposed to denote as the *wide sense properties* of a random function $X(t)$ all the properties which are actually inherent in the normal function $X^{(0)}(t)$ having the same first and second moments as $X(t)$. This suggestion is clearly compatible with the meaning of the term "wide sense stationarity".

It should be emphasized that conditions (1.26) and (1.27) implicitly suggest that the random variable $X(t)$ has finite variance, while conditions (1.23) or (1.24) are unrelated to this limitation. Hence the function $X(t)$, which satisfies conditions (1.23) or (1.24) (i.e. is strictly stationary), may, in principle, be non-stationary in the wide sense. For us, however, the last circumstance will not be essential, because in this book the existence of first and second moments of the random variables $X(t)$ will always beimplied a priori (and will not even be stipulated).

[9]The proof of the Chebyshev inequality (1.39) is very simple. If $\tilde{F}(x)$ is the distribution function of the random variable $X - X_n$, then

$$\langle (X - X_n)^2 \rangle = \int_{-\infty}^{\infty} x^2 d\hat{F}(x) \;\geqslant\; (\int_{-\infty}^{-\epsilon} + \int_{\epsilon}^{\infty})\, x^2 d\,\hat{F}(x)$$

$$\geqslant\; \epsilon^2 (\int_{-\infty}^{-\epsilon} + \int_{\epsilon}^{\infty})\, d\hat{F}(x) = \epsilon^2 P\{|X - X_n| \geqslant \epsilon\}.$$

The next, more general result is proved quite similarly: *If the condition $\langle |Y|^\alpha \rangle < \infty$, where $\alpha > 0$, is satisfied for a random variable Y, then*

$$(1.7') \qquad P\{|Y| > \epsilon\} \;\leqslant\; \frac{\langle |Y|^\alpha \rangle}{\epsilon^\alpha}.$$

Inequality (1.7') is sometimes called the *Markov inequality.*

[10]See, e.g., the discussion of the various kinds of convergence for sequences of random variables in the books by Parzen (1960), Gnedenko (1962), Loève (1963), Papoulis (1965, 1984), Cramér and Leadbetter (1967), and Shiryaev (1980).

[11]See, e.g. Loève (1963), Sec. 9.4, Gihman and Skorohod (1969), Sec. II.5, or Shiryaev (1980), Sec. II.10. The statement formulated here actually refers to mathematical analysis (more precisely, to functional analysis) rather then to probability theory; it is often called the Fischer and Riesz theorem in analysis.

[12]It has already been mentioned in Note 1 to Chap. 1 that in order to find the continuity conditions for all the realizations $x(t)$ of the process $X(t)$ (more precisely, the conditions of continuity of $x(t)$ with probability one, i.e. the continuity of all the curves of $x(t)$ with the exception of some "exclusive realizations", whose total probability is equal to zero) we must modify the very definition of the random process. Indeed, it is easy to show that the probability of the continuity of $x(t)$ is not determined by the finite-dimensional distributions (1.3). For proving this it is sufficient to consider the following simple example. Let $X(t)$ be a random process on the interval $0 \leqslant t \leqslant 1$ having only continuous realizations $x(t)$ (e.g, $X(t) = \cos(\omega t + \Phi)$, where Φ is a fixed random variable). Suppose now that Z is a random variable distributed uniformly on the interval $[0,1]$ and that $X_1(t) = X(t)$ for $t \neq Z$, but $X_1(t) = X(t) + 1$ for $t = Z$. Since $P\{Z = t\} = 0$ for any fixed t and, what is more, $P\{Z = t_1,$ or $Z = t_2,$ ..., or $Z = t_n\} = 0$ for any n fixed numbers $t_1, t_2, ..., t_n$, we have

$$P\{X(t_1) = X_1(t_1),\; X(t_2) = X_1(t_2),\; ...,\; X(t_n) = X_1(t_n)\} = 1$$

for any $t_1, t_2, ..., t_n$. Hence all the finite-dimensional distributions (1.3) are identical for processes $X(t)$ and $X_1(t)$. However,

$$P\{x(t) \text{ is continuous}\} = 1, \quad P\{x_1(t) \text{ is continuous}\} = 0$$

(because $x_1(t)$ is clearly discontinuous at the point $t = z$, where z is the realization of Z).

We see that the conditions imposed on the finite-dimensional probability distributions of the process can only guarantee, at best, the existence of a process which has such finite-dimensional distributions and all the realizations of which are continuous (but not the continuity of all the realizations for every random process having such finite-dimensional distributions). This is the exact sense of all the statements in this and the following Notes concerning the conditions of the continuity (or, e.g., differentiability) of realizations with probability one. Note, however, that the last remark is immaterial from the practical point of view. Indeed, if $X(t)$ is a random process having continuous (or, say, differentiable) realizations, then all the "non-regular" processes with the same finite-dimensional distributions will usually represent some "mathematical pathologies" never met in applications (the above-mentioned process $X_1(t)$ is a good example of such a "pathological process"). These "pathological processes" can also be fully excluded from consideration if we require the fulfillment of some very general (and always valid in real practice) regularity condition, which is not imposed on the finite-dimensional distributions of a process $X(t)$, but is based on the other specification of random processes. See, in this respect, the book by Cramér and Leadbetter (1967), Sec. 3.6 and Chap. 4, or the more complicated mathematical treatises by Doob (1953), Loève (1963), and Gihman and Skorohod (1969, 1974).

The simplest and most important of the known continuity conditions for realizations of random processes was found by A.N. Kolmogorov in the middle 1930s (and first published in Slutsky's paper, 1937). It states that *realizations of a process* $X(t)$ *given in an interval* $a \leqslant t \leqslant b$ *are continuous functions with probability one, provided the following inequality holds for some* $\alpha > 0$, $\delta > 0$, $C < \infty$ *and for all sufficiently small* h *(say, for* $h \leqslant h_0$, *where* h_0 *is a fixed number):*

$$(1.8') \qquad \langle |X(t + h) - X(t)|^\alpha \rangle < C|h|^{1+\delta}.$$

(The condition (1.8') clearly imposes limitations only on two-dimensional probability distributions of the process $X(t)$.) We shall not give a strict proof of the italicized statement (which is not very simple for the above-indicated reasons), and shall restrict ourselves to some considerations which make it almost obvious.

Note that for a "random telegraph signal" the quantity $\langle |X(t + h) - X(t)| \rangle$ is very small for *any fixed* t and sufficiently small h. Nevertheless, if we sample *all the possible values* of t, it would inevitably be large at some t. To prove that the condition (1.8') excludes such a possibility, we divide the interval $[a,b]$ into 2^n equal parts (of length $(b - a)/2^n$ each) by points $t_k = a + (b - a)k/2^n$, $k = 1, 2, ..., 2^n - 1$; $t_0 = a$, $t_{2^n} = b$. Let n be so large that $(b - a)/2^n < h_0$. It is easy to show that for any random events $A_1, A_2, ..., A_{2^n}$,

$$P(A_1 + A_2 + ... + A_{2^n}) \leqslant P(A_1) + P(A_2) + ... + P(A_{2^n}),$$

where $A_1 + A_2 + ... + A_{2^n}$ is an event consisting of the occurrence of at least one of the events A_1, A_2, ..., A_{2^n}.   (The inequality turns into an equality when the events A_1, A_2, ..., A_{2^n} are disjoint.) Let A_k, $k = 1, 2, ...,$ 2^n, be an event for which the inequality $|X(t_k) - X(t_{k-1})| < 1/n^2$ holds. If we apply the inequality (1.7') to the variable $Y = X(t_k) - X(t_{k-1})$ putting $\epsilon = 1/n^2$, then, by virtue of (1.8'), we get

$$P\{|X(t_k) - X(t_{k-1})| > 1/n^2 \text{ for at least one } k\}$$

$$(1.9') \qquad < 2^n \, \frac{C(b - a)^{1+\delta} \, n^{2\alpha}}{2^{n(1+\delta)}} = C(b - a)^{1+\delta} \, \frac{n^{2\alpha}}{2^{n\delta}} \, .$$

But $n^{2\alpha}/2^{n\delta} \to 0$, as $n \to \infty$, for any $\alpha > 0$ and $\delta > 0$.   Hence, assuming that n is very large, we obtain that under condition (1.8') the increments of $X(t)$ will be extremely small in absolute value (less than $1/n^2$) with probability very near to one on all 2^n equal parts of the interval $[a,b]$ of length $(b - a)/2^n$. This conclusion shows that a situation similar to the one depicted in Fig. 11 (related to a random telegraph signal) cannot take place in the case concerned and that the realization $x(t)$ must apparently be continuous here with probability one.

Note that the assumption about the stationarity of the process $X(t)$ has not been used anywhere in the above reasoning. Indeed, this assumption is not required in this case. In the particular case of a normal (Gaussian) process $X(t)$ it is not hard to show that

$$\langle |X(t + h) - X(t)|^{m\alpha} \rangle = C_{m,\alpha} \langle |X(t + h) - X(t)|^{\alpha} \rangle^m$$

where $C_{m,\alpha}$ is a constant depending on m and α. Therefore Kolmogorov's condition will hold for a Gaussian process, provided

$$\langle |X(t + h) - X(t)|^{\alpha_1} \rangle < C_1 |h|^{\beta_1} \text{ for some } \alpha_1 > 0, \beta_1 > 0, \text{ and } C_1 < \infty.$$

A strict proof of Kolmogorov's condition and of some other conditions for the continuity of realizations generalizing and supplementing Kolmogorov's result can be found in the book by Cramér and Leadbetter (1967), Secs. 4.2, 7.3, and 9.2–9.5. See also more sophisticated books by Loève (1963) and Gihman and Skorohod (1969, 1974).

[13]Note that the existence of $B''(\tau)$ implies the validity of Kolmogorov's condition (1.8') with $\alpha = 2$ and $\delta = 1$. Therefore, *the realizations* x(t) *of a process* X(t), *that is differentiable in the mean (square), are necessarily continuous.* However the derivative $x'(t)$ may not exist here at some points t (as the realizations of a process

continuous in the mean could have points of discontinuity). It is important to note, however, that if a differentiable in the mean process $X(t)$ satisfies some very general regularity condition, which is always fulfilled in practical applications, then its realization $x(t)$ will have, with probability one, a derivative $x'(t)$ for almost all values of t (i.e. for all t, maybe with the exception of some set of "exclusive values of t", whose total "length" is equal to zero). The derivative $x'(t)$ will not necessarily be continuous, but it will necessarily be integrable and such that $x(t)$ plus a constant coincides with the indefinite integral of $x'(t)$. Moreover, the value of the derivative $x'(t)$ at fixed t (which depends on the choice of a realization, i.e. which is a random variable) will coincide, with probability one, with the value of the mean square derivative $X'(t)$ of the process $X(t)$ (see Doob, 1953, p. 536). Therefore, in practical applications the realization of the mean square derivative $X'(t)$ can always be equated to the derivative of the realization of the process $X(t)$.

Note also that *if the process* X(t) *is twice differentiable in the mean* (which requires the existence of a continuous fourth derivative $B^{IV}(\tau)$), *then its realizations* x(t) *will have a continuous derivative* x'(t) *evertwhere.* This result is due to Slutsky (1937); see also the discussion of the conditions for differentiability of realizations in the book by Cramér and Leadbetter (1967), Secs. 4.3, 7.3, and 9.2–9.5.

[14]The above-mentioned proof can be found, for instance, in the book by Gihman and Skorohod (1974), Sec. IV.3. See also Doob (1953), pp. 62–63 and 518, where results are proved which imply the result stated in this book.

[15]Similar to the real case, to prove this it will suffice to show that for *any complex positive definite function* B(τ) *there exists a Gaussian, also complex, random function* X(t) (specified by its finite-dimensional probability distributions), *whose correlation function coincides with* B(τ). The construction of such a function is more cumbersome than the similar construction in the real case (because here, in place of n-dimensional probability distributions, one has to consider more complicated $2n$-dimensional distributions), but it is just as simple in principle. Note that now the values of $\langle X(t) \rangle = m$ and $\langle X(t)\overline{X(s)} \rangle = B(t - s))$ no longer uniquely define the required $2n$-dimensional Gaussian probability distributions, the specification of which requires the knowledge of all the moments $\langle X_j(t)X_k(s) \rangle$, $j = 1, 2$; $k = 1, 2$, rather than just one combination (1.57) of them. Therefore additional conditions must be imposed on the random function $X(t)$. For instance if $\langle X(t) \rangle = 0$ we can require the fulfillment of the relation $\langle X(t)X(s) \rangle = 0$ for all t, s. In this respect see, e.g., Doob (1953), Sec. II.3, or Loève (1963), Sec. 34.1, where the desired proof is given for the general case of an arbitrary positive definite kernel $B(t,s)$, not only for a kernel of the form $B(t - s)$.

[16]The   Cauchy–Buniakovsky–Schwarz   inequality   for   complex   random variables X and Y has the form: $|\langle X\bar{Y}\rangle| \leqslant \langle |X|^2\rangle^{1/2}\langle |Y|^2\rangle^{1/2}$.

It implies that $\langle |X + Y|^2\rangle = \langle |X|^2\rangle + \langle X\bar{Y}\rangle + \langle \bar{X}Y\rangle + \langle |Y|^2\rangle \leqslant \langle |X|^2\rangle + 2\langle|X|^2\rangle^{1/2}\langle|Y|^2\rangle^{1/2} + \langle|Y|^2\rangle = (\langle|X|^2\rangle^{1/2} + \langle|Y|^2\rangle^{1/2})^2$.   Geometrically, this last inequality means that in the space H the length of the side $X + Y$ of the triangle does not exceed the sum of the lengths of its two other sides X and Y.

[17]There are plenty of texts on functional analysis and operator theory containing parts devoted to the study of the Hilbert space; see, e.g., Stone (1932), Halmos (1951), Riesz and Sz.-Nagy (1955), Akhiezer and Glazman (1980), Reed and Simon (1972).   As a typical example of geometric theorems referring to such a space we give the formulation of the following *theorem on the perpendicular: In Hilbert space* H, *one can drop a unique perpendicular, from any point* Y, *to a given linear subspace* H$_1$, *and the length of this perpendicular is the shortest distance between the point* Y *and points of the subspace* H$_1$.* This theorem has important applications in the theory of stationary random functions (see, e.g., Yaglom, 1962; Rozanov, 1967; Papoulis, 1965, 1984; or Lambert and Poskitt, 1983).

In considering random variables as points of the Hilbert space H we must assume that the variables X and Y coincide if $\|X - Y\|^2 = \langle |X - Y|^2\rangle = 0$ (i.e. if $X = Y$ with probability one).   From the probabilistic point of view such identification of X and Y is, of course, quite natural.   We also wish to emphasize that for H to be considered a Hilbert space in the sense adopted in functional analysis it is important that the space H is complete in the sense that from the condition $\lim\limits_{n\to\infty,m\to\infty} \|X_n - X_m\|^2 = 0$

follows the existence of $X = \lim\limits_{n\to\infty}X_n$ (cf. Note 11 above).   Also, in defining the linear subspace H_1 one should in fact require not only that all the combinations $c_1X_1 + c_2X_2$ appear in H_1 together with X_1 and X_2, but also that the element $X = \lim\limits_{n\to\infty} X_n$ necessarily appears in H_1 together

with each converging sequence of elements $X_1, X_2, ..., X_n, ...$ (i.e. if condition $\lim\limits_{n\to\infty,m\to\infty} \|X_n - X_m\|^2 = 0$ holds).

[18]Strictly speaking, in order to be able to talk about $X(t)$ as a curve, we must also require that the following continuity condition hold: The distance between the points $X(t)$ and $X(t')$ of the curve must approach zero as $t' \to t$.   It is easy to see that this condition is equivalent to the condition of mean square continuity of the process $X(t)$.   In general, the mean square convergence of the sequence of

---

*A linear subspace H$_1$ is a set of vectors containing, together with any pair of vectors X$_1$ and X$_2$, also all their linear combinations c$_1$X$_1$ + c$_2$X$_2$ and closed in the sense which will be explained a little later.

random variables $X_1, X_2, ..., X_n, ...$ to the variable X has a very simple geometric interpretation: The distance between the point X of the space H and the variable point X_n tends to zero as $n \to \infty$.

The notion of a random process as a curve in the Hilbert space was first introduced by Kolmogorov (1940a). The advantage of the utilization of the theory of Hilbert spaces in investigating random functions was clearly demonstrated in Kolmogorov's classical paper (1941a); see also, e.g., Karhunen (1947), Yaglom (1962), and Rozanov (1967).

Chapter 2

[1]See Anderson (1971), Sec. 5.7.1, where it is proved that if $B(\tau) = 0$ for $|\tau| > n$, then numbers b_0, b_1, ..., b_n can be found which satisfy (2.5). Let us now consider the more general case of the complex-valued function $B(\tau)$. Assuming that $b_k = 0$ for $k > n$ and $k < 0$, we can write the corresponding generalization of the result (2.5) in the form

$$(2.1') \qquad B(t - s) = \sum_{k=-\infty}^{\infty} b_{k+t-s}\,\bar{b}_k = \sum_{j=-\infty}^{\infty} b_{t-j}\,\bar{b}_{s-j}.$$

The last relationship implies the possibility of representing the random sequence $X(t)$ in the form (2.4) by virtue of the so-called theorem on generalized spectral representation of random functions (see the closing part of Note 17 below).

Particular examples of moving average sequences (2.4) were considered by Yule (1921) and Slutsky (1927a). The general case was apparently first discussed in 1938 in the first edition of Wold's book (1954).

[2]A simple description of the classes of stationary random sequences $X(t)$ representable as two-sided or, respectively, one-sided moving averages of an infinite order was given by Kolmogorov (1941a); see also Doob (1953), Secs. X.8 and XII.4, or Anderson (1971), Secs. 7.5.2 and 7.6.3. This description is based on the spectral representation of the correlation function $B(\tau)$ of stationary sequences $X(t)$ considered in Sec. 13 of this book. According to Kolmogorov's results, a sequence $X(t)$ can be represented as a two-sided moving average of the form (2.8) if, and only if, it has a spectral density $f(\omega)$ (i.e. when its correlation function $B(\tau)$ can be represented as a Fourier integral of the form (2.67)). For $X(t)$ to be representable as a one-sided moving average (2.6) the spectral

density $f(\omega)$ must satisfy the condition $\int_{-\pi}^{\pi} |\log f(\omega)|\, d\omega < \infty$ (so that, in particular, $f(\omega)$ cannot vanish on any subinterval of the segment $[-\pi, \pi]$, no matter how small this subinterval is). These results are proved quite similarly to the proof of related results for moving average stationary processes considered in Sec. 26.2 of this book.

[3]If all the roots of (2.14) are larger than 1 in absolute value, then

$$(2.2') \qquad (1 + a_1 z + \dots + a_m z^m)^{-1} = \sum_{j=0}^{\infty} \gamma_j z^j, \quad \gamma_0 = 1,$$

where the series on the right-hand side converges for $|z| \leqslant 1$. It is easy to verify now that the sequence

$$X(t) = c \sum_{j=0}^{\infty} \gamma_j E(t - j)$$

will satisfy (2.13) and will be the unique stationary solution of this difference equation. Moreover, the solution of (2.13) corresponding to any initial values x_{t_0}, x_{t_0-1}, ..., x_{t_0-m+1} will tend to the stationary

solution as $t - t_0 \to \infty$ (see Anderson, 1971, Sec. 5.2.1; cf. also S. Goldberg, 1958, Brand, 1966, and Gel'fond, 1967).

Particular examples of autoregressive sequences (of orders $m \leqslant 4$) were considered by Yule (1927). The general autoregressive model of an arbitrary order was defined by G. Walker (1931), who extended Yule's approach.

[4]Short tables of random numbers can be found, e.g., at the end of the book by Buslenko et al. (1966); there exist also much more extensive tables. At present, however, sequences $e(1)$, $e(2)$, ..., imitating the realization of a sequence of independent random variables (so-called "pseudorandom" numbers) are much more often calculated anew each time according to definite rules on computers; see, e.g., Buslenko et al. (1966), or Rubinstein (1981).

[5]Equations (2.15) $-$ (2.15b) were first obtained (in a slightly different form) by Kendall (1944); see also Bartlett (1978), Sec. 5.2, Kendall and Stuart (1968), Sec. 47.19, Anderson (1971), Sec. 5.2.2, Box and Jenkins (1970), Sec. 3.2.4, Priestley (1981), Sec. 3.5.3. [In some of the indicated sources only the equations for $R(\tau) = B(\tau)/B(0)$ are given; the most usual method for deriving these equations is based on the use of the Yule–Walker equations (2.16) $-$ (2.16a).] A very simple method of the derivation of equations (2.15) $-$ (2.15b) is considered in Sec. 15 of this book.

[6]Suppose that γ_j are again determined by (2.2'). It is then easy to verify that

$$(2.3') \qquad X(t) = c \sum_{j=0}^{\infty} \sum_{k=0}^{n} \gamma_j b_k E(t - j - k), \quad b_0 = 1,$$

will be a stationary solution of the difference equations (2.17).

Using elementary arguments from the theory of linear difference equations it is also easy to show that the stationary solution of (2.17) is unique and that any initial value solution of this equation tends to the stationary solution, as $t - t_0 \to \infty$, no matter what the choice of the initial values is (see Anderson, 1971, Sec. 5.8.1; cf. also the textbooks on difference equations referred to in Note 3).

[7]Note that a process with a discrete spectrum can also be defined as a stationary random process whose correlation function has the form (2.39).   Indeed, it was shown by Slutsky (1938) that every stationary random process $X(t)$ with a correlation function of the

form (2.39) can be represented in the form (2.38) where $\langle X_k \bar{X}_l \rangle = 0$ for $k \neq l$.   This result of Slutsky is a special case of the general theorem on the spectral representation of stationary processes, which is treated in Sec. 8 of this book.

It should also be mentioned that in the case of a real process $X(t)$ the spectrum is sometimes defined as the whole set of numbers ω_k (which always includes $-\omega_k$ along with each ω_k), and sometimes as only the set of nonnegative values of ω_k (appearing in the real representation of $X(t)$ in a form similar to (2.35)).   To avoid misunderstanding, the set of all values of ω_k can be called a *two-sided spectrum* of the process $X(t)$, and the set of nonnegative values of ω_k is its *one-sided* spectrum.

[8]The variables E_n and E_m with $m \neq n$ are independent; therefore, $\langle E_n E_m \rangle = m_E^2$ for $m \neq n$ and $\langle E_n E_m \rangle = \langle E_n^2 \rangle = m_E^2 + \sigma_E^2$ for $m = n$. Also, the numbers of Poisson points t_k on two disjoint intervals are independent variables, and therefore $\langle dN(s)dN(s_1) \rangle = \langle dN(s) \rangle \langle dN(s_1) \rangle = \lambda^2 ds ds_1$ for $s_1 \neq s$. Moreover the contribution of the diagonal $s_1 = s$ to the value of the double integral over $ds ds_1$ of the bounded function of s_1 and s_2 is clearly equal to zero. Hence, it is easy to show that

$$\begin{aligned}
\langle X(t + \tau)X(t) \rangle &= \sum_n \sum_m \langle E_n E_m \rangle \langle \Gamma(t + \tau - t_n)\Gamma(t - t_m) \rangle \\
&= (m_E^2 + \sigma_E^2)\langle \sum_n \Gamma(t + \tau - t_n)\Gamma(t - t_n) \rangle \\
&\quad + m_E^2 \langle \sum_n \sum_{m \neq n} \Gamma(t + \tau - t_n)\Gamma(t - t_m) \rangle \\
&= (m_E^2 + \sigma_E^2) \int_{-\infty}^{\infty} \Gamma(t + \tau - s)\Gamma(t - s)\langle dN(s) \rangle \\
&\quad + m_E^2 \int_{-\infty}^{\infty}\int_{-\infty}^{\infty} \Gamma(t + \tau - s)\Gamma(t - s_1)\langle dN(s)dN(s_1) \rangle \\
&= \lambda(m_E^2 + \sigma_E^2) \int_{-\infty}^{\infty} \Gamma(t + \tau - s)\Gamma(t - s)ds \\
&\quad + \lambda^2 m_E^2 \int_{-\infty}^{\infty}\int_{-\infty}^{\infty} \Gamma(t + \tau - s)\Gamma(t - s_1)ds ds_1.
\end{aligned}$$

The obtained equation is equivalent to (2.44).

Equations (2.43) and (2.46) for the case where $\sigma_E^2 = 0$ are due to N. Campbell; see, e.g., Campbell (1909a,b) or Rice (1944). Formulae (2.43) to (2.46) are the simplest results of the correlation theory of random pulse processes, which is covered by an extensive literature (see, e.g., Middleton, 1960, Chap. II; Stratonovich, 1967, Sec. 6, Rytov, 1976, Chap. II; or Levin, 1974, Chap. II, where many additional references can be found).

[9]Let us mention one more example of radio engineering importance. In the physical theory of vacuum tubes it is shown that the total current $X(t)$ flowing through such a tube can, in many cases, be represented by (2.41), where $E_n = 1$ for all n (i.e. $m_E = 1$ and $\sigma_E = 0$) and $\Gamma(t)$ is described reasonably well by the expression

$$\Gamma(t) = \begin{cases} \alpha t & \text{for } 0 \leqslant t \leqslant T_0, \\ 0 & \text{for } t < 0 \text{ or } t > T_0. \end{cases}$$

(Here, $\alpha = 2\epsilon/T_0^2$, where $-\epsilon$ is the charge of an electron and T_0 is the time taken by an electron to travel from cathode to anode; see, e.g., Davenport and Root, 1958, Sec. 7.1. The time T_0 is, of course, quite short, usually of the order of 10^{-9} seconds.) Now, using (2.43), (2.45) and (2.46), we obtain

$$m = \lambda \alpha T_0^2/2 \ (= \lambda \epsilon), \quad \sigma_X^2 = \lambda \alpha^2 T_0^3/3 \ (= 4\lambda \epsilon^2/3T_0),$$

$$b(\tau) = \begin{cases} \dfrac{\lambda \alpha^2 T_0^3}{3} \left(1 - \dfrac{3}{2} \dfrac{|\tau|}{T_0} + \dfrac{1}{2} \dfrac{|\tau|^3}{T_0^3}\right) & \text{for } |\tau| \leqslant T_0, \\ 0 & \text{for } |\tau| > T_0 \end{cases}$$

where λ is the intensity of the Poisson point system $\{t_n\}$, i.e. the mean number of electrons emitted by the cathode per unit time.

Let us also give one simple generalization of (2.43) and (2.45). Consider a random pulse process $X(t)$ of the form

$$(2.4') \qquad X(t) = \sum_n E_n \Gamma \ (t - t_n; Z_n),$$

where $\{E_n\}$ and $\{t_n\}$ are the same random sequences as in (2.41), and $\{Z_n\}$ is another sequence of mutually independent, identically distributed random variables that are independent both of all the amplitudes $\{E_n\}$ and of the Poisson points $\{t_n\}$. In other words, we again consider the Poisson sequence of pulses, but now we assume that the shape of the nth pulse depends on the random parameter Z_n, i.e. the shape itself is random. Then, repeating the reasoning that led us to (2.43) $-$ (2.46), but including in it the additional averaging over the probability distribution of variables Z_n (which we shall, for simplicity, assume to have a probability density $p(z)$), we find that in this case

$$(2.5') \qquad m = \langle X(t) \rangle = \lambda m_E \langle \int_{-\infty}^{\infty} \Gamma(u;Z)du \rangle = \lambda m_E \int_{-\infty}^{\infty} \int_{-\infty}^{\infty} \Gamma(u;z)p(z)du\,dz,$$

$$(2.6') \quad \begin{aligned} b(\tau) &= \lambda \langle E^2 \rangle \langle \int_{-\infty}^{\infty} \Gamma(u + \tau;Z)\Gamma(u;Z)du \rangle \\ &= \lambda \langle E^2 \rangle \int_{-\infty}^{\infty} \int_{-\infty}^{\infty} \Gamma(u + \tau;z)\Gamma(u;z)p(z)du\,dz. \end{aligned}$$

Let us assume, e.g., that all the pulses are rectangular, of random width Z (i.e. $\Gamma(u;z) = 1$ for $-z \leqslant u \leqslant 0$ and $\Gamma(u;z) = 0$ for all the other values of u) and Z has an exponential probability distribution (i.e. $p(z) = \alpha e^{-\alpha z}$ for $z \geqslant 0$ and $p(z) = 0$ for $z < 0$). Then

$$\int_{-\infty}^{\infty} \Gamma(u;z)du = z, \quad \langle \int_{-\infty}^{\infty} \Gamma(u;Z)du \rangle = \langle Z \rangle = \alpha^{-1},$$

so that $m = \lambda \alpha^{-1} m_E$. Moreover,

$$\int_{-\infty}^{\infty} \Gamma(u + \tau;z)\Gamma(u;z)du = \begin{cases} z - |\tau| & \text{for } |\tau| \leqslant z, \\ 0 & \text{for } |\tau| > z, \end{cases}$$

so that

$$\langle \int_{-\infty}^{\infty} \Gamma(u + \tau;Z)\Gamma(u;Z)du \rangle = \alpha \int_{|\tau|}^{\infty} e^{-\alpha z} (z - |\tau|)dz = \alpha^{-1}e^{-\alpha|\tau|},$$

$$b(\tau) = \lambda \alpha^{-1}(m_E^2 + \sigma_E^2)e^{-\alpha|\tau|}.$$

Thus we obtain one more example of a random process having an exponential correlation function (cf. Sec. 2 and (2.48)).

A class of random pulse processes more general than processes (2.4') is considered in the paper by J. Rice (1977) containing many related references.

[10]This follows also from the easily verifiable statement that any function of the form (2.52) is positive definite (cf. Note 3 to the Introduction, especially (0.2'), (0.8'), and (0.9')). Moreover, the same result is implied by the statement related to (1.34) (see Vol. I, p. 60) and the fact that eq. (2.24) describes a correlation function.

[11]Example (2.54) and eq. (2.55) are due to Khinchin (1934), who used (2.55) to prove that any function of the form (2.53) is a correlation function of some stationary random process $X(t)$.

[12]It is well known (see, e.g., Wiener, 1933; or Titchmarsh, 1948; or Bochner, 1959; or R. Goldberg, 1961) that for any (even an unbounded) function $\Gamma(u)$, such that $\int_{-\infty}^{\infty} \Gamma^2(u)du < \infty$, there exists a Fourier transform

$$\gamma(\omega) = \frac{1}{\sqrt{2\pi}} \int_{-\infty}^{\infty} \Gamma(u)e^{-i\omega u}du$$

(where the integral must sometimes be understood in some special sense for unbounded functions $\Gamma(u)$). Note also that $\gamma(-\omega) = \overline{\gamma(\omega)}$, and $\int_{-\infty}^{\infty} |\gamma(\omega)|^2 d\omega < \infty$. Moreover, the following Parseval's formula is valid:

$$\int_{-\infty}^{\infty} \Gamma(u + \tau)\Gamma(u)du = \int_{-\infty}^{\infty} e^{i\omega\tau}|\gamma(\omega)|^2 d\omega = 2\int_{0}^{\infty} \cos\omega\tau|\gamma(\omega)|^2 d\omega.$$

Hence, (2.45) can be rewritten as

$$b(\tau) = \int_{-\infty}^{\infty} e^{i\omega\tau} dF(\omega) = \int_{0}^{\infty} \cos\omega\tau\, dG(\omega),$$

where

$$F(\omega) = \lambda(\,m_E^2 + \sigma_E^2)\int_{-\infty}^{\omega} |\gamma(\omega')|^2 d\omega'; \quad G(\omega) = 2F(\omega).$$

It is seen that (2.45) can always be rewritten in the form (2.52) or (2.53), where $F(\omega)$ and $G(\omega)$ are monotone nondecreasing functions.

[13]The formulated Khinchin's theorem is evidently a simple consequence of the Bochner theorem (see Note 3 to the Introduction) and of the fact that the correlation function is necessarily positive definite.

[13a]An amazing early note by Einstein (1914) (a summary of his talk at a meeting of the Swiss Physical Society in February 1914) became known to the author after he had finished writing this book. The note shows that Einstein in 1914 was already aware of the concepts of a stationary time series (random process) and its correlation function and that he at that time could prove that the Fourier transform of a correlation function $b(\tau)$ (if it exists) is necessarily nonnegative. This is the most important special case of the Khinchin (or Wiener-Khinchin) theorem; therefore there are grounds to call this theorem the *Einstein–Wiener–Khinchin theorem*.
The contents of Einstein's note (1914) is not exhausted by the statement above. In this note Einstein anticipated some important results of the modern spectral theory of stochastic processes; see Note 41 to Chap. 3 and also the recent comments to Einstein's paper (Yaglom 1985, 1986a,b). (All these comments begin with the text of Einstein's note.)

[14]Almost periodic functions $x(t)$ representable in the form

$$x(t) = \sum_{k} a_k e^{i\omega_k t}$$

are treated, in particular, in Secs. 23 and 24 of Wiener's book (1933); see also Riesz and Sz.-Nagy (1955), pp. 254–256. These functions are characterized by the following property: *For every $\epsilon > 0$ it is* *possible to find a positive number* $T_0 = T_0(\epsilon)$ *such that every interval of the real axis of length* T_0 *contains at least one number* t_0 *such that* $|x(t + t_0) - x(t)| < \epsilon$ *for any t.* Therefore it is clear that an almost periodic function $x(t)$ cannot go to zero as $t \to \infty$.

[15]See Wiener (1930) and (1933), Chap. IV. Representation of the function $x(t)$ (satisfying the condition (2.56)) in the form of superposition of complex harmonics $e^{i\omega t}$ can be written as

$$(2.7') \qquad x(t) = \lim_{\epsilon \to \infty} \lim_{T \to \infty} \int_{-T}^{T} e^{i\omega t} \frac{z(\omega + \epsilon) - z(\omega - \epsilon)}{2\epsilon} d\omega,$$

where the limit, as $T \to \infty$, is understood as the functional mean square limit. The function $z(\omega)$ here has the following relation to the "spectrum" $F(\omega)$:

$$(2.8') \qquad F(\omega_2) - F(\omega_1) = \lim_{\epsilon \to \infty} \int_{\omega_1}^{\omega_2} \frac{|z(\omega + \epsilon) - z(\omega - \epsilon)|^2}{2\epsilon} d\omega$$

(cf. Doob, 1949). Equations (2.7') and (2.8') can also be written in several other forms; see, e.g., Bass (1962) or Brillinger (1975), Sec. 3.9.

[16]A classical example of an important applied work using such a representation of a "noise" is the well-known paper by S. O. Rice (1944-45).

[17]First proofs of the spectral representation theorem were given by Kolmogorov (1940a, 1941a), Cramér (1942), Loève (1945-46), Blanc–Lapierre and Fortet (1946a,b), Karhunen (1947), and Maruyama (1949). There are also numerous books containing such a proof, e.g., Doob (1953), Blanc-Lapierre and Fortet (1953), Yaglom (1962), Loève (1963), Lévy (1965), Cramér and Leadbetter (1967), Rozanov (1967), Gihman and Skorohod (1969, 1974), Wentzell (1981), Lamperti (1977), Priestley (1981), Rosenblatt (1985), and many others. The most direct method leading to the proof of the indicated theorem is apparently based on the use of the inversion of the Fourier–Stieltjes integral on the right-hand side of (2.61). It is known that, at least in some cases, there exists an inversion formula for Fourier–Stieltjes intergrals of numerical (nonrandom) functions (see, e.g., the inversion formula (0.6') for characteristic functions in Note 3 to the Introduction). By analogy with this formula we can suppose that if the representation (2.61) holds, then the function $Z(\omega)$ is given in terms of $X(t)$ by the formula

$$(2.9') \qquad Z(\omega) = \lim_{T \to \infty} \frac{1}{2\pi} \int_{-T}^{T} \frac{e^{-i\omega t} - 1}{-it} X(t) dt + Z_0$$

(where Z_0 is an arbitrary constant, which can be equated to zero). Now let $X(t)$ be a stationary random process and define the random function $Z(\omega)$ by (2.9'). In order to make this equation meaningful, we still have to show that the integral from $-T$ to T appearing in (2.9') exists in the mean square sense (this can be easily done by using the results of Sec. 4) and that, as $T \to \infty$, this integral converges in the mean to a definite random variable $Z(\omega)$. This last assertion can be proved with the aid of Khinchin's formula (2.52) for the correlation function $B(\tau)$. In fact, from (2.52) and (1.55), generalized to the case of complex $f(t)$ and $X(t)$, it follows that for $T' > T$,

$$\left\langle \left| \int_{-T'}^{T'} \frac{e^{-i\omega t} - 1}{-2\pi i t} X(t)dt - \int_{-T}^{T} \frac{e^{-i\omega t} - 1}{-2\pi i t} X(t)dt \right|^2 \right\rangle = \left\langle \left| \int_{T<|t|<T'} \frac{e^{-i\omega t} - 1}{-2\pi i t} X(t)dt \right|^2 \right.$$

$$= \int_{T<|t|<T'} \int_{T<|s|<T'} \frac{e^{-i\omega t} - 1}{-2\pi i t} \frac{e^{i\omega s} - 1}{2\pi i s} \langle X(t)\overline{X(s)} \rangle dt\,ds$$

$$(2.10') \qquad = \int_{T<|t|<T'} \int_{T<|s|<T'} \int_{-\infty}^{\infty} \frac{e^{-i\omega t} - 1}{-2\pi i t} \frac{e^{i\omega s} - 1}{2\pi i s} e^{i\omega'(t-s)} dt\,ds\,dF(\omega')$$

$$= \int_{-\infty}^{\infty} \left| \int_{T<|t|<T'} \frac{e^{-i\omega t} - 1}{-2\pi i t} e^{i\omega' t} dt \right|^2 dF(\omega').$$

For simplicity, we shall assume temporarily that the function $F(\omega)$ is continuous. Then, by virtue of (2.10') the proof of the existence of the limit (2.9') is a consequence of the existence of the following limit involving only numerical functions:

$$\Psi(\omega',\omega) = \lim_{T\to\infty} \frac{1}{2\pi} \int_{-T}^{T} \frac{e^{i\omega' t} - e^{i(\omega'-\omega)t}}{it} dt$$

$$(2.11') \qquad = \frac{1}{\pi} \int_{0}^{\infty} \frac{\sin\omega' t}{t} dt - \frac{1}{\pi} \int_{0}^{\infty} \frac{\sin(\omega'-\omega)t}{t} dt$$

or, more precisely, a consequence of the fact that $\Psi(\omega',\omega)$ exists and satisfies the relation

$$(2.12') \qquad \lim_{T\to\infty} \int_{-\infty}^{\infty} |\Psi(\omega',\omega) - \Psi_T(\omega',\omega)|^2 dF(\omega') = 0,$$

where

$$\Psi_T(\omega',\omega) = \frac{1}{2\pi} \int_{-T}^{T} \frac{e^{i\omega' t} - e^{i(\omega'-\omega)t}}{it} dt.$$

However it is well known that

$$(2.13') \qquad \frac{1}{\pi} \int_{0}^{\infty} \frac{\sin\omega t}{t} dt = \begin{cases} 1/2 & \text{for } \omega > 0, \\ 0 & \text{for } \omega = 0, \\ -1/2 & \text{for } \omega < 0. \end{cases}$$

It follows from this that the limit (2.11') exists and is equal to

$$(2.14') \qquad \Psi(\omega',\omega) = \begin{cases} 1 & \text{for } 0 < \omega' < \omega, \\ 1/2 & \text{for } \omega' = \omega > 0 \text{ or } \omega' = 0, \omega > 0, \\ 0 & \text{for } \omega' > 0, \omega < \omega', \text{or } \omega' = \omega = 0, \text{or } \omega' < 0, \omega > \omega, \\ -1/2 & \text{for } \omega' = \omega < 0, \quad \text{or } \omega' = 0, \omega < 0, \\ -1 & \text{for } 0 > \omega' > \omega, \end{cases}$$

and, as is easy to see, relation (2.12') is also fulfilled. Hence, (2.9') actually defines a random function $Z(\omega)$.

The spectral representation theorem will be proved if we show that the function $Z(\omega)$ has the properties (a') and (b') and satisfies relation (2.61). If $\langle X(t)\rangle = 0$ and Z_0 is chosen to have zero mean, the fact that $Z(\omega)$ has the property (a') is quite obvious. Let us verify now that it has the property (b') and satisfies (2.61). Arguing as we did in deriving (2.10') and using (2.11'), we can easily show that

$$(2.15') \quad \begin{aligned} &\langle [Z(\omega_2) - Z(\omega_1)]\overline{[Z(\omega_4) - Z(\omega_3)]}\rangle \\ &= \int_{-\infty}^{\infty} \Psi(\omega - \omega_1, \omega_2 - \omega_1)\overline{\Psi(\omega - \omega_3, \omega_4 - \omega_3)}\,dF(\omega). \end{aligned}$$

It follows from (2.15') and (2.14') that for $\omega_2 > \omega_1$

$$(2.16') \quad \langle |Z(\omega_2) - Z(\omega_1)|^2\rangle = F(\omega_2) - F(\omega_1)$$

and for $\omega_1 < \omega_2 \leqslant \omega_3 < \omega_4$

$$(2.17') \quad \langle [Z(\omega_2) - Z(\omega_1)]\overline{[Z(\omega_4) - Z(\omega_3)]}\rangle = 0,$$

i.e., $Z(\omega)$ has the property (b'). Moreover, again using (2.10'), (2.11'), and (2.52), we readily find that

$$\langle X(t)\overline{[Z(\omega_2) - Z(\omega_1)]}\rangle = \int_{-\infty}^{\infty} \Psi(\omega_2 - \omega, \omega_2 - \omega_1)e^{i\omega t}\,dF(\omega).$$

Hence, for $\omega_2 > \omega_1$

$$(2.18') \quad \langle X(t)\overline{[Z(\omega_2) - Z(\omega_1)]}\rangle = \int_{\omega_2}^{\omega_1} e^{i\omega t}\,dF(\omega).$$

Using (2.16') and (2.17') it is easy to show that the integral on the right-hand side of (2.61) (understood as is explained in (2.62)) does exist. Then, from (2.18') and (2.62), we obtain

$$(2.19') \quad \langle X(t)\overline{\int_{-\infty}^{\infty} e^{i\omega s}\,dZ(\omega)}\rangle = \lim_{T\to\infty} \int_{-T}^{T} e^{i\omega(t-s)}\,dF(\omega) = B(t - s),$$

and it also follows from (2.16'), (2.17'), and (2.62) that

$$(2.20') \quad \langle \int_{-\infty}^{\infty} e^{i\omega t}\,dZ(\omega)\overline{\int_{-\infty}^{\infty} e^{i\omega s}\,dZ(\omega)}\rangle = \int_{-\infty}^{\infty} e^{i\omega(t-s)}\,dF(\omega) = B(t - s).$$

Equations (2.19'), (2.20'), and (2.52) imply the relation

$$(2.21') \quad \langle \left| X(t) - \int_{-\infty}^{\infty} e^{i\omega t}\,dZ(\omega)\right|^2\rangle = 0,$$

which is clearly equivalent to (2.61). This completes the proof of the existence of the spectral representation (2.61) in the case where the function $F(\omega)$ is continuous.

If the function $F(\omega)$ is not continuous, but contains jumps (which are the only possible discontinuities of the nondecreasing function

$F(\omega)$), then, by virtue of discontinuity of the function (2.14'), some of the above relations become inapplicable to discontinuity points of $F(\omega)$. In this case we must define $Z(\omega)$ by using (2.9') only at the points of continuity of $F(\omega)$, while at the point of discontinuity of $F(\omega)$ we set $Z(\omega) = [Z(\omega - 0) + Z(\omega + 0)]/2$, where $Z(\omega - 0)$ and $Z(\omega + 0)$ are defined in the usual way. It is not hard to verify that then all the formulae written above remain in force. Therefore, our proof of the existence of the spectral representation goes through in this case too.

There are also many other proofs of the spectral representation theorem for stationary random processes and almost all such proofs have their specific merits. Since the theorem is very important, we shall consider one more way to prove it. The new proof does not include the derivation of an explicit formula for $Z(\omega)$, but in principle it is very simple and permits an important generalization, which will also be outlined at the end of this Note.

Begin again with the case where $F(\omega)$ is a continuous function of ω. Let us use Khinchin's formula (2.52) for the correlation function $B(\tau)$ and consider, together with the Hilbert space H_X introduced at the end of Sec. 5, another Hilbert space $L_2(F)$ consisting of all complex-valued functions $\varphi(\omega)$, $-\infty < \omega < \infty$, satisfying condition

$$(2.22') \qquad \int_{-\infty}^{\infty} |\varphi(\omega)|^2 dF(\omega) < \infty.$$

The norm $\|\varphi(\omega)\|_L$ of the vector $\varphi(\omega)$ from $L_2(F)$ is defined as the square root of the left-hand side of (2.22'), and the scalar product of two such vectors $\varphi(\omega)$ and $\psi(\omega)$ is defined correspondingly as

$$(2.23') \qquad (\varphi(\omega), \psi(\omega))_L = \int_{-\infty}^{\infty} \varphi(\omega)\overline{\psi(\omega)} dF(\omega).$$

We shall set up a one-to-one isometric correspondence between these two Hilbert spaces and deduce from this the existence of the spectal representation.

We begin with the subset $H_X^{(1)}$ of H_X consisting of all finite linear combinations of the variables $X(t)$ and the subset $L_2^{(1)}(F)$ of $L_2(F)$ consisting of all trigonometric polynomials of the form

$$\sum_{k=1}^{m} c_k e^{it_k \omega}.$$

Define the correspondence between them by letting

$$(2.24') \qquad \sum_{k=1}^{m} c_k X(t_k) \longleftrightarrow \sum_{k=1}^{m} c_k e^{it_k \omega}$$

(so that $X(t) \longleftrightarrow e^{it\omega}$). It is easily verified, by using (2.52), that the correspondence (2.24') is isometric, i.e. the scalar product of any two elements of $H_X^{(1)}$ is equal to the scalar product of the corresponding

elements of $L_2^{(1)}(F)$:

$$\left\langle\left[\sum_{k=1}^{m} c_k X(t_k)\right]\overline{\left[\sum_{l=1}^{n} d_l X(t_l)\right]}\right\rangle = \int_{-\infty}^{\infty}\left[\sum_{k=1}^{n} c_k e^{it_k\omega}\right]\overline{\left[\sum_{l=1}^{n} d_l e^{it_l\omega}\right]}dF(\omega).$$

It is also clear that (2.24') is a linear mapping of $H_X^{(1)}$ onto $L_2^{(1)}(F)$. It follows at once that if $Y_1, Y_2, ..., Y_n, ...$ is a sequence of elements of $H_X^{(1)}$ converging (in the mean), as $n \to \infty$, to a vector Y of H_X, then the corresponding sequence $\varphi_1(\omega), \varphi_2(\omega), ..., \varphi_n(\omega)$ of elements of $L_2^{(1)}(F)$ converges, as $n \to \infty$, to some element $\varphi_Y(\omega)$ of $L_2(F)$ (i.e. $\|\varphi_Y(\omega) - \varphi_n(\omega)\|_L^2 \to$ 0 as $n \to \infty$). The element $\varphi_Y(\omega)$ of $L_2(F)$ depends only on Y but not on the choice of the sequence $Y_1, ..., Y_n, ...$ in $H_X^{(1)}$; therefore, it is natural to extend the correspondence (2.24') to the limit elements by putting

(2.25') $Y \longleftrightarrow \varphi_Y(\omega).$

It is clear that the extended correspondence (2.25') is a linear and isometric mapping of the whole Hilbert space H_X— the closure of the set $H_X^{(1)}$ — onto the closure of the set $L_2^{(1)}(F)$ of trigonometric polynomials in $L_2(F)$. However, it is easy to see that the closure of $L_2^{(1)}(F)$ in $L_2(F)$ coincides with the whole space $L_2(F)$.* Hence (2.25') is a one-to-one isometric correspondence between the Hilbert spaces H_X and $L_2(F)$.

In the first proof of the spectral representation theorem discussed in this Note we used the form (2.61) of the spectral representation. Now it will be more convenient for us to use the equivalent form (2.58) of this representation and to introduce the random interval function $Z(\Delta\omega)$ instead of the random point function $Z(\omega)$. Let $\chi_{\Delta\omega}(\omega)$ be the so-called *indicator function* of the interval $\Delta\omega$, i.e. $\chi_{\Delta\omega}(\omega) = 1$ for $\omega_1 < \omega \leqslant \omega_2$ and $\chi_{\Delta\omega}(\omega) = 0$ for $\omega \leqslant \omega_1$ or $\omega > \omega_2$, where ω_1 and ω_2 are the left and right endpoints of $\Delta\omega$. It is clear that the function $\chi_{\Delta\omega}(\omega)$ belongs to $L_2(F)$ and so it must correspond to some element of H_X. Let us denote this element by $Z(\Delta\omega)$.** To prove the spectral representation theorem we only need to show that the random interval function $Z(\Delta\omega)$ has the properties (a), (b), and (c) and satisfies the relation (2.58).

Since all the elements of H_X are random variables with mean value zero (if $\langle X(t)\rangle = 0$), it is obvious that $Z(\Delta\omega)$ has the property (a). The

---

*This follows from the completeness of the set of trigonometric polynomials in $L_2(F)$, i.e. from the non-existence of an element $\psi(\omega)$ of $L_2(F)$ such that

$$\int_{-\infty}^{\infty} e^{-it\omega}\psi(\omega)dF(\omega) = 0 \text{ for all t and } \int_{-\infty}^{\infty}|\psi(\omega)|^2 \neq 0.$$

**Hence to determine $Z(\Delta\omega)$ we must only find the sequence

$$\sum_{k=1}^{m_n} c_k^{(n)} e^{it_k^{(n)}\omega}, \quad n = 1,2, ... ,$$

of trigonometric polynomials tending to $\chi_{\Delta\omega}(\omega)$ in $L_2(F)$. Then obviously $Z(\Delta\omega)$

$$= \lim_{n\to\infty} \sum_{k=1}^{m_n} c_k^{(n)} X(t_k^{(n)}) \text{ where the limit is understood in the mean square sense.}$$

property (c) of $Z(\Delta\omega)$ follows from the linearity of the correspondence $(2.24') - (2.25')$ and from the fact that $X_{\Delta_1\omega}(\omega) + X_{\Delta_2\omega}(\omega) = X_{\Delta_1\omega+\Delta_2\omega}(\omega)$

for any pair of adjacent intervals $\Delta_1\omega$ and $\Delta_2\omega$. Finally, the isometric property of $(2.25')$ implies the result

$$(2.26') \quad \langle Z(\Delta_1\omega)\overline{Z(\Delta_2\omega)}\rangle = \int_{-\infty}^{\infty}X_{\Delta_1\omega}(\omega)\overline{X_{\Delta_2\omega}(\omega)}dF(\omega),$$

which is valid for any two intervals $\Delta_1\omega$ and $\Delta_2\omega$. Letting $\Delta_1\omega$ and $\Delta_2\omega$ be disjoint, we arrive at property (b). If, however, $\Delta_1\omega = \Delta_2\omega = \Delta\omega$ then $(2.26')$ takes the form

$$(2.27') \quad \langle|Z(\Delta\omega)|^2\rangle = F(\omega_2) - F(\omega_1)$$

where ω_2 and ω_1 are the endpoints of $\Delta\omega$.
   It also follows from the linearity of the considered correspondence that

$$(2.28') \quad \sum_k a_k Z(\Delta_k\omega) \longleftrightarrow \sum_k a_k X_{\Delta_k\omega}(\omega).$$

If we now choose $\Delta_1\omega$, ..., $\Delta_n\omega$ to be a partition of the whole axis $-\infty < \omega < \infty$ and equate $a_1, ..., a_n$ to the values of a given continuous function $\varphi(\omega)$ at some points of the intervals $\Delta_1\omega$, ..., $\Delta_n\omega$, then $\sum_{k=1}^{n} a_k Z(\Delta_k\omega)$ will be an integral sum corresponding to $\int_{-\infty}^{\infty}\varphi(\omega)Z(d\omega)$, and $\sum_{k=1}^{n} a_k X_{\Delta_k\omega}(\omega)$ will be a step-function approximation to $\varphi(\omega)$. Hence it is clear that

$$(2.29') \quad \int_{-\infty}^{\infty}\varphi(\omega)Z(d\omega) \longleftrightarrow \varphi(\omega)$$

for any continuous function $\varphi(\omega)$ from $L_2(F)$.* Choosing the function $e^{it\omega}$ as $\varphi(\omega)$ in $(2.29')$ and taking into account that $X(t) \longleftrightarrow e^{it\omega}$ by virtue of $(2.24')$, we obtain the desired spectral representation of $X(t)$:

$$\int_{-\infty}^{\infty}e^{it\omega}Z(d\omega) = X(t).$$

---

*In accordance with the definition of the integral $\int_{-\infty}^{\infty}\varphi(\omega)Z(d\omega)$ by equations of the form (2.59) and (2.60), it is reasonable to begin with the partition $\Delta_1\omega$, ..., $\Delta_n\omega$ of the finite interval $[a,b]$. Then we obtain the relation $\int_a^b \varphi(\omega)Z(d\omega) \longleftrightarrow X_{[a,b]}(\omega)\varphi(\omega)$, where $X_{[a,b]}(\omega)$ is the indicator function of $[a,b]$. Letting $a \to -\infty$, $b \to \infty$ in the last relation we obtain $(2.29')$.

   Note that in fact relation $(2.29')$ also holds for discontinuous functions from $L_2(F)$, but this fact is of no importance to us.

This completes the proof of the spectral representation theorem when $F(\omega)$ is a continuous function.

If the function $F(\omega)$ has discontinuity points ("jumps"), all the preceding arguments relating to $Z(\Delta\omega)$ remain valid for $\Delta\omega$ being continuity intervals of F, both ends of which are continuity points of $F(\omega)$. Thus, we can assume that the values of $Z(\Delta\omega)$ are determined for all continuity intervals $\Delta\omega$ and now we must only determine these values for intervals $\Delta\omega$ in the cases where at least one endpoint of $\Delta\omega$ coincides with the discontinuity point of F. This is very easy to do by applying a procedure similar to that applied above in this Note (at the end of the first proof of the theorem) to the point function $Z(\omega)$. Note also that we can even restrict ourselves to specification of $Z(\Delta\omega)$ for continuity intervals of F, agreeing that the limit on the right-hand side of (2.60) is taken only with respect to partitions $a = \omega_0 < \omega_1 < ... < \omega_{n-1} < \omega_n = b$, where none of the points ω_k, $k = 0$, 1, ..., n, coincide with the discontinuity point of $F(\omega)$.

To conclude this Note we shall consider briefly the possibility of extending the second proof of the spectral representation theorem for stationary random processes which we have just outlined to some other classes of random functions. Let us consider a random function $X(t)$ defined on an arbitrary set T of points t and having mean value zero and a bounded correlation function $B(t,s) = \langle X(t)\overline{X(s)}\rangle$ representable in the form

$$(2.30')\quad B(t,s) = \int_A \varphi(t,a)\overline{\varphi(s,a)}F(da).$$

Here, $\varphi(t,a)$ is a function (generally speaking, complex-valued) of the point t in the set T and of an additional parameter a in another set A, and F is a *measure* on A. This means, of course, that F is a numerical nonnegative function of subsets Δa of the set A, defined for a sufficiently wide class of such subsets and additive, i.e. such that $F(\Delta_1 a) + F(\Delta_2 a) = F(\Delta_1 a + \Delta_2 a)$, where $\Delta_1 a + \Delta_2 a$ is a union of disjoint subsets $\Delta_1 a$ and $\Delta_2 a$. In this case, under appropriate (sufficiently wide) regularity conditions, the integral (2.30') with respect to the measure $F(da)$ can be defined similarly to the usual definition of the Stieltjes integral.* Now we can consider the isometric correspondence

$$X(t) \longleftrightarrow \varphi(t,a)$$

between the elements $X(t)$ of the Hilbert space H_X and the elements $\varphi(t,a) = \varphi_t(a)$ of the Hilbert space $L_2(F)$ of the function $\varphi(a)$ on A such

---

*We shall not spend time on the formulation of the quite general conditions on the set A and the measure F making the concept of an integral over A with respect to F(da) meaningful, and on the related mathematical subtleties (see, e.g., Halmos, 1950, or Munroe, 1953). Let us only emphasize that in principle this concept is rather simple and intuitively clear.

that $\int_A |\varphi(a)|^2 F(da) < \infty$.   Then we extend this correspondence in the usual way, linearly and continuously, to an isometry between the set of all linear combinations of random variables $X(t)$ and all the (mean square) limits of the sequences of such combinations — which is the whole Hilbert space H_X — and the set of all linear combinations of the functions $\varphi_t(a) = \varphi(t,a)$ and their limits in $L_2(F)$.   Let us assume, first of all, that by using linear combinations of the form $\sum_{k=1}^{m} c_k \varphi(t_k,a)$ we can approximate, to any degree of accuracy (in the sense of mean square with respect to the measure $F(da)$) the indicator function $\chi_{\Delta a}(a)$ of any subset Δa from the class of the subsets used in the construction of the integrals of the form (2.30'). Then our second proof of the spectral representation theorem can be repeated without any difficulty for the random function $X(t)$ with a correlation function (2.30'). When applied to such a function $X(t)$, it turns into the proof of the existence of a *generalized spectral representation* of the form

$$(2.31')\quad X(t) = \int_A \varphi(t,a) Z(da).$$

Here, Z is a complex-valued random additive function of the subset Δa such that $\langle Z(\Delta a)\rangle = 0$, $\langle Z(\Delta_1 a)\overline{Z(\Delta_2 a)}\rangle = 0$ for disjoint subsets $\Delta_1 a$ and $\Delta_2 a$, and $\langle |Z(\Delta a)|^2\rangle = F(\Delta a)$.   The integral with respect to $Z(da)$ is defined similarly to the integral (2.58) with respect to $Z(d\omega)$. Moreover, it is also possible to show, with the aid of a simple special method, that the possibility of approximating all the indicator functions $\chi_{\Delta a}(a)$ by linear combinations of the form $\Sigma c_k \varphi(t_k,a)$ is by no means necessary for the existence of the representation (2.31').* Thus, the *representability of*

---

*The above-mentioned method is as follows: If the linear combinations of the function $\varphi(t,a)$ of a are not sufficient for approximating all the functions $\chi_{\Delta a}(a)$, then we supplement these functions by some additional functions $\psi(r,a)$, where r runs through some new set R. These functions are selected in such a way that the linear combinations $\Sigma_{k=1}^n c_k \varphi(t_k,a) + \Sigma_{j=1}^m d_j \psi(r_j,a)$ are already sufficient for approximating

all the functions $\chi_{\Delta a}(a)$ and $\int_A \psi(r,a)\overline{\varphi(t,a)}F(da) = 0$ for all r in R and t in T.

After this we determine a new random function $X_1(r)$ on R such that $\langle X_1(r)\rangle = 0$, $\langle X_1(r_1)\overline{X_1(r_2)}\rangle = \int_A \psi(r_1,a)\psi(r_2,a)F(da)$ and $\langle X_1(r)\overline{X(t)}\rangle = 0$ for all r in R and t in T. It is easy to show that such a function $X_1(r)$ always exists (and can even be chosen to be Gaussian). Therefore, the proof of the existence of the representation (2.31') for the random function X(t) on T can be reduced to the similar proof related to the random function on the larger set T + R (the union of T and R) which takes values X(t) on T and values $X_1(r)$ on R, and for this function the stated condition of the approximability of functions $\chi_{\Delta a}(a)$ is already fulfilled.

the correlation function B(t,s) of X(t) *in the form* (2.30') *always implies the possibility of representing the random function* X(t) *in the form* (2.31') (where, however, the values of the random measure $Z(\Delta a)$ do not necessarily belong to the Hilbert space H_X). The italicized statement is due to Karhunen (1947) and is also proved in the books by Grenander and Rosenblatt (1956a), Sec. 1.4, and Gihman and Skorohod (1974), Sec. IV.5; see also M. M. Rao (1985). This statement is sometimes called *the theorem on generalized spectral representation* (or *orthogonal representation) of random functions.* The set A appearing in (2.30') and (2.31') may also be discrete, and then the integrals on the right-hand sides of these equations naturally turn into ordinary sums. (We have already referred to this particular case of Karhunen's theorem in Note 1 to this chapter, where it was used to prove that the representability of the correlation function $B(t,s) = B(t - s)$ in the form (2.1') with $b_k = 0$ for $k > n$ and $k < 0$ implies the representability of the corresponding random sequence $X(t)$ in the form (2.4).)

[18]Fourier integrals are treated, e.g., in the monographs by Wiener, 1933; Titchmarsh, 1948; Bochner, 1959; R. Goldberg, 1961, and in many other books (including a number of advanced calculus textbooks). In the case at hand, however, many theorems on Fourier integrals acquire a simpler-than-usual form, because the functions $B(\tau)$ are continuous and positive definite, and therefore they cannot be too irregular.

[19]Let us recall that any bounded monotone nondecreasing function $F(\omega)$ can always be represented as a sum of the form $F(\omega) = \overset{\smile}{F}_I(\omega) + F_{II}(\omega) + F_{III}(\omega)$, where $\overset{\smile}{F}_I(\omega)$ is absolutely continuous (i.e. $\overset{\smile}{F}_I(\omega) = \int_{-\infty}^{\omega} f(\omega')d\omega'$, $f(\omega') \geqslant 0$), $F_{II}(\omega)$ is a step-function, and $F_{III}(\omega)$ is a "singular component" (i.e. a continuous, but not absolutely continuous, monotone function whose derivative vanishes at almost all points); see Note 2 to the Introduction. Hence, a continuous component $F_I(\omega)$ of $F(\omega)$ is, generally speaking, represented by the sum $\overset{\smile}{F}_I(\omega) + F_{III}(\omega)$. It is not difficult to show that

$$B_I(\tau) = \int_{-\infty}^{\infty} e^{i\tau\omega} dF_I(\omega)$$

necessarily tends to zero as $|\tau| \to \infty$ if $F_I(\omega) = \overset{\smile}{F}_I(\omega)$ (i.e., if a singular component is missing); see, e.g., Cramer, 1962, Chap. IV. Moreover, the nonzero singular components are in fact highly pathological and they never appear in spectral distribution functions relating to applied problems of practical interest.

[20]The *discrete spectrum* of the process $X(t)$ is sometimes called the *point spectrum*, or *line spectrum*. A process $X(t)$ having both discrete

and continuous spectrum components is usually said to have a *mixed spectrum.*   Note also that in the case of a real $X(t)$ we must distinguish a *two-sided spectrum* — the subset of the whole axis $-\infty < \omega < \infty$ symmetric with respect to the point $\omega = 0$ — and a *one-sided spectrum*, which is strictly nonnegative.

[21]As is well known, the first example of a nowhere differentiable function, which attracted the attention of the scientists, was constructed by the German mathematician K. Weierstrass in the 1870s.* This example caused general confusion at first; many mathematicians insisted for a long time that it was incorrect, while others denied that nowhere differentiable functions had the right to be called functions. Gradually, however, the mathematicians got used to the fact that nowhere differentiable functions do exist, though the physicists refused to admit it for a long time and perceived such functions as an ugly product of mathematical fantasy which had nothing to do with the real universe (i.e. they proceeded from the principle that "all functions are differentiable in physics").

Nowhere differentiable functions apparently first appeared in a physical problem in the Brownian motion theory when Wiener proved that the Einstein–Smoluchowski model of such a motion implies that the trajectory of a Brownian particle is nowhere differentiable with probability one.   However, one could assume that non-differentiability in this case is the consequence of the inadmissible idealization (because a more exact theory of Uhlenbeck and Ornstein, which takes into account the inertia of Brownian particles, leads to the conclusion that their trajectories are everywhere differentiable).   Therefore, it is worth noting that in the spectral theory of stationary random processes, nowhere differentiable functions $Z(\omega)$ arise quite naturally and one can get rid of them only by abandoning the stationarity condition, which has a clear physical meaning and alone makes the theory under consideration simple and suitable for use in practice.

A number of other, rather unexpected, examples of the application of nowhere differentiable functions were indicated recently by Mandelbrot (1977, 1982). This author showed that in fact nowhere differentiable functions give natural models for many spatial patterns of applied origin related not only to various parts of physics but also to such fields as geography and geomorphology.

[22]The derivation of (2.91) in Vol. I is not rigorous since, in principle, the component $B_{\mathrm{I}}(\tau)$ of $B(\tau)$ does not necessarily tend to zero as $|\tau| \to \infty$. (This has already been mentioned on p. 109 of Vol. I.) To obtain a rigorous proof let us note that in the real case, where $B(\tau) = B(-\tau)$,

---

*A similar example had been in fact given by the Czech mathematician and philosopher B. Bolzano in a paper written not later then 1830 (but not published until 1930).

$$\frac{1}{T}\int_0^T B(\tau)d\tau = \frac{1}{2T}\int_{-T}^T B(\tau)d\tau = \frac{1}{2T}\int_{-T}^T\int_{-\infty}^\infty e^{i\tau\omega}dF(\omega)d\tau$$

$$= \frac{1}{2T}\int_{-\infty}^\infty \frac{e^{iT\omega} - e^{-iT\omega}}{i\omega}\,dF(\omega) = \int_{-\infty}^\infty \varphi_T(\omega)dF(\omega),$$

where $\varphi_T(\omega) = \sin\omega T/\omega T$. It is clear, that $\varphi_T(0) = 1$, $|\varphi_T(\omega)| \leqslant 1$ for all ω and T and $|\varphi_T(\omega)| \to 0$, as $T \to \infty$, provided that $\omega \neq 0$. Moreover, $\Delta F(0) = \lim_{h\to 0}[F(h) - F(-h)]$, where $F(h) - F(-h)$ is a monotonically

nondecreasing function of h. Now we choose h so that $F(h) - F(-h) < \Delta F(0) + \epsilon$ and assume

$$\int_{-\infty}^\infty \varphi_T(\omega)dF(\omega) = \int_{-\infty}^{-h}\varphi_T(\omega)dF(\omega) + \int_{-h}^h \varphi_T(\omega)dF(\omega) + \int_h^\infty \varphi_T(\omega)dF(\omega)$$

$$= I_T^{(1)} + I_T^{(2)} + I_T^{(3)}.$$

For any $\epsilon > 0$ there obviously exists $T_0 = T_0(h,\epsilon)$ such that $|\varphi_T(\omega)| < \epsilon$ for $T > T_0$ and $|\omega| > h$. Since $\int_{-\infty}^{-h}dF(\omega) < B(0)$ and $\int_h^\infty dF(\omega) < B(0)$, it is clear that $I_T^{(1)} \to 0$ and $I_T^{(3)} \to 0$ as $T \to \infty$.

In addition, $I_T^{(2)} < F(h) - F(-h) < \Delta F(0) + \epsilon$, since $\varphi_T(\omega) < 1$. On the other hand, for any T and h there exists an $h_T < h$ such that $|\varphi_T(\omega)| > 1 - \epsilon$ for $|\omega| < h_T$. Hence, $I_T^{(2)} > (1 - \epsilon)[F(h_T) - F(-h_T)] \geqslant (1 - \epsilon)\Delta F(0)$. Thus, $I_T^{(2)} \to \Delta F(0)$ as $T \to \infty$.

It is easy to prove in a similar manner that

$$(2.32')\quad \frac{1}{T}\int_{-T/2}^{T/2}X(t)dt \to \Delta Z(0) \text{ as } T \to \infty,$$

where $\Delta Z(0)$ is a jump of a random function $Z(\omega)$ at the point $\omega = 0$. The quantity $\Delta Z(0)$ necessarily includes the nonrandom term $m = \langle X(t)\rangle$; it may or may not include also a purely random term (with mean value zero), depending on whether the jump $\Delta \overset{\circ}{F}(0)$ is positive or equal to zero, where $\overset{\circ}{F}(\omega)$ is the spectral distribution function of the centered process $\overset{\circ}{X}(t) = X(t) - \langle X(t)\rangle$.

[23]In fact, since $B(\tau)$ is a correlation function, $B^2(\tau)$ is also a correlation function. (See Sec. 4, where it is shown that a product of two correlation functions is always a correlation function.) According to the well-known "convolution theorem" of the theory of Fourier integrals, if $F_1(\omega)$ and $F_2(\omega)$ are the spectral distribution functions corresponding to correlation functions $B_1(\tau)$ and, respectively, $B_2(\tau)$, then the spectral distribution function

$$(2.33')\quad F_{12}(\omega) = \int_{-\infty}^\infty F_1(\omega - \omega')dF_2(\omega')$$

corresponds to the correlation function $B_1(\tau)B_2(\tau)$. (To prove the "convolution theorem" it is sufficient to check that $\int_{-\infty}^{\infty} e^{i\tau\omega} dF_{12}(\omega) = B_1(\tau)B_2(\tau)$ if $\int_{-\infty}^{\infty} e^{i\tau\omega} dF_i(\omega) = B_i(\tau)$ for $i = 1, 2$.) Hence, to the correlation function $B^2(\tau)$ corresponds the spectral distribution function

$$F^{(2)}(\omega) = \int_{-\infty}^{\infty} F(\omega - \omega') dF(\omega') = \int_{-\infty}^{\infty} [F_I(\omega - \omega')$$
$$+ F_{II}(\omega - \omega')][dF_I(\omega') + dF_{II}(\omega')]$$

where $F_I(\omega)$ and $F_{II}(\omega)$ are continuous and discrete (having the form of a step-function) components of $F(\omega)$. Equation (2.91) now implies that $\beta_1 = \Delta F^{(2)}(0)$.

It is clear that $\int_{-\infty}^{\infty} F_I(\omega - \omega') dF_I(\omega')$ is a continuous function of ω which contributes nothing to the value of $\Delta F^{(2)}(0)$. Moreover,

$$(2.34')\quad \int_{-\infty}^{\infty} F_I(\omega - \omega') dF_{II}(\omega') = \sum_k f_k F_I(\omega - \omega_k),$$

where the ω_k's are discontinuity points of $F_{II}(\omega)$ and f_k is the jump of $F_{II}(\omega)$ at $\omega = \omega_k$. The function $(2.34')$ is also a continuous function, and $\int_{-\infty}^{\infty} F_{II}(\omega - \omega') dF_I(\omega') = \int_{-\infty}^{\infty} F_I(\omega - \omega') dF_{II}(\omega')$ since both sides of the last equation represent the Fourier transform of $B_I(\tau)B_{II}(\tau)$. Thus, $\Delta F^{(2)}(0)$ coincides with the jump at $\omega = 0$ of the function $\int_{-\infty}^{\infty} F_{II}(\omega - \omega') dF_{II}(\omega')$, i.e. it is determined by the discrete component $F_{II}(\omega)$ of the spectral distribution function $F(\omega)$.

[24]The "uncertainty principle" is illustrated in the simplest way by the fact that the transition from the function $B(\tau)$ to the "a times contracted" function $aB(a\tau)$ * leads, in terms of Fourier transformations, to a transition from the function $f(\omega)$ to the "a times expanded function" $f(\omega/a)$. Mathematically, this principle can be written in the form of the inequality $\Delta B \Delta f \geqslant \mu$, where ΔB and Δf are the "widths" of the function $B(\tau)$ and, of its Fourier transform, $f(\omega)$, respectively (these widths can be defined in different ways), and μ is some constant depending, of course, on the definitions of the widths ΔB and Δf (see, e.g., Kharkevich, 1960, Landau and Pollak, 1961, and

---

*The factor a is added to preserve the area under the graph of the function, i.e. the value of its Fourier transform at zero frequency.

DeBruijn, 1967). The term "uncertainty principle" emphasizes the fact (first discovered by H. Weyl) that one of the simplest forms of the indicated inequality is exactly equivalent to the famous Heisenberg uncertainty principle in quantum mechanics.

[25]See, e.g., Chandrasekhar (1943), Chap. II; Doob (1949), Sec. 5; Davenport and Root (1958), Secs. 7.4 and 9.4; Middleton (1960), Chaps. 10 and 11.

Schottky's formula can also be easily deduced from the equation for the shot noise correlation function given in Note 9 to Chap. 2.

In fact, this equation implies that $f(\omega) \approx f(0) = (1/\pi)\int_0^\infty B(\tau)d\tau =$

$\lambda\epsilon^2/2\pi = \epsilon i_0/2\pi$ for all values of ω such that $\omega T_0 \ll 1$.

"White noise" is the simplest (and also the most important) example of the so-called *generalized stationary random processes* considered in Chap. 4 of this book (see Sec. 24).

[26]This result can also be easily derived from the "convolution theorem" (see Note 23 above). In fact, let $F_1(\omega)$ be the indefinite integral of the spectral density (2.95) and $F_2(\omega)$, the spectral distribution function of the form (2.99) but with $C = 1$. Then the spectral distribution function $F(\omega) = F_{12}(\omega)$ corresponding to correlation function (2.101) will be of the form (2.33') by virtue of the convolution theorem. Since $F_2(\omega)$ is a step-function having only two jumps of magnitude $1/2$ at points $\omega = -\omega_0$ and $\omega = \omega_0$, the right-hand side of (2.33') is equal to $[F_1(\omega + \omega_0) + F_1(\omega - \omega_0)]/2$ (cf. (2.34')). Hence,

$$f(\omega) = \frac{dF(\omega)}{d\omega} = \frac{C\alpha}{2\pi}\left\{ \frac{1}{\alpha^2 + (\omega + \omega_0)^2} + \frac{1}{\alpha^2 + (\omega - \omega_0)^2} \right\}.$$

[27]For instance, such a form has a correlation function of the amplitude-modulated oscillation $X(t) = A(t)\cos(\omega_0 t + \Phi)$, where the frequency ω_0 is fixed, Φ is a random variable uniformly distributed from 0 to 2π, and $A(t)$ is a random function, which is independent of Φ and has an exponential correlation function. See, e.g., Chap. 5 of Bendat's book (1958) especially devoted to correlation functions of the form (2.101).

[28]Calculation of the correlation function of an averaged pressure field and its approximation by the function (2.124) is due to Yaglom (1955c), who used these results for forecasting experiments. In place of atmospheric pressure at the Earth's surface the height of the 700 millibar isobaric surface was used in the calculations; this height represents the pressure at an altitude of about 3 km and is indicated on regularly issued weather charts. The averaging of the data over a large area was carried out in order to pass from the characteristic of the rapidly changing local weather conditions to that of the

"average weather over a large area", which has a relatively slowly decreasing correlation function.

Note that in solving certain problems on the average pressure field, where high accuracy is not needed, one can use a cruder approximation and replace the initial portion of the solid curve in Fig. 26 by the function of the form (2.101) (or even (2.94); see, e.g., Leith, 1973).

[29]See, e.g., Thompson (1957), p. 291. The inadmissibility of Thompson's model of the correlation function was indicated by Novikov (1959) who noted that the Fourier transform of the selected function $B(\tau)$ has alternating signs and proposed to replace it by a "Gaussian function" of the form (2.126).

A function of the form

$$B(\tau) = \begin{cases} C(2T^3 - 3|\tau|T^2 + |\tau|^3) & \text{for } |\tau| < T, \\ 0 & \text{for } |\tau| > T, \end{cases}$$

where $C > 0$ and $T > 0$, is also used sometimes as a model correlation function (see, e.g., Note 9). It is shown in Note 9 that this function can be represented in the form (2.45); hence, we can be sure that it is positive definite.

[29a]The fact that for $0 < m \leqslant 1$ the function (2.125a) is a possible correlation function (i.e., it is positive definite) follows at once from Pólya's theorem; see example 10 in Sec. 10. On the other hand, substituting $U = X(T + h) - X(T)$ and $V = X(h) - X(0)$ in inequality (1.42) we obtain

$$|2B(T) - B(T + h) - B(T - h)| \leqslant 2|B(0) - B(h)|.$$

Hence, if $B(\tau)$ is a correlation function, then the last inequality must be valid for any $T > 0$ and $h > 0$. However, if $B(\tau)$ is the function (2.125a), then the left-hand side of the inequality has the same order of magnitude as h for small values of h while the right-hand side is of the order of h^m. Hence the inequality does not hold in this case for small enough values of h if $m > 1$.

[30]The "Gaussian correlation function" (2.126) possesses the following curious property: the correlation function $B(\tau)$ and the spectral density $f(\omega)$ have the same shape (i.e. their graphs are similar and differ only in scale parameters). Other examples of the same phenomenon also exist; e.g.,

$$(2.35') \quad B(\tau) = \frac{C}{\cosh \alpha\tau}, \quad f(\omega) = \frac{C}{2\alpha \cosh(\pi\omega/2\alpha)}$$

(cf. Feller, 1966, p. 476-477).

[31]Wiener (1949) was the first to show that many important

problems of the theory of stationary processes (particularly, the linear extrapolation and linear filtering problems) have simple explicit solutions in the case where the spectral density is a rational function of ω. (A simpler derivation of Wiener's results was given by Yaglom, 1962a; see also Yaglom, 1955a, where some additional problems are treated). The solution of various linear approximation problems (e.g., extrapolation, interpolation, filtering and so on) for stationary processes with rational spectral densities is discussed at length in a great number of books; see, e.g., mathematical texts by Rozanov (1967) and Gihman and Skorohod (1969, 1974), or the books with an engineering flavor by Laning and Battin (1956), Bendat (1958), Davenport and Root (1958), Middleton (1960), Solodovnikov (1960), and Pugachev (1965). The simplicity of solution of statistical problems involving linear approximations is associated with the specific Markov-type properties of the class of stationary processes with rational spectral densities. These properties were studied by Doob (1944) and Yaglom (1965).

[32]The case of a general rational spectral density is discussed at greater length in the books by Doob (1953), Sec. XI.10, and Rozanov (1967), Sec. I.10.

[33]Pólya's theorem is given in the paper by Pólya (1949) but, without an explicit formulation, it has been in fact proved and applied in establishing some particular results by the same author much earlier (cf. Pólya, 1949, footnote on p. 117; see also the next Note). The proof of this theorem can also be found in the books by Feller (1966), p. 279 and 282, and by Lukacs (1970), Sec. 4.3, and in a number of papers; see, e.g., Fuchs (1968), where some additional references are presented and it is proved that a spectral density corresponds to any correlation function satisfying Pólya's theorem. Askey (1975) recently proved a new interesting theorem of Pólya's type:

Askey's theorem. *If* $B(\tau)$ *is a bounded even function of* τ *which tends to zero as* $|\tau| \to \infty$, *is differentiable for* $\tau > 0$ *and has the property that the graph of the function* $-B'(\tau)$ *in* $0 \leqslant \tau < \infty$ *is a concave curve, then* $B(\tau)$ *is a correlation function to which corresponds the spectral density* $f(\omega)$, *having its maximum at* $\omega = 0$ *and decreasing monotonically for both positive and negative* ω.

It is clear that the existence of the nonpositive derivative $B'''(\tau)$ for all $\tau > 0$ is sufficient for concavity of the graph of $-B'(\tau)$ in $0 \leqslant \tau < \infty$. Therefore, since the exponential correlation function (2.94) satisfies the relation $B'''(\tau) = -C\alpha^2\exp(-\alpha\tau) < 0$ for $\tau > 0$, Askey's theorem implies, in particular, that the spectral density must correspond to this correlation function and this spectral density must decrease monotonically on both sides of the point $\omega = 0$. (This, of course, agrees with the shape of the graph in Fig. 19(b).) The theorem also implies that a similar behavior is exhibited by the spectral density corresponding to the correlation function of a form $B_a(\tau) = C\exp\{-a|\tau| - |\tau|^3\}$, where $C > 0$ and $a \geqslant 3$. (It follows from Pólya's theorem that $B_a(\tau)$ is a correlation function for all $a \geqslant (1.5)^{1/3} \approx 1.15$.) Another consequence of Askey's theorem will be mentioned in Note 35.

[34]The history of the attempts to prove the positive definiteness of functions (2.132) for $0 < m \leqslant 2$ is rather long. It is said that in the nineteenth century Cauchy was aware of the positivity of the Fourier transforms of all these functions, but had no strict proof of it except for the simplest cases where $m = 1$ or $m = 2$. Apparently Polya was the first to prove in 1920 that at least for $0 < m \leqslant 1$ the functions (2.132) are positive definite; his proof was in fact based on the use of Pólya's theorem (which he did not formally state at the time). The positive definiteness of functions (2.132) for all m in the interval $0 < m \leqslant 2$ was proved quite differently by Lévy (1925); see the next Note and the books by Loève (1963), Sec. 23.4, Feller (1966), Sec. XVII.4, Ibragimov and Linnik (1971), Sec. II.1, and Zolotarev (1983). The proof of an even more general statement is also sketched in Note 50 to Chap. 4 of this book.

[35]The class of normalized correlation functions $R(t) = B(t)/B(0)$ evidently coincides with that of characteristic functions $\psi(t)$ of probability distributions (see Note 3 to the Introduction). The functions $\psi(t) = \exp(-\alpha|t|^m)$, $0 < m \leqslant 2$, belong to the important class of *stable characteristic functions* (characteristic functions of *stable probability distributions*) introduced by Lévy (1925); namely, they are characteristic functions of *symmetric stable distributions*. Hence, the Fourier transforms of $\exp(-\alpha|t|^m)$, $0 < m \leqslant 2$, give the spectral densities corresponding to the correlation functions (2.132) for $C = 1$ and simultaneously the probability densities of symmetric stable distributions (*symmetric stable densities*). Similarly, the integrated Fourier transforms of $\exp(-\alpha|t|^m)$ determine both the spectral distribution functions corresponding to the correlation functions (2.132) and the *symmetric stable probability distribution functions*. This implies that the results of numerous investigations concerning stable probability distributions (see, e.g., the books by Feller, 1966, Sec. XVII.6; Ibragimov and Linnik, 1971, Chap. II; and Zolotarev, 1983; the papers by Zolotarev, 1954 and 1964; Fama and Roll, 1968; and Worsdale, 1975) can be directly applied to the study of spectral densities and spectral distribution functions corresponding to correlation functions (2.132).

The spectral density corresponding to the correlation function (2.132) can be expressed in a closed form in terms of elementary functions only if $m = 1$ or $m = 2$. However, there are also cases where the spectral density $f(\omega)$ can be expressed in terms of some higher mathematical functions. In particular, Zolotarev (1954) showed that

$$f(\omega) = C\left(\frac{3}{\pi}\right)^{1/2}|\omega|^{-1}e^{-2\alpha^3/27\omega^2}W_{1/2,1/6}\left(\frac{4\alpha^3}{27\omega^2}\right)$$

for $m = 2/3$ and

$$f(\omega) = \frac{C\alpha}{2\sqrt{2\pi}|\omega|^{3/2}}\left[\cos\frac{\alpha^2}{4|\omega|}\{1 - 2C\left(\frac{\alpha}{2\sqrt{|\omega|}}\right)\} + \sin\frac{\alpha^2}{4|\omega|}\{1 - 2S\left(\frac{\alpha}{2\sqrt{|\omega|}}\right)\}\right]$$

for $m = 1/2$, where $W_{\nu,\mu}(x)$, $C(x)$ and $S(x)$ are Whittaker's function and two Fresnel's integrals defined, e.g., in Gradshteyn and Ryzhik's book (1980).* The densities $f(\omega)$ are, of course, given by the equation

$$f(\omega) = \frac{C}{\pi}\int_0^\infty e^{-\alpha|\tau|^m}\cos\omega\tau\,d\tau, \quad 0 < m \leqslant 2,$$

and this equation can easily be used to derive the representation of the spectral distribution function $F(\omega)$ in the form of a definite integral of elementary functions. Quite a different representation of $F(\omega)$ in the form of a definite integral was given by Zolotarev (1964); the numerical values of these functions for a number of values of m, some special selection of the "scale parameter" α, and when $C = 1$ where tabulated by Fama and Roll (1968) and Worsdale (1975).

It is easy to see that Askey's theorem (formulated in Note 33) implies that for $0 < m \leqslant 1$ the spectral density $f(\omega)$ corresponding to the correlation function (2.132) necessarily decreases monotonically on both sides of its maximum at the point $\omega = 0$ (note that this density is symmetric, i.e. $f(-\omega) = f(\omega)$). Askey's theorem cannot be applied to cases where $1 < k \leqslant 2$. Nevertheless, using more complicated arguments, one can show that in fact the density $f(\omega)$ behaves similarly for all admissible values of m (see Wintner, 1936, or the books by Ibragimov and Linnik, 1971, Sec. II.5, and Zolotarev, 1983, Sec. 2.7).

[36]For information concerning higher mathematical functions see, e.g., the excellent reference books by Erdélyi et al. (1953), Abramowitz and Stegun (1964), and Gradshteyn and Ryzhik (1980), and the very useful treatise by Watson (1958), devoted to the theory of Bessel functions.

[37]Equation (2.133) and/or the more general equation (2.136) can be found in all the books cited in the preceding Note. See, in particular, Erdélyi et al. (1953), Vol. 2, eq. 7.12(7); Abramowitz and Stegun (1964), eq. 9.1.20; Gradshteyn and Ryzhik (1980), eqs. 3.753.2 and 8.411.2.

---

*Note that the definition of two Fresnel's integrals $C(x)$ and $S(x)$ by Gradshteyn and Ryzhik (1980) slightly differs from both the definitions used in Erdélyi et al. (1953) and in Abramowitz and Stegun (1964).

[38]See either Erdélyi et al. (1953), Vol. 2, eq. 7.12(27), or Abramowitz and Stegun (1964), eq. 9.6.25, or Gradshteyn and Ryzhik (1980), eq. 3.771.2. An equivalent equation can also be found in Watson (1958).

[39]The model (2.139) of the correlation function $B(\tau)$ has the property that at $\omega \gg \alpha$ the corresponding spectral density $f(\omega)$ behaves as a power function of the frequency (namely, $f(\omega) \sim \omega^{-(2\nu+1)}$ at $\omega \gg \alpha$). Therefore such a model can be useful in describing fluctuations $X(t)$ having a spectral density which is proportional to some power of the frequency ω over a wide frequency range. Power densities $f(\omega) \sim \omega^{-(2\nu+1)}$ are typical for noises in vacuum tube circuits generated by fluctuations in cathode emission intensity (so called "flicker noises") and also for many types of electrical noises in semiconductors (see, e.g., MacDonald, 1962, Van der Ziel, 1959, 1970, 1976, Robinson, 1974, Wolf, 1978, and Buckingham, 1983). Equations (2.139) and (2.140), as applied to such noises, were used in particular by Blanc-Lapierre (cf. Blanc-Lapierre and Fortet, 1953, p. 453). In turbulence theory the case of the power density $f(\omega) \sim \omega^{-(2\nu+1)}$, where $2\nu + 1 = 5/3$, is especially important (see, e.g., Monin and Yaglom, 1975, Chap. 8). The use of (2.139) with $\nu = 1/3$ to describe the correlation function of turbulent velocity fluctuations was first suggested by von Kárman (1948), and later was accepted by many authors; equations (2.139) and (2.140) are mentioned in this connection, in particular, in the books by Hinze (1975), p. 247, and by Monin and Yaglom (1975), p. 9.

[40]See, e.g., the books by Lukacs (1970) and Ramachandran (1967), containing many additional references. Numerous examples of equations for $B(\tau)$ can also be found by examining the tables of cosine Fourier transforms (e.g., that by Erdélyi et al., 1954) and selecting pairs of reciprocal transforms such that at least one function of the pair is everywhere nonnegative. (This function can be taken as $f(\omega)$ and its Fourier transform gives the expression for $B(\tau)$.) The tables of Laplace transforms (e.g., that by Erdélyi et al., 1954) can also be useful since any function of the form

$$B(t) = \int_0^\infty e^{-p|\tau|} \varphi(p)\,dp$$

is clearly a correlation function if $\varphi(p) \geqslant 0$ for all nonnegative p (cf. eq. (1.33) in Sec. 4). Finally, the most convenient source of examples of correlation functions is provided by extensive tables of the characteristic functions and the corresponding probability densities compiled by Oberhettinger (1973).

[41]See, e.g., the books by W. Brown (1963), Zadeh and Desoer (1963), and Kailath (1980) on the linear system theory or the books on linear filters cited in Note 43. As for the contents of numerous mathematical monographs on the theory of linear operators (Stone,

1932, and Akhiezer and Glazman, 1980, are typical examples), they cover a different area from the contents of both the engineering books on linear systems (filters) and Secs. 11—12 of the present book.

[42]The transfer function $H(\omega)$ of a physically realizable transformation $\mathfrak{L}$ is specified by the relation

$$H(\omega) = \int_0^\infty e^{-i\omega u}\, h(u)du.$$

The function $\exp(-i\omega u)$ evidently falls off rapidly, as $u \to \infty$, if $\mathrm{Re}\,\omega < 0$ (i.e. for all ω from the lower half-plane of the complex plane). Therefore, if the integral determining $H(\omega)$ converges for all real ω, it will be all the more convergent for complex ω in the lower half-plane. Moreover, it is clear that in the vicinity of each complex point ω from the lower half-plane the integral specifies an analytic function of ω. It is easy to deduce from this that *the class of transfer functions* H(ω) *of physically realizable transformations* $\mathfrak{L}$ *can be specified as the class of functions* H(ω) *of the real variable* ω *which are boundary values on the real axis of functions* H(ω) *of the complex variable, analytic in the lower half-plane.* In other words, all the transfer functions $H(\omega)$ considered can be extended analytically into the entire lower half-plane of the complex variable ω .

One can easily characterize the class of gains $A(\omega) = |H(\omega)|$ of physically realizable linear transformations $\mathfrak{L}$. According to the well-known result by Paley and Wiener (1934), theorem 12, *the nonnegative function* A(ω) *whose square is integrable from* $-\infty$ *to* ∞, *can be the gain of a physically realizable linear transformation* $\mathfrak{L}$ *if, and only if,*

$$(2.36') \qquad \int_{-\infty}^{\infty} \frac{|\log A(\omega)|}{1 + \omega^2}\, d\omega < \infty.$$

[43]See, e.g., the books by Jackson (1950), Cauer (1958), Christian and Eisenmann (1967), and Temes and Mitra (1973) containing many additional references.

Note that according to the condition (2.36') the gain of any physically realizable filter cannot vanish identically on any frequency interval (no matter how narrow it is). Therefore, the high-pass, low-pass, and band-pass ideal filters, having a zero gain in some frequency intervals, are always physically unrealizable. It follows that in practice one always has to deal with non-ideal filters, whose gain $A(\omega)$ is only small (but is not exactly zero) outside the pass band (and within this band is usually not strictly constant).

The following arguments can be adduced to show that in principle there exist physically realizable filters which are arbitrarily close to the given ideal band-pass filter. If the filter is ideal, then $H(\omega) = 1$ for $\omega_1 \leqslant \omega \leqslant \omega_2$ and $H(\omega) = 0$ otherwise. According to (2.154) the corresponding weighting function $h(u)$ is $[\exp(i\omega_2 u) - \exp(i\omega_1 u)]/2\pi i u$ and, hence, the ideal filter is specified by the equation:

$$(2.37')\quad \mathscr{L}\{X(t)\} = \frac{1}{2\pi}\int_{-\infty}^{\infty}\frac{e^{i\omega_2 u} - e^{i\omega_1 u}}{iu}\,X(t-u)du.$$

The filter (2.37') is clearly physically unrealizable. We can, however, use, in place of (2.37'), the physically realizable filter

$$(2.38')\quad\begin{aligned}\mathscr{L}_T\{X(t)\} &= \frac{1}{2\pi}\int_0^{\infty}\frac{e^{i\omega_2(u-T)} - e^{i\omega_1(u-T)}}{i(u-T)}\,X(t-u)du \\ &= \frac{1}{2\pi}\int_{-T}^{\infty}\frac{e^{i\omega_2 u'} - e^{i\omega_1 u'}}{iu'}\,X(t+T-u')du'.\end{aligned}$$

By choosing T to be sufficiently large we can ensure that the result of the application of the filter (2.38') to the process $X(t)$ approximates, to any preassigned degree of accuracy, the result obtained by applying the ideal filter (2.37') to the "time-shifted" process $X(t+T)$. It is clear, however, that such a time shift cannot play any role.

The filters (2.37') and (2.38') are inconvenient since they have complex weighting functions. If we are interested only in real filters, we should define an ideal band-pass filter as a filter for which $H(\omega) = 1$ when $\omega_1 \leqslant \omega \leqslant \omega_2$ or $-\omega_2 \leqslant \omega \leqslant -\omega_1$ and $H(\omega) = 0$ for all other ω. Equation (2.37') is then replaced by

$$(2.37a')\quad \mathscr{L}\{X(t)\} = \frac{1}{\pi}\int_{-\infty}^{\infty}\frac{\sin\omega_2 u - \sin\omega_1 u}{u}\,X(t-u)du,$$

and (2.38') is changed similarly.

[44]See, e.g., Olson (1958) and Chap. 5 (written by R.A. Johnson) in Temes and Mitra (1973), where additional references can be found.

[45]In order to apply electrical filters to turbulent velocity fluctuations, these fluctuations $X(t)$ must first be converted into fluctuations of electrical current. This conversion is carried out by using a *hot-wire anemometer*, a device whose basic element is a thin conducting wire located in the turbulent flow. The change in velocity of the flow past the wire produces changes in the heat exchange of the wire, thereby changing its temperature and resistance, and ultimately the current in the wire. See, e.g., Bradshaw (1971) and Hinze (1975) for more details.

[46]A detailed presentation of the proof, which does not assume, in particular, that the function $B(\tau)$ falls off rapidly at infinity, can be found in the paper by Blanc-Lapierre and Fortet (1947a).

[47]The derivation of the spectral-representation theorem based on the theory of linear filters is due to Blanc-Lapierre and Fortet (1946a,b); see also their book (1953), Chap. VIII.

[48]The applications of the theory of stationary processes to analysis of physical measurements are considered, e.g., in the survey paper by McCombie (1953), the Russian book by Nemirovsky (1964), and in many scientific papers and special sections in books on measuring instruments. As examples of papers devoted to particular instruments we shall mention the papers by Hall (1950) and by Kaganov and Yaglom (1976) on meteorological instruments, the paper by Uberoi and Kovasznay (1953) on turbulent measurements, and the paper by Milatz and van Zolingen (1953) on the electrometers (described by a third-order differential equation).

[49]See, e.g., the description of the measurements of small-scale turbulent spectra in the paper by Alekseev (1965) or the paper by Owens (1978) on the transformations (2.181) and (2.181a), where a number of geophysical applications is mentioned.

[50]Using the residue theorem of the theory of functions of complex variables, it is easy to show that *if* $P(i\omega)$ *has no roots lying in the lower half-plane of the complex variable* ω, *then*

$$h(u) = (2\pi)^{-1} \int_{-\infty}^{\infty} e^{i\omega u}[P(i\omega)]^{-1} du$$

vanishes for $u < 0$ *and is equal to* i *multiplied by the sum of all the residues of the function* $\exp(i\omega u)/P(i\omega)$ *at its poles* (lying in the upper half-plane) *for* $u > 0$. This result implies that the function $h(u)$ falls off exponentially as $u \to \infty$. Moreover, all the particular solutions of the homogeneous equation corresponding to the non-homogenous equation (2.187) die out as $t \to \infty$, and $h(u)$ coincides with the Green function of the indicated equation. It follows from this that the solution of the problem with the initial values at $t = t_0$ for equation (2.187) tends to the solution (2.190a) as $t_0 \to -\infty$.

[51]The form

$$(2.39') \quad dY(t) + aY(t)dt = dW(t)$$

of (2.191a) was used, in particular, by Doob (1942). Integrating all the terms of (2.39'), we obtain

$$(2.39a') \quad Y(t) = Y(t_0) - a \int_{t_0}^{t} Y(s)ds + \int_{t_0}^{t} dW(s).$$

Stochastic integral equations (i.e. integral equations including random processes) similar to equation (2.39a') are widely used in modern probability theory to specify some classes of random processes (see, e.g., McKean, 1969, or Gihman and Skorohod, 1979, Chaps. II-III). Doob (1944, 1949, 1953) also used the following form of (2.191a):

$$(2.39b') \quad \int f(t)dY(t) + a \int f(t)Y(t)dt = \int f(t)dW(t),$$

where $f(t)$ is an arbitrary, sufficiently smooth function, which dies
out rapidly enough at infinity. (Doob also considered a similar
form of higher-order equations and gave much attention to the
interpretation of the equations as ones referring to individual
realization $y(t)$ and $w(t)$ of the random processes $Y(t)$ and $W(t)$.)
Equation (2.39b') is close to the modern interpretation of processes
$cE(t)$ and $Y(t)$ entering into (2.191a) as generalized random processes;
see Sec. 24 in Chap 4 of this book. Rigorous mathematical
investigation of the solutions of general linear differential equations
(of arbitrary order) with constant coefficients and a "white noise"
$cE(t)$ on the right-hand side is carried out by Dym (1966).

[52]See Uhlenbeck and Ornstein (1930) and also Doob (1942),
Chandrasekhar (1943), Wang and Uhlenbeck (1945), and Nelson
(1967). The Langevin equation (first proposed by Langevin in 1908)
is considered in many books; see, e.g., Middleton (1960), Chap. X,
and Rytov (1976), Sec. 28.

[53]The term "the Ornstein-Uhlenbeck process" is due to Doob (1942).

[54]General processes with a rational spectral density can be obtained
as stationary solutions of idealized linear differential equations of
the form

$$(2.40') \quad \frac{d^n Y(t)}{dt^n} + a_1 \frac{d^{n-1}Y(t)}{dt^{n-1}} + ... + a_n Y(t) = c\left[\frac{d^m E(t)}{dt^m} + b_1 \frac{d^{m-1}E(t)}{dt^{m-1}} + ... + b_m E(t)\right]$$

where $a_1, ..., a_n, b_1, ..., b_m$ are constant coefficients, or as solutions of
systems of linear differential equations with constant coefficients of
the form

$$(2.40a') \quad \frac{dY_i(t)}{dt} = \sum_{j=1}^{n} a_{ij} Y_j(t) + \sum_{k=1}^{m} b_{ik} E_k(t), \quad i = 1,2,...,n,$$

where $E_1(t), ..., E_m(t)$ are uncorrelated white noises. Equations of the
form (2.40a') describe, in particular, the Brownian motion of a
system of coupled harmonic oscillators (and also the Brownian
motion of an arbitrary mechanical system in the vicinity of its
equilibrium position) or the electrical fluctuations in an arbitrary
linear network of n meshes due to thermal noises. For information
on this subject, see, e.g., the survey papers by Wang and Uhlenbeck
(1945) and Doob (1949), and the books by James, Nichols, and
Phillips (1947), Chaps. II and IV, Livshits and Pugachev (1963), Vol.
1, Chap. 5, Åström (1970), Chap. 4, and many others. Since the
processes described by equations (2.40') and (2.40a') are often
encountered in applications, the theory of stationary random
processes with rational spectral densities is of great practical value.
Special mathematical properties of stationary processes with
rational spectral densities derived from the fact that these processes
are solutions of equations of the form (2.40a') are investigated, e.g.,

in the papers by Zakai and Snyders (1970), and Erickson (1971).

[55]The first proof of the Herglotz theorem was given in the paper by Herglotz (1911); the proof of this theorem can also be found, e.g., in the books by Doob (1953), Sec. III.3; Loève (1963), Sec. 14.1; Lamperti (1977), Sec. 3.3; Shiryaev (1980), Sec. VI.1; Lambert and Poskitt (1983), Chap. 6; or Rosenblatt (1985), Sec. I.2. The application of this theorem to the proof of the spectral representation theorem for correlation functions of stationary sequences was indicated by Wold in 1938 (in the first edition of Wold, 1954).

[56]The spectral representation theorem for stationary sequences was first proved by Kolmogorov (1941a) in terms of the geometry of the Hilbert space H (see Sec. 4). The proof of this theorem can also be found, e.g., in the books by Doob (1953), Sec. III.4; Rozanov (1967), Chap. I; Gihman and Skorohod (1969), Sec. V.4, and (1974), Sec. IV.5; Wentzell (1981), Sec. 4.2; Shiryaev (1980), Sec. VI.3; Lambert and Poskitt (1983), Chap. 7; or Rosenblatt (1985), Sec. I.3.

In fact, all proofs of the spectral representation theorem for stationary processes can easily be carried over to the case of stationary sequences. To accomplish this, only some obvious changes in the proofs must be made which usually even simplify the arguments. Thus, in the case of the first proof presented in Note 17 (see pp. 33–36), we must, instead of using (2.9'), determine $Z(\omega)$ as follows:

$$(2.41')\qquad Z(\omega) = \frac{1}{2\pi}\left\{ \omega X(0) - \sum_{t\neq 0} \frac{e^{-i\omega t}}{it} X(t)\right\}.$$

Then all the reasoning presented in Note 17 in relation to formula (2.9') can be applied to formula (2.41') with a few rather trivial changes.*

The carrying over of the second proof given in Note 17 (pp. 36–39) in the case of stationary sequences also requires only some minor modifications. In particular, instead of the Hilbert space $L_2(F) = L_2(F;-\infty,\infty)$ of all functions $\varphi(\omega)$, $-\infty < \omega < \infty$, satisfying condition

$\int_{-\infty}^{\infty}|\varphi(\omega)|^2 dF(\omega) < \infty$, we must now consider the similar Hilbert space

$L_2(F; -\pi,\pi)$ of all functions $\varphi(\omega)$, $-\pi \leqslant \omega \leqslant \pi$, satisfying the condition

$\int_{-\pi}^{\pi}|\varphi(\omega)|^2 dF(\omega) < \infty$. This replacement of $L_2(F;-\infty,\infty)$ by $L_2(F;-\pi,\pi)$

---

*For example, instead of using (2.13') we must now use the formula

$$\frac{1}{\pi}\left[\frac{\omega}{2} + \sum_{t=1}^{\infty} \frac{\sin\omega t}{t}\right] = \begin{cases} 1/2 & \text{for } 0 < \omega \leqslant \pi, \\ 0 & \text{for } \omega = 0, \\ -1/2 & \text{for } 0 > \omega \geqslant -\pi \end{cases}$$

(cf. Gradshteyn and Ryzhik, 1980, eq. 1.441.1).

simplifies somewhat all the related reasoning of Note 17.

Finally, we can also apply the generalized spectral representation theorem stated on pp. 40–41 to stationary sequences $X(t)$. According to this theorem the spectral representation (2.204) of a stationary random sequence $X(t)$ follows at once from the spectral representation (2.206) of its correlation function $B(\tau)$ (i.e., in fact, from Herglotz' theorem).

[57]The formulated statement can easily be checked for stationary sequences with a discrete spectrum (i.e. of the form $X(t) = \sum_k X_k \exp(i\omega_k t)$). In the general case, one must substitute into the left-hand side of (2.211) the spectral representation (2.206) of the function $B(\tau)$. Then the proof of the equality of the left-hand side of (2.211) to the jump of the function $F(\omega)$ at the point $\omega = \omega_k$ can be based on simple arguments similar to those applied in Note 22 on pp. 42–43 to the case of continuous t and $\omega_k = 0$. (However, now the function

$$\varphi_T(\omega) = \frac{1}{2T}\int_{-T}^{T} e^{i\omega\tau}\,d\tau = \frac{\sin\omega T}{\omega T},$$

$-\infty < \omega < \infty$, must be replaced by the function

$$\varphi_T^{(1)}(\omega - \omega_k) = \frac{1}{2T+1}\sum_{\tau=-T}^{T} e^{i(\omega-\omega_k)\tau} = \frac{\sin[(T+1/2)(\omega - \omega_k)]}{(2T+1)\sin[(\omega-\omega_k)/2]}, \quad -\pi \leqslant \omega \leqslant \pi.$$

[58]Equation (2.215) has actually been known for a long time to mathematicians studying Fourier coefficients of the form (2.206); see, e.g., Hausdorff (1923). It can also be found, e.g., in the books by Doob and Rozanov mentioned in Note 56.

[59]Equations equivalent to (2.231a) were first given in 1938 in the first edition of Wold's book (1954), p. 70, where only the case of a real process $X(t)$ and of the real form of spectral representation of the functions $B(\tau)$ and $B_\Delta(\tau)$ were considered. The importance of these equations in estimating the spectral density of a continuous process $X(t)$ from observational data at equidistant time points was emphasized by Blackman and Tukey (1959). An accurate mathematical derivation of (2.231) − (2.231a) requires a slight (and very simple) refinement of the arguments in Vol. I of the book (which is needed to prove that if $F(\omega)$ is right-continuous, i.e. $F(\omega + h) \to F(\omega)$, as $h \to 0$, then $F_\Delta(\omega)$ is also bounded and right-continuous, and $B_\Delta(\tau)$ in the first equation (2.231) coincides with $B(\tau\Delta)$ at integral values of τ); cf., e.g., Prousa (1960).

It should also be noted that the relation $f_\Delta(\omega) = \Delta^{-1}\sum_k f([\omega + 2\pi k]/\Delta)$

is very close to the mathematical theorem obtained by S. Poisson as far back as the beginning of the nineteenth century. Indeed, suppose that $\Delta = 1$, and that the function $B(\tau)$, $-\infty < \tau < \infty$, has a Fourier transform $f(\omega)$. Then, using (2.213), we can rewrite the indicated relation in the form

$$(2.42') \quad \sum_{k=-\infty}^{\infty} f(\omega + 2k\pi) = \frac{1}{2\pi} \sum_{\tau=-\infty}^{\infty} e^{-i\omega\tau} \int_{-\infty}^{\infty} e^{i\omega'\tau} f(\omega') d\omega'.$$

The relation (2.42') (whose derivation does not require, in fact, that the function $f(\omega)$ be nonnegative) is nothing other than one of the forms of the well-known *Poisson summation formula*; see, e.g., Courant and Hilbert, 1953, Sec. II.5.5.

[60]For ordinary (nonrandom) band-limited functions $X(t)$ representable in the form of a Fourier (or Fourier–Stieltjes) integral, restricted to a finite frequency band $-\pi/\Delta \leqslant \omega \leqslant \pi/\Delta$, an interpolation formula of the form (2.234) was obtained by Whittaker (1915). (A related result had been known, in fact, to Cauchy in the middle of the nineteenth century; see Lloyd, 1959, where, however, it is erroneously asserted that the exact form of equation (2.234) can be found in Cauchy's paper published in 1841.) Independently of the mathematicians' works, this formula was derived (also for ordinary functions) by Kotel'nikov (1933), who emphasized its great importance for communication engineering; see Kharkevish, 1960, Sec. 14, or Levin, 1974, Appendix VI. Later Shannon (1948, 1949) explained that formula (2.234) plays an important part in information theory; therefore, in the engineering literature this formula is often attributed to Shannon or to Kotel'nikov. Many papers and sections in monographs are devoted to discussion of the sampling formula (2.234), to its applications and various extensions and modifications pertaining to numerical (nonrandom) band-limited functions $X(t)$; see, e.g., the survey paper by Jerri (1977) containing an extensive reference list.

Let us now pass on to the case where $X(t)$ is a stationary random process. The interpolation problem of the best approximation of the random variable $X(t)$ by the mean-square limit $X^*(t)$ of a sequence of

variables $X_n^*(t)$, $n = 1, 2, ...$, of the form $X_n^*(t) = \sum_{k=-n}^{n} a_k^{(n)}(t) X(k\Delta)$ was

studied by Yaglom (1949). The general expression for $\sigma_\Delta^2(t) = \min \langle |X(t) - X^*(t)|^2 \rangle$ in terms of the spectral distribution function $F(\omega)$ of the process $X(t)$ was found in this paper. This expression implies, in particular, that $\sigma_\Delta^2(t) = 0$ for any t if the distance between any two points of the frequency spectrum of $X(t)$ differs from any multiple of $2\pi/\Delta$. The last condition is, of course, satisfied if the whole frequency spectrum Λ of $X(t)$ lies within the segment $-\pi/\Delta < \omega < \pi/\Delta$. Moreover, the same result also implies that $\sigma_\Delta^2(t) = 0$ for any t if all the points of the frequency spectrum of $X(t)$ belong to the closed interval $-\pi/\Delta \leqslant \omega \leqslant \pi/\Delta$, but its two endpoints do not belong simultaneously to the discrete spectrum of $X(t)$; cf. also Lloyd (1959). From Yaglom's results (1949) it is also easy to deduce, for such processes $X(t)$, the sampling formula (2.234), where the infinite sum is interpreted as the mean square limit (see Yaglom, 1955c, Sec. 18, or Lloyd, 1959, where similar arguments are used). However, the first published proof of formula (2.234) for stationary processes $X(t)$ is due to Balakrishnan (1957), who assumed that the whole spectrum of $X(t)$ lies in the interval $-\pi/\Delta \leqslant \omega \leqslant \pi/\Delta$, and neither of its end points is a discontinuity point of $F(\omega)$. Later on, other proofs of the same theorem were proposed, in particular by Lloyd (1959), who found the necessary and sufficient conditions for the validity of (2.234) and investigated the convergence properties of the series entering this formula, and by Beutler (1961b), who assumed that one of the endpoints of the segment $-\pi/\Delta \leqslant \omega \leqslant \pi/\Delta$ may be a discontinuity point of $F(\omega)$. A very simple proof of (2.234), outlined in Vol. I, was proposed by Parzen (see Jerri, 1977) and is presented neatly (under the same conditions as were adopted by Balakrishnan) in the book by Rozanov (1967), Sec. I.6; see also Gihman and Skorohod (1969, 1974). Quite another proof of the same equation was given by Belyaev (1959) and was also outlined in the book by Khurgin and Yakovlev (1971), Sec. 6. This proof uses the assumption that the whole spectrum $X(t)$ belongs to a closed segment lying within the segment $-\pi/\Delta < \omega < \pi/\Delta$, but interprets the convergence of the series on the right-hand side of (2.234) not as mean square convergence, but as almost-sure convergence. More general necessary and sufficient conditions for almost-sure convergence of the series on the right-hand side of (2.234) was given by Gaposhkin (1977a).

[61]If the coefficients h_j fail to satisfy the condition (2.242), but satisfy (2.242a), the Fourier expansion (2.243) of the function $H(\omega)$ must be interpreted in some special sense. However, summation transformations $\mathfrak{L}$, whose weighting sequence does not satisfy (2.242), are rarely encountered in practice. Linear transformations $\mathfrak{L}$, which cannot be specified by a weighting sequence at all, are also rare but, nevertheless, there are problems where one has to deal with such transformations. For instance, the transformation $\mathfrak{L}$ having a transfer function $H(\omega) = [1 - \exp(-i\omega)]^{-1}$ is applied to the stationary sequence $X(t)$ with a spectral density of the form $f(\omega) = A\sin^2(\omega/2)$ in the early paper by Kozulyaev (1941) and in some subsequent works (cf., e.g., Hannan, 1970, Sec. III.2, where more general related transformations are also considered). The transfer function $H(\omega) = [1 - \exp(-i\omega)]^{-1}$ cannot be expanded into a convergent Fourier series, and the corresponding transformation $\mathfrak{L}$ cannot be represented in the form (2.240). However, condition (2.236) is satisfied here and therefore $\mathfrak{L}$ can be represented in the form (2.235): according to Hannan (1970), p. 131,

$$(2.43') \qquad \mathfrak{L}\{X(t)\} = \lim_{n \to \infty} \sum_{j=0}^{n} (1 - \frac{j}{n})X(t - j)$$

in this case.

[62]Digital filters (both linear and nonlinear) are covered by an enormous and rapidly growing literature (see, in particular, a special bibliography by Helms, Kaiser, and Rabiner, 1975). Of the many recent books devoted to digital filters we shall only mention rather elementary texts by Hamming (1977) and Otnes and Enochson (1972, 1978), a somewhat more complicated book by Oppenheim and Schafer (1975), an advanced manual by Rabiner and Gold (1975), and a recent handbook by F. Taylor (1983).

[63]Let us restrict ourselves, for simplicity, to summation transformations of the form (2.240), though in fact this restriction is not necessary. Then we find easily that if $\mathcal{L}$ is a physically realizable transformation, its transfer function $H(\omega)$ is the boundary value on the unit circle $z = \exp(-i\omega)$, $-\pi \leqslant \omega \leqslant \pi$, of the function $H(z) = \Sigma h_j z^j$, which is analytic within the unit circle. The theory of such boundary values is presented, e.g., in the books by Privalov (1950), and Hoffman (1962), see also Doob (1953), Sec. IV.6. The results from any of these books imply, in particular, that the class of gain functions $A(\omega)$ of physically realizable discrete linear transformation $\mathcal{L}$ coincides with the class of functions satisfying the condition

$$(2.44') \qquad \int_{-\pi}^{\pi} |\log A(\omega)| d\omega < \infty .$$

(Cf. also Kolmogorov, 1941a, where, in quite a different connection, the condition (2.44') is also proved for the absolute value of the boundary value on the unit circle of the function $H(z)$ analytic in the unit circle.)

[64]See discussion of this problem by Moran (1950), p. 277–278. Cf. also the related remark on the behavior of the output process of the band-pass filter with a narrow pass band in Sec. 11 (p. 156 in Vol. I).

[65]The "sinusoidal limit theorem" and some of its generalizations were studied by Slutsky (1927a,b), Romanovsky (1932, 1933), Moran (1949, 1950), and Kedem (1984); see also Kendall and Stuart (1968), Sec. 47.15. Let C be the simplest moving average specified by the equation $C\{X(t)\} = X(t) + X(t-1)$ and $\mathcal{D}$ be the first-order differencing. Apply the transformation $\mathcal{D}^m C^n$ to the "purely random" sequence $E(t)$ and assume that $n \to \infty$, $m \to \infty$, $m/n \to \lambda$, where $0 < \lambda < 1$. Then, according to Slutsky (1927a,b), any finite section of the sequence $\mathcal{D}^m C^n E(t)$ will tend asymptotically to a section of a sinusoid of angular frequency $\omega_0 = \cos^{-1}[(1 - \lambda)/(1 + \lambda)]$.

Thus Slutsky's theorem played, in its day, an important role, since it helped to show that apparent cyclical behavior of observed time series does not necessarily imply the existence of a strictly periodic mechanism generating this series. Today, however, the spectral theory of stationary processes makes the theorem not surprising at all. A simple spectral proof of Slutsky's theorem was given by

Moran (1949); see also Kendall and Stuart (1968), and Kedem (1984). The proof is based on the fact that the transfer function of the transformation $\mathcal{D}^m \, C^n$ is $H(\omega) = [1 - \exp(-i\omega)]^m [1 + \exp(-i\omega)]^n$. Thus, this transformation leads to multiplication of the spectral density by $|H(\omega)|^2 = 2^{m+n}(1 - \cos\omega)^m(1 + \cos\omega)^n$. Let us multiply $|H(\omega)|^2$ by the normalizing constant $A = A(m,n)$ determined by the condition

$A\int_{-\pi}^{\pi}|H(\omega)|^2 d\omega = 2$. Then it is not hard to show that $A|H(\omega)|^2 \to 0$ for $\omega \neq$

$\pm\omega_0$, and $A|H(\omega)|^2 \to \infty$ for $\omega = \omega_0$ or $\omega = -\omega_0$, as $n \to \infty$, $m \to \infty$, $n/m \to \lambda$, where $\omega_0 = \cos^{-1}[(1 - \lambda)/(1 + \lambda)]$, i.e. $A|H(\omega)|^2 \to \delta(\omega - \omega_0) + \delta(\omega + \omega_0)$, where δ is Dirac's δ-function. This easily implies Slutsky's result stated above.

Two simple generalizations of Slutsky's theorem were proposed by Romanovsky (1932, 1933); both of them can also be easily proved by using spectral arguments (see Moran, 1949). Later Moran (1950), using the same method, studied the asymptotic behavior, as $n \to \infty$, of the sequence $\mathcal{L}^n X(t)$, where $\mathcal{L}$ is a general moving average transformation and $X(t)$ is an arbitrary stationary sequence having a strictly positive spectral density. The results of this work can also be considered as an extension of Slutsky's "sinusoidal limit theorem". Quite another approach to the proof of the "sinusoidal limit theorem" and some of its generalizations were considered recently by Kedem (1984).

[66]See, e.g., Sec. 7 in the book by R.G. Brown (1962) devoted to economic applications of time-series theory. The use of the filter specified by (2.256) is also discussed in a number of journal articles.

[67]The correlation function $B(\tau)$, $\tau = 0,\pm1,\pm2, ...$, of a stationary sequence, having a spectral representation (2.206), obviously coincides with the set of values at integral values of τ, of the following correlation function of a band-limited stationary process:

$$B(\tau) = \int_{-\pi}^{\pi} e^{i\tau\omega}dF(\omega), \quad -\infty < \tau < \infty.$$

Moreover, there are also many other correlation functions of stationary processes which, when sampled at unit time intervals, generate the same correlation sequence $B(\tau)$, $\tau = 0,\pm1,\pm2, ...$. In fact, if $F_c(\omega)$ and $F_d(\omega)$ are the spectral distribution functions of a stationary process and a stationary sequence, while $B_c(\tau)$ and $B_d(\tau)$ are the corresponding correlation functions, then, according to (2.231a), $B_c(\tau) = B_d(\tau)$ for $\tau = 0,\pm1,\pm2, ...$, if and only if

$$(2.45') \quad F_d(\omega) = \sum_{k=-\infty}^{\infty} [F_c(\omega + 2k\pi) - F_c((2k - 1)\pi)], \quad -\pi \leq \omega \leq \pi.$$

In particular, if the correlation functions $B_c(\tau)$ and $B_d(\tau)$ are specified by spectral densities $f_c(\omega)$ and $f_d(\omega)$, then the condition

(2.45') takes the form

$$(2.46') \qquad f_d(\omega) = \sum_{k=-\infty}^{\infty} f_c(\omega + 2k\pi), \quad -\pi \leqslant \omega \leqslant \pi.$$

[68]The relationship between the spectral densities (2.124) and (2.264) can be generalized considerably. Let $B_d(\tau)$ be an arbitrary correlation function of a stationary sequence. Priestley (1963) showed that by plotting all the points $(\tau, B_d(\tau))$, $\tau = 0, \pm1, \pm2, \ldots$, in a plane and joining the consecutive points by straight lines, we always obtain a plot of the correlation function $B_c(\tau)$ of a stationary process. Obviously, $B_c(\tau) = B_d(\tau)$ when τ is an integer and if $B_d(\tau)$ is a correlation function (2.263), then $B_c(\tau)$ is a function of the form (2.123), where $T = m$, $C = m^{-1}$. Moreover, if the spectral density $f_d(\omega)$ corresponds to the correlation function $B_d(\tau)$, then, according to Priestley, the correlation function $B_c(\tau)$, $-\infty < \tau < \infty$, will be specified by a spectral density

$$(2.47') \qquad f_c(\omega) = \left(\frac{\sin\omega/2}{\omega/2} \right)^2 f_d(\omega), \quad -\infty < \omega < \infty,$$

where $f_d(\omega)$ is now considered as a periodic function of period 2π. This is the generalization of the above-mentioned relationship; the validity of (2.46') is here also guaranteed by (2.265).

[69]Identity (2.269) can be derived from Poisson's summation formula (2.42'); this derivation is very close to the one given in the present book. This identity can be found, in a slightly different formulation, in Jolley (1925), pp. 176-177.

[70]Indeed, any function $f(\omega)$ rational in $\exp(i\omega)$ is the ratio of two polynomials of $\exp(i\omega)$. Factorizing both these polynomials, we find that

$$(2.48') \qquad f(\omega) = C' e^{ik\omega} \frac{(e^{i\omega} - \beta_1') \ldots (e^{i\omega} - \beta_{n_1}')}{(e^{i\omega} - \alpha_1') \ldots (e^{i\omega} - \alpha_{m_1}')},$$

where k is an integer (positive, negative, or zero), all the complex numbers β_j' and α_k' are nonzero, and none of the β_j' coincides with any of the α_k'. Since the spectral density $f(\omega)$ is a positive integrable function, it is clear that no α_k' is equal to 1 in absolute value, and any β_j' of absolute values 1 appear in the numerator of (2.48') an even number of times. Further, since $f(\omega) = \overline{f(\omega)}$,

$$C' e^{ik\omega} \frac{(e^{i\omega} - \beta_1') \ldots (e^{i\omega} - \beta_{n_1}')}{(e^{i\omega} - \alpha_1') \ldots (e^{i\omega} - \alpha_{m_1}')} = \overline{C'} e^{-ik\omega} \frac{(e^{-i\omega} - \overline{\beta_1'}) \ldots (e^{-i\omega} - \overline{\beta_{n_1}'})}{(e^{-i\omega} - \overline{\alpha_1'}) \ldots (e^{-i\omega} - \overline{\alpha_{n_1}'})}$$

$$= \overline{C'} e^{-i(k+n_1-m_1)\omega} \frac{\overline{\beta_1'} \ldots \overline{\beta_{n_1}'} (1/\overline{\beta_1'} - e^{i\omega}) \ldots (1/\overline{\beta_{n_1}'} - e^{i\omega})}{\overline{\alpha_1'} \ldots \overline{\alpha_{m_1}'} (1/\overline{\alpha_1'} - e^{i\omega}) \ldots (1/\overline{\alpha_{m_1}'} - e^{i\omega})}.$$

Hence to each β_j', corresponds a root $\beta_k' = 1/\overline{\beta_j'}$, and to each α_j'

corresponds a root $\alpha_k' = 1/\overline{\alpha_j'}$. But it is easy to see that, if $\gamma \neq 0$,

$$|e^{i\omega} - 1/\overline{\gamma}| = |\gamma|^{-1}|\overline{\gamma}e^{i\omega} - 1| = |\gamma|^{-1}|\overline{\gamma} - e^{-i\omega}| = |\gamma|^{-1}|e^{i\omega} - \gamma|.$$

Moreover, $f(\omega) = |f(\omega)|$ since $f(\omega)$ is real and positive, and hence
(2.48') can be rewritten as

$$(2.49') \quad f(\omega) = C\frac{|(e^{i\omega} - \beta_1) \ldots (e^{i\omega} - \beta_n)|^2}{|(e^{i\omega} - \alpha_1) \ldots (e^{i\omega} - \alpha_m)|^2},$$

where $C > 0$, $0 < |\beta_j| \leqslant 1$ and $0 < |\alpha_k| < 1$. (Here β_j, $j = 1, \ldots, n$, are
all the roots β_j', $j = 1, \ldots, n_1$, which have absolute values less than 1
and half of those roots which have absolute values equal to 1, while
α_k, $k = 1, \ldots, m$, are those values of α_k', $k = 1, \ldots, m_1$, which have
absolute values less than 1.) Thus, (2.273) is proved.

If $X(t)$ is a real stationary sequence, then $f(\omega) = f(-\omega)$. Therefore,
in this case to each complex root β_j corresponds a complex conjugate
root $\overline{\beta}_j$, and to each complex root α_k, a root $\overline{\alpha}_k$. Consequently, all the
complex roots β_j and α_k are divided into pairs of complex conjugate
numbers. This implies that all the coefficients $b_1, \ldots, b_n$ and $a_1, \ldots,$
a_m in (2.273) can be considered real.

Stationary random sequences whose spectral densities are rational
functions of $\exp(i\omega)$ are quite similar to stationary processes with a
density $f(\omega)$ that is a rational function of ω. Such sequences possess
the same particular properties that were mentioned in Note 31 on
pp. 46–47 with reference to stationary processes, and many
important statistical problems in the theory of stationary random
functions can also be solved especially simply for such sequences.
See, in this connection, e.g., Doob (1944), Wiener (1949), Yaglom
(1962a), Rozanov (1967), Chap. 3, and also the book by Box and
Jenkins (1970), where random sequences $X(t)$ are considered which
are somewhat more general than stationary sequences with a rational
spectral density in $\exp(i\omega)$.

[71]See, e.g., Hannan (1970), Sec. VI.8, and Phadke and Wu (1974).
Let $B(\tau)$, $-\infty < \tau < \infty$, be a correlation function either of an
autoregressive process of order m, where $m > 1$, or of a more general
process with the rational spectral density of the form (2.129), where
$m > n \geqslant 1$. It is shown in the above mentioned works that then the
values of $B(\tau)$ at integer-valued τ (or at τ or the form $\tau = k\Delta$, where
the k's are integers) will form a correlation function of a stationary
sequence with the particular spectral density of the form (2.273)
where m is the same as above and $n = m - 1$.

Chapter 3

[1]The class of all possible consistent estimates of a given parameter c is very extensive and it includes a number of particular practically important estimates thoroughly studied in many statistical texts (see, e.g., Cramér, 1946; Wilks, 1962; Kendall and Stuart, 1966; Silvey, 1970; Zacks, 1971). To compare two different unbiased consistent estimators of c, say, C_N^* and C_N^{**}, the *relative efficiency* $e_r(C_N^{**}, C_N^*)$ or C_N^{**} as compared with C_N^* is sometimes evaluated by the formula $e_r(C_N^{**}, C_N^*) = \sigma^2(C_N^*)/\sigma^2(C_N^{**})$. Clearly, C_N^{**} is preferable to C_N^* if, and only if, $e_r(C_N^{**}, C_N^*) > 1$. Note also that under some general conditions the lower bound $\sigma_N^2(c)$ to the variance $\sigma^2(C_N^*)$ of any unbiased estimator C_N^* of a parameter c can be determined, and hence the absolute *efficiency* $e(C_N^*) = \sigma^2(C_N^*)/\sigma_N^2(c)$ of C_N^* can be evaluated. The estimator C_N^* (and the estimate c_N^*) is called *efficient* if $e(C_N^*) = 1$ and *asymptotically efficient* if $e(C_N^*) \to 1$ as $N \to \infty$.

[2]Note also that the estimate m_N^* of the mean value m is computed rather easily and that in many important cases (e.g., when X has the normal or Poisson distribution) m_N^* is an efficient (i.e., the most accurate) estimate of the parameter m.

[3]If $b(\tau) \to 0$, as $\tau \to \infty$, then for any $\epsilon > 0$ there exists such $T_0 = T_0(\epsilon)$ that $|b(\tau)| < \epsilon/2$ for $T > T_0$. Since $|b(\tau)| \leqslant b(0)$ for any τ, in the case of continuous time we have

$$\left| \frac{1}{T} \int_0^T b(\tau)d\tau \right| \leqslant \frac{1}{T} \int_0^T |b(\tau)|d\tau \leqslant \frac{b(0)T_0}{T} + \frac{\epsilon(T - T_0)}{2T} \leqslant \frac{b(0)T_0}{T} + \frac{\epsilon}{2}$$

for $T > T_0$. Thus,

$$\left| \frac{1}{T} \int_0^T b(\tau)d\tau \right| < \epsilon \text{ for } T > T_1(\epsilon) = 2b(0)T_0/\epsilon,$$

i.e. $\lim\limits_{T\to\infty} \dfrac{1}{T} \int_0^T b(\tau)d\tau = 0$.   (The discrete time case is treated quite similarly.) Moreover, if $b(\tau)$ is the sum of a function decaying to zero at infinity and several periodic summands of the form $g_k \cos\omega_k \tau$, then the condition (3.10) or (3.10a) will also be fulfilled. (This is easy to check directly and can also be deduced from the equivalence of Slutsky's condition to the condition $\Delta \overset{\circ}{F}(0) = 0$; see p. 220 in Vol. I.)

Slutsky's theorem was first proved (for the continuous time case) in Slutsky's paper (1938).   However, even before that, Khinchin (1933, 1934) proved the following *law of large numbers* for stationary random functions: *For any stationary function* X(t) *the time averages* M_T^* *necessarily tend* (*in the mean square*) *to a definite limit as* T → ∞.   (It was shown later that this limit is equal to $\Delta Z(0)$; see eq. (2.32′) on p. 43.) Khinchin's law of large numbers is, in turn, very similar to the so-called *quasi-ergodic theorem* of Neumann (1932a) referring to general mechanics (theory of dynamic systems).

[4]In the case of continuous t, it follows from (1.55) that

$$\sigma^2(M_T^*) = \frac{1}{T^2} \int_0^T\!\int_0^T b(t - s)\,dt\,ds.$$

Changing now the variables t, s to $\tau = t - s$, $\tau_1 = t + s - T$ and noting that $b(-\tau) = b(\tau)$, we obtain the next to last or the last expression (3.11a) for $\sigma^2(M_T^*)$ depending on whether we first integrate with respect to τ and then to τ_1 or vice versa.   The results (3.11) are obtained quite similarly, but we must take into account that in the discrete time case the moment $\tau = 0$ is exceptional since $-\tau$ coincides with τ only at this τ.

[5]It is easy to see that the left-hand side of (3.10) or (3.10a) coincides with $\langle (M_T^* - m)(X(0) - m) \rangle$.   Therefore, if $\sigma^2(M_T^*) = \langle (M_T^* - m)^2 \rangle \to 0$ as $T \to \infty$, then this left-hand side also tends to zero, as $T \to \infty$, by virtue of Cauchy–Buniakovsky–Schwarz inequality (1.42).

In proving the converse result that Slutsky's condition implies the convergence of $\sigma^2(M_T^*)$ to zero, as $T \to \infty$, we shall assume, for definiteness, that the time t is continuous.   Then Slutsky's condition has the form (3.10a) and it implies that for any $\epsilon > 0$ there exists $T_0 = T_0(\epsilon)$ such that

$$\int_0^{\tau_1} b(\tau)d\tau \leqslant \epsilon\tau_1 \text{ for } \tau_1 > T_0.$$

On the other hand, since $|b(\tau)| \leqslant b(0)$,

$$\int_0^{\tau_1} b(\tau)d\tau \leqslant b(0)\tau_1 \text{ for any } \tau_1.$$

Therefore, if $T > T_0$

$$\sigma^2(M_T^*) = \frac{2}{T^2} \int_0^T \int_0^{T_1} b(\tau)d\tau d\tau_1 \leqslant \frac{2}{T^2} \left[\int_0^{T_0} b(0)\tau_1 d\tau_1 + \int_{T_0}^T \epsilon \tau_1 d\tau_1 \right]$$

$$= \frac{b(0)T_0^2}{T^2} + \frac{\epsilon(T^2 - T_0^2)}{T^2} < \frac{b(0)T_0^2}{T^2} + \epsilon.$$

Consequently, $\sigma^2(M_T^*)$ can be made arbitrarily small by choosing T to be sufficiently large. Hence, $\sigma^2(M_T^*) \to 0$, as $T \to \infty$. The case of discrete time t can be considered quite similarly.

Equation (3.11a) appears for the first time in quite a different connection in G. I. Taylor's famous paper (1921) on turbulent diffusion. Equation (3.11) is due to Slutsky (1929a,b).

[6]See, e.g., S. Vilenkin (1959, 1967), where it is also shown that under some general conditions (imposed on $b(\tau)$) an integer N can be found for any given $T > 0$ such that the mean-square error of the discrete estimator $M_{TN}^* = [1/(N + 1)]\Sigma_{k=0}^N X(kT/N)$ of m will be lower than the

mean square error of the integral estimator $M_T^* = (1/T)\int_0^T X(t)dt$. (Some

particular examples illustrating this phenomenon were also adduced by Fine and Johnson, 1965, and Morris and Ebey, 1984.) However, the optimal value of N depends on the exact form of $b(\tau)$ which is rarely known in applied problems, and besides, the difference between the mean square errors of the estimator M_T^* and the best estimator M_{TN}^* is small in all cases.

Many explicit formulae for the variances $\sigma^2(M_T^*)$ and $\sigma^2(M_{TN}^*)$ related to particular forms of $b(\tau)$ can be found, e.g., in the papers by Zernike (1932), Yule (1945), Bayley and Hammersley (1946), Frenkiel (1953), Kharybin (1957), Venchkovsky (1962), Soloviev (1970), Olberg (1972), and in the books by Livshits and Pugachev (1963), Korn (1966), Vilenkin (1967), Sveshnikov (1968), Mirsky (1972), Max (1981), and Bendat and Piersol (1986). These formulae are sometimes accompanied in the indicated sources by tables, graphs, and nomograms. The climatological application of (3.11a) is discussed by Leith (1973).

Accurate calculations of the variances $\sigma^2(M_T^*)$ and $\sigma^2(M_{TN}^*)$ by (3.11) and (3.11a), of course, have some sense only if the function $b(\tau)$ is known exactly. If, however, $b(\tau)$ is unknown (which happens most often in practice), then the true values of $b(\tau)$ must be replaced on the right-hand sides of (3.11) and (3.11a) by the approximate estimates of $b(\tau)$ (which are discussed in Sec. 17). Therefore, only rather crude estimates of $\sigma^2(M_T^*)$ and $\sigma^2(M_{TN}^*)$ can be obtained in cases where $b(\tau)$ is unknown. See in this connection the relevant discussion at the end of Yule's paper (1945) and also the papers by Hannan (1957) and Moran (1975), where some

approximate expressions for $\sigma^2(M_T^*) = \sigma^2(T^{-1}\Sigma_{t=1}^T X(t))$ are suggested which depend only on the variables $X(1), ..., X(T)$.

We also note that if the function $b(\tau)$ is known exactly, it is possible, in principle, to find the *best* (i.e. minimum variance) *unbiased linear estimator* $\hat{M}_T$ of $m = \langle X(t) \rangle$ which is more accurate than the estimator M_T^* (or M_{TN}^*). For this we must consider the whole class of linear unbiased estimators of m of the form $M_T^{(a)} = \sum_{t=1}^{T} a(t)X(t)$ (or $M_T^{(a)} = \int_0^T a(t)X(t)dt$), where $\sum_{t=1}^{T} a(t) = 1$ (or $\int_0^T a(t)dt = 1$

and $a(t)$ can also be a generalized function). Then the optimal values of $a(t)$ can be determined from the condition $\sigma^2(M_T^{(a)}) = $ minimum (see, e.g., Grenander, 1950; Davenport, Johnson, and Middleton, 1952; Yaglom, 1955a). The optimal function $a(t)$ depends, of course, on the form of $b(\tau)$; e.g., if $b(\tau) = C\exp(-\alpha|\tau|)$ and the time t is continuous, then the best choice is $a(t) = [\delta(t) + \delta(t - T) + \alpha]/(2 + \alpha T)$, i.e. $\hat{M}_T = [X(0) + X(t) + \alpha\int_0^T X(t)dt]/(2 + \alpha T)$.

Moreover, in most cases the difference between mean square errors of the estimators M_T^* and $\hat{M}_T$ is slight and tends to zero, as $T \to \infty$ (see, e.g., the error calculations for the process $X(t)$ with the correlation function $b(t) = C\exp(-\alpha|\tau|)$ presented in Davenport, Johnson, and Middleton, 1952; and Fine and Johnson, 1965). However, if the spectral density $f(\omega)$ of $\mathring{X}(t) = X(t) - m$ vanishes at $\omega = 0$, then the difference $\sigma^2(M_T^*) - \sigma^2(\hat{M}_T)$ can be quite significant even at large values of T (cf. Vitale, 1973; Adenstedt, 1974; Rasulov, 1976; and Chistyakov, 1986).

[7]Formula (3.18) was first obtained by G. I. Taylor (1921) in quite another connection. The applications of formulae (3.18) and (3.18a) in the investigation of atmospheric turbulence are indicated, e.g., by Lumley and Panofsky (1964), Sec. 1.15, and Sreenivasan, Chambers, and Antonia (1978).

[8]It is easy to show that if $f(\omega) \sim \omega^\beta$, $-1 < \beta < 1$, in the vicinity of zero frequency (i.e. if $f(\omega) = \omega^\beta f_0(\omega)$, where $f_0(\omega)$ is continuous and different from zero at small positive values of ω), then $\sigma^2(M_T^*) \sim T^{-1-\beta}$ (see Leonov 1960, 1964; cf. also Cox, 1981, Sec. 2.3). If $B(\tau) = -B_0''(\tau)$ for continuous τ or $B(\tau) = -B_0(\tau - 1) + 2B_0(\tau) - B_0(\tau + 1)$ for integer-valued τ, where $B_0(\tau)$ is also a correlation function (in particular, if $f(\omega) \sim \omega^\beta$, $\beta > 1$, at small values of ω), then and only then $\sigma^2(M_T^*) \sim T^{-2}$ (E. Robinson, 1960; Leonov, 1961). Of course, an increase of the averaging time guaranteeing the given accuracy of the mean value estimation with an increase of the intensity of the components of $X(t)$ with longest periods (i.e. with the increase of $f(\omega)$ at small ω) is natural from the physical point of view. Recall also that unbounded spectral densities $f(\omega)$ proportional to $\omega^{-\alpha}$, $\alpha \lesssim 1$, at small values of ω are typical for many electrical noises (see Note 39 to Chap. 2 and the literature there).

Hence, the mean value estimation in the presence of such noises requires the averaging over rather a long time interval.

[9]For information on the *central limit theorem* for stationary random functions $X(t)$ see, e.g., the books by Rozanov (1967), Sec. IV.11; Hannan (1970), Sec. IV.4; Ibragimov and Linnik (1971), Chap. XVIII; Brillinger (1975), Sec. 4.4; Rosenblatt (1985), Chap. III; and Eberlein and Taqqu (1986), Sec. 2; also, and the survey papers by Lumley (1972) and Rosenblatt (1972), intended for physicists. (These sources also contain many additional references.) The conditions guaranteeing the validity of the theorem have the form of rather natural requirements for the asymptotic independence of the remote future of the function $X(t)$ from its remote past. These conditions usually cannot be checked directly, but they are very natural and they can safely be assumed valid in most of the practical situations. Specific examples of theorems on asymptotic normality of M_T^* can be found, e.g., in Hannan (1970), Sec. IV.4; Anderson (1971), Sec. 8.4.1; and Brillinger (1975), Sec. 4.4. Examples of another type, illustrating that there are situations where the asymptotic normality is not obtained, are given by Rosenblatt (1985), Sec. III.5; see also Eberlein and Taqqu (1986).

Construction of confidence intervals for a given estimator is considered in all statistical texts referred to in Note 1; the case of the estimator M_T^* was specially studied by Brillinger (1979).

[10]In the mathematical literature the statement on the almost sure convergence (i.e. convergence with probability one) of the time averages M_T^* (i.e. of their realizations m_T^*) to m, as $T \to \infty$, is usually called the *strong law of large numbers* for stationary functions $X(t)$. Sufficient conditions for the validity of this law were first found by Loeve (1945) and Blanc-Lapierre and Brard (1946); their results are stated also in the books by Doob, 1953, Secs. X.6 and XI.6; Loeve, 1963, Sec. 34.7; Cramér and Leadbetter, 1967, Secs. 5.5 and 7.10; and Hannan, 1970, Sec. IV.3. These results imply, in particular, that *the strong law of large numbers necessarily holds if ϵ > 0 exists such that the condition* (3.10) *or* (3.10a) *remain valid after the replacement of the factor* $1/T$ *on the left-hand side by the factor* $1/T^{1-\epsilon}$ (cf. Verbitskaya, 1964, 1966). Moreover, Verbitskaya also proved that in fact *it is sufficient to require only that the factor* $1/T$ *in* (3.10) *and* (3.10a) *should be replaceable by* $(\log T)^a/T$, *where a* > 3. However, the original Slutsky's condition (3.10) or (3.10a) is not sufficient for the validity of the strong law of large numbers. This last assertion follows from the existence of examples of stationary functions $X(t)$ for which Slutsky's condition (and hence the ordinary law of large numbers) is valid, but the strong law is not (see Blanc-Lapierre and Tortrat, 1968; Gaposhkin, 1973, 1977b).

Loève's and Blanc-Lapierre and Brard's condition for the validity of the strong law of large numbers implies that *this law is necessarily valid if* $C > 0$ *and* $\alpha > 0$ *exist such that* $|b(\tau)| < C|\tau|^{-\alpha}$ *for all sufficiently large* τ (i.e. *if* $b(\tau)$ *decreases not slower than* $\tau^{-\alpha}$ *as* $\tau \to \infty$). Verbitskaya's condition shows that it is sufficient to require *that*

$b(\tau)$ *decrease not slower than* $(\log \tau)^{-a_1}$, *where* $a_1 > 3$, *as* $\tau \to \infty$. However, the strongest result of such a type was proved by Gaposhkin (1973, 1977b) who showed that *the strong law of large numbers is necessarily valid if* $b(\tau)$ *decreases at large values of* τ *not slower than* $(\log \log \tau)^{-2-\epsilon}$, *where* ϵ *is any positive number, but that this law can be violated if* $b(\tau)$ *decreases only as* $(\log \log \tau)^{-2}$. Later Gaposhkin (1981a) estimated also the rate of the almost sure convergence of m_T^* to m for stationary processes $X(t)$ with a given rate of decrease of $b(\tau)$ as $\tau \to \infty$.

Note that all the examples of stationary functions $X(t)$ for which Slutsky's condition is fulfilled, but m_T^* does not tend almost surely to m, as $T \to \infty$, (i.e. the ordinary law of large numbers is valid, but the strong law is not), are rather complex and artificial. This is due, to a great extent, to the fact that such functions $X(t)$ must necessarily belong to the class of random functions, which are stationary only in the wide, but not in the strict, sense. (It has been already indicated in this book that the functions of this class are almost never encountered in applications; cf. the end of Sec. 3 in Vol. I.) For strictly stationary random functions the almost sure convergence of the time averages m_T^* as $T \to \infty$, was proved under very general conditions as far back as the early 1930s. Namely, Birkhoff proved in 1931 an important ergodic theorem for dynamic systems, and it was indicated by Khinchin almost immediately after (in 1932) that by virtue of Birkhoff's theorem $\lim_{T \to \infty} m_T^*$ *must exist with*

probability one for any strictly stationary random function X(t) *satisfying the condition* $\langle |X(t)| \rangle < \infty$ (and, if t is continuous, still another very general regularity condition which is always fulfilled in practice). The proof of this famous *Birkhoff–Khinchin ergodic theorem* can now be found in many books (see, e.g., Khinchin, 1949; Doob, 1953; Gnedenko, 1962; Loève, 1963; Cramér and Leadbetter, 1967; Rozanov, 1967; Gihman and Skorohod, 1969, 1974; Lamperti, 1977, or Shiryaev, 1980). Note that this theorem does not even require the existence of the correlation function $B(\tau)$ (i.e. the validity of the condition $\langle X^2(t) \rangle < \infty$). If, however, $X(t)$ is a strictly stationary function such that $\langle X^2(t) \rangle < \infty$ and the function $b(\tau) = B(\tau) - m^2$ satisfies Slutsky's condition (3.10) or (3.10a), then for this $X(t)$ the variable M_T^* converges to m in the mean square, as $T \to \infty$, and simultaneously the time averages M_T^* also converge to a certain limit with probability one. It readily follows that *for any strict-sense stationary random function satisfying Slutsky's condition* M_T^* *converges to* m *with probability one, i.e. the strong law of large numbers necessarily holds.*

The Birkhoff–Khinchin ergodic theorem is associated with some other important results of the theory of strictly stationary random processes, which should be briefly mentioned here. In Sec. 3 (on p.

51 in Vol. I) we have already emphasized that if $X(t)$ is a strictly stationary random function, the function $Y(t) = g(X(t))$, where $g(x)$ is an arbitrary function of one variable, is also strictly stationary. It is proved quite similarly that in the case at hand the function $Y_\Phi(t) = \Phi(X(t),X(t + \tau_1), ..., X(t + \tau_{n-1}))$, where $\Phi(x_1,x_2, ..., x_n)$ is an arbitrary function of n variables, is also strictly stationary. Therefore, if a finite mean value $\langle\Phi(X(t),X(t + \tau_1), ..., X(t + \tau_{n-1}))\rangle = m_\Phi(\tau_1, ..., \tau_{n-1})$ exists for the function Φ, then, by virtue of the Birkhoff–Khinchin theorem, the average of $\Phi(x(t),x(t + \tau_1), ..., x(t + \tau_{n-1}))$, where $x(t)$ is a realization of $X(t)$, over a time interval of length T (we denote this average by $m_T^{(\Phi)}$) must converge, with probability one, to a definite

limit $m^{(\Phi)} = m^{(\Phi)}(\tau_1, ..., \tau_{n-1})$ as $T \to \infty$. Of special interest, of course, are those stationary functions $X(t)$, for which $m^{(\Phi)}(\tau_1, ..., \tau_{n-1}) = m_\Phi(\tau_1, ..., \tau_{n-1})$ for any function $\Phi(x_1,x_2, ..., x_n)$ such that $\langle\Phi(X(t),X(t + \tau_1), ..., X(t + \tau_{n-1}))\rangle < \infty$. These functions $X(t)$ are usually called *ergodic* (or *metrically transitive*) stationary random functions. Note that a particular case of the function $\Phi(X(t),X(t + \tau_1), ..., X(t + \tau_{n-1}))$ with a finite mean value is $\chi_{a_1}[X(t)]\chi_{a_2}[X(t + \tau_1)] ... \chi_{a_n}[X(t + \tau_{n-1})]$ (see Note 1 to Chap. 1 on pp.

14–16). Therefore, it is clear that for a realization of the stationary ergodic process $X(t)$ the relation (1.1') will necessarily be fulfilled with probability one. (Here $F_{t_1,...,t_n}(x_1, ..., x_n)$ is the n-dimensional

distribution function (1.3).) It is also easy to see that the validity of the relations (1.1') in fact implies the ergodicity of the process $X(t)$.

Naturally, not every stationary random function is ergodic. Generally speaking, the ergodicity of $X(t)$ is very hard to check (the only exception being Gaussian functions $X(t)$, which are treated on pp. 233–234 in Vol. I and in Note 20 on p. 76 of this volume). Therefore, in applications the ergodicity is often assumed without proof, referring to physical intuition. Note also that if a stationary function $X(t)$ is not ergodic, then *under very general regularity conditions* (which can always be assumed valid in practice) *the set of all its realizations* x(t) *can be decomposed into some nonintersecting subsets, within each of which all the time averages* $m_T^{(\Phi)}$ *almost surely converge, as* T → ∞, *to limiting values which are equal to the conditional mean values of* $\Phi(X(t),X(t + \tau_1), ..., X(t + \tau_{n-1}))$ *relative to the condition that realization* x(t) *belongs to the corresponding subset.* This important fact was proved by Neumann (1932b) for a rather general (but not the most general) class of dynamic systems and it was later generalized considerably by Rokhlin (1949), who also stated his results in terms of general dynamics. The proof of the same result formulated in terms of the theory of stationary processes can be found, e.g., in the book by Rozanov (1967), Sec. IV.8. The formulated statement imparts an exact meaning to the treatment on pp. 215–217 in Vol. I of the general (non-ergodic) stationary random function $X(t)$ as a mixture of some ergodic functions related to "statistical subensembles" of realizations that are parts of the ensemble corresponding to the function $X(t)$.

[11]Equation (3.19) can easily be derived with the aid of the general equation (0.16'), which expresses the moments of a multidimensional probability distribution in terms of the corresponding characteristic function. According to (1.6'), in the case of a Gaussian function $X(t)$ with $\langle X(t) \rangle = 0$, the characteristic function of the two-dimensional vector $\{X(t), X(t + \tau)\}$ is equal to

$$\psi(\theta_1, \theta_2) = \exp\{- [b(0)(\theta_1^2 + \theta_2^2) + 2b(\tau)\theta_1\theta_2]/2\} \ .$$

Hence $\langle X^2(t + \tau)X^2(t) \rangle = \partial^4\psi(\theta_1, \theta_2)/\partial\theta_1^2\partial\theta_2^2\big|_{\theta_1=\theta_2=0} = 2b^2(\tau) + b^2(0).$

The general Isserlis formula for the arbitrary even-order central moment of a multidimensional Gaussian distribution can be derived in a similar way. (All the odd-order central moments of such a distribution are obviously equal to zero.) According to Isserlis (1918), *if the vector* $\mathbf{X} = (X_1, ..., X_n)$ *has a multidimensional normal distribution, and* $\langle X_i \rangle = m_i$, $\langle X_i X_j \rangle = b_{ij}$, $k_1 + k_2 + ... + k_n = 2K$, *then*

$$(3.1')\quad \langle(X_1 - m_1)^{k_1}(X_2 - m_2)^{k_2} ... (X_n - m_n)^{k_n}\rangle = \Sigma b_{i_1 i_2} b_{i_3 i_4} \cdots b_{i_{2K-1} i_{2K}}$$

where the sum on the right-hand side is taken over all $(2K)!/(2^K K!)$ *partitions of the 2K indices* $1,1, ..., 1; 2,2, ..., 2; ...; n,n, ...; n$ *(where 1 is repeated* k_1 *times, ..., n is repeated* k_n *times) into K pairs* (i_1, i_2), (i_3, i_4), *...,* (i_{2K-1}, i_{2K}). In particular, for $2K = 4$ we get

$$(3.2')\quad \langle(X_1 - m_1)(X_2 - m_2)(X_3 - m_3)(X_4 - m_4)\rangle = b_{12}b_{34} + b_{13}b_{24} + b_{14}b_{23}.$$

When $X_1 = X_2 = X(t + \tau)$, $X_3 = X_4 = X(t)$ the last formula turns into (3.19). Formula (3.1') follows readily from (0.16'), where the function $\psi(\theta_1, ..., \theta_n)$ must now be replaced by $\exp\{- (1/2)\sum_{i,j=1}^{n} b_{ij}\theta_i\theta_j\}$ (cf. (0.20')) and expanded in powers of $\theta_1, ..., \theta_n$.

[12]Let us assume, for definiteness, that t is continuous. Substituting the values $a = 0$, $b = T$, $f(\tau) = b(\tau)$, $g(\tau) = 1/T$ into the well-known Buniakovsky–Schwarz inequality

$$\left|\int_a^b f(\tau)g(\tau)d\tau\right|^2 \leqslant \int_a^b f^2(\tau)d\tau \ \int_a^b g^2(\tau)d\tau,$$

we see at once that condition (3.20) implies (3.10a).

[13]Let $X(t) = \overset{\circ}{X}(t) + m$, where $m = \langle X(t) \rangle$. Then it is easy to see that

$$b_T^*(0) = \frac{1}{T}\int_0^T [X(t) - \frac{1}{T} \int_0^T X(s)ds]^2 dt = \frac{1}{T}\int_0^T \left[\overset{\circ}{X}(t) - \frac{1}{T}\int_0^T \overset{\circ}{X}(s)ds \right]^2 dt$$

$$(3.3')\qquad = \frac{1}{T}\int_0^T [\overset{\circ}{X}(t)]^2 dt - \left[\frac{1}{T} \int_0^T \overset{\circ}{X}(t)dt\right]^2$$

(the time t is assumed continuous for definiteness). Hence,

$$\delta(b_T^*(0)) = \langle b_T^*(0)\rangle - b(0) = - \langle\Big[\frac{1}{T}\int_0^T X(t)dt - m\Big]^2\rangle,$$

i.e. the bias of $b_T^*(0)$ is negative and equal in absolute values to the variance of M_T^*. Recall now that usually $\sigma^2(M_T^*) \sim T^{-1}$ at large values of T. Therefore, $\delta^2(b_T^*(0))$ is of the order of T^{-2}, whereas the variance of the estimate $b_T^*(0)$ in the case where $\langle X(t)\rangle$ is known is usually of the order of T^{-1} (see (3.21)). The variance of $b_T^*(0)$ when m is unknown is equal to the sum of the variances of two terms on the right-hand side of (3.3$'$) minus twice their covariance. It is possible to show that under general conditions (which are mostly fulfilled in practice and, in particular, are valid in the case of a Gaussian function $X(t)$) the main contribution to $\sigma^2(b_T^*(0))$ at large values of T is made by the variance of the first term which coincides with the variance of $b_T^*(0)$ when m is known (see, e.g., Veselova and Gribanov (1970a) where only the case of a Gaussian $X(t)$ is considered).

Note also that even when m is known the mean square error

$$\Delta^2(b_T^*(0)) = \sigma^2(b_T^*(0)) + \delta^2(b_T^*(0)) \text{ of the estimator } b_T^*(0) = (1/T)\int_0^T [X(t)$$

$$- (1/T)\int_0^T X(s)ds]^2 dt \text{ is often smaller than the mean square error (i.e.}$$

the variance) of the simpler estimator $b_T^{**}(0) = (1/T)\int_0^T [X(t) - m]^2 dt$.

Therefore, even if $\langle X(t)\rangle$ is known (e.g., if it is known that $\langle X(t)\rangle = 0$), in estimating the variance of $X(t)$ it is often expedient to begin with subtraction of the time average m_T^* from all the observed values $x(t)$ (see, e.g., German, Morozov, and Svalov, 1974). Such subtraction, however, cannot increase the accuracy of the estimate $b_T^*(0)$ substantially, except in the case of a short observation interval T (in comparison with the correlation time T_1), where the whole realization $x(t),\, 0 \leqslant t \leqslant T$, is often placed on one side of $\langle X(t)\rangle$ (i.e. the difference $x(t) - m$ does not change sign).

Many examples of mean square error calculations for the estimate $b_T^*(0)$ (and for the related discrete estimate $b_{TN}^*(0)$ similar to the estimate (3.13b)) of the variance of a Gaussian stationary random function $X(t)$ can be found in the available literature. These examples refer to the cases of both a known and an unknown mean value $\langle X(t)\rangle$ and of both discrete and continuous time t; see, e.g., the papers by Bayley and Hammersley (1946), Kharybin (1957), Venchkovsky (1962), S. Veselova and Gribanov (1968), Soloviev (1970), Olberg (1972), German, Morozov, and Svalov (1974), and the books by Livshits and Pugachev (1963), Vilenkin (1967), Mirsky (1972), Max (1981), and Bendat and Piersol (1986). The applied value of such calculations is naturally greatly diminished by the fact that the function $b(\tau)$ appearing in the equations for the mean square error is usually unknown and must be estimated from the data. Cf. in this connection, e.g., the paper by Brillinger (1979), where the construction of confidence intervals for the correlation function is discussed.

[14]Recall that in the case where $\langle X(t) \rangle = 0$ the validity of Slutsky's condition (3.10a) is equivalent to the continuity of the spectral distribution function $F(\omega)$ of $X(t)$ at the point $\omega = 0$, and the validity of condition (3.20) is equivalent to the continuity of $F(\omega)$ at all points ω. Moreover, according to the convolution theorem of the theory of Fourier integrals, the nth power $b^n(\tau)$ of $b(\tau)$, where n is an integer, is also a correlation function corresponding to the spectral distribution function $F^{(n)}(\omega)$ which is equal to an n-fold convolution of $F(\omega)$ with itself, i.e.

$$F^{(n)}(\omega) = \int_{-\infty}^{\infty} \cdots \int_{-\infty}^{\infty} F(\omega - \omega_1 - \cdots - \omega_{n-1}) dF(\omega_{n-1}) \cdots dF(\omega_1).$$

It is easy to see that if $F(\omega)$ is a continuous function, then all its successive convolutions $F^{(n)}(\omega)$, $n = 2, 3, \ldots$, are also continuous. Hence, $\Delta F^{(n)}(0) = 0$ for all integers n and this implies that

$$(3.4') \qquad \lim_{T \to \infty} \frac{1}{T} \int_0^T b^n(\tau) d\tau = 0, \quad n = 2, 3, \ldots,$$

and also

$$(3.4a') \qquad \lim_{T \to \infty} \frac{1}{T} \int_0^{\infty} b_{x^{2k}}(\tau) d\tau = 0, \quad k = 2, 3, \ldots .$$

[15]General equations for the mean square errors of the estimators $M_T^{(2k)}$ in the Gaussian case are given, e.g., by Alekseev (1970). The asymptotic formulae (3.22) and (3.23) can be found, e.g., in the books by Lumley and Panofsky (1964), p. 38, and Lumley (1970), p. 72. Some applications of these formulae to the estimation of the accuracy of the measured moments of turbulent fluctuations in the atmosphere are discussed by Sreenivasan et al. (1978).

[16]See Alekseev (1970).

[17]There are many special models of non-Gaussian random processes $X(t)$, for which it is possible to evaluate the variances of the estimators $M_T^{(n)}$; see, e.g., Peschel (1961), Alekseev (1970), Geranin (1972), Vishnyakov, Vishnyakova et al. (1974), Vishnyakova, Geranin et al. (1975), Vishnyakova, Gaevoi et al. (1976), Vishnyakova, Gartstron et al. (1978), and Vishnyakova and Geranin (1978), where some additional references can be found. (In most of these papers only the estimator $M_T^{(2)}$ of $\mu^{(2)}$, which coincides with $b_T^*(0)$ if $\langle X(t) \rangle = 0$, is considered. The only exception is the paper by Alekseev (1970), where the accuracy of the estimators $M_T^{(2)}, M_T^{(3)}$, and $M_T^{(4)}$ is studied.) The models of the processes $X(t)$ analyzed in the indicated papers include, in particular, the non-centered and centered Poisson pulse process (0.42) with constant amplitudes E_n and several specific pulse shapes $\Gamma(t)$ (Peschel; Vishnyakova, Geranin et al; Vishnyakova, Gartstron et al.); random

telegraph signal (Vishnyakova, Geranin et al.); centered processes of the form $|Y(t)|$, or $Y(t) - c|Y(t)|$, c = const., or $\exp\{Y(t)\}$, where, as everywhere in this Note, $Y(t)$ is a Gaussian random process (Alekseev; Vishnyakova and Geranin); processes of the form $Y^2(t)$ or $Y(t)Y_1(t)$, where $Y(t)$ and $Y_1(t)$ are independent Gaussian processes (Vishnyakov, Vishnyakova et al.; Vishnyakova, Geranin et al.), and modulated oscillations of the form $Y(t)\cos(\omega_0 t + \Phi)$ with a Gaussian amplitude $Y(t)$ and a uniformly distributed random phase Φ (Geranin; Vishnyakov, Vishnyakova et al.). A survey of 14 different models is given by Vishnyakova and Geranin (1978).

As an example of the obtained results, we point out, first of all, the asymptotic formula

$$(3.5') \qquad \frac{\sigma^2(b^*_T(0))}{[b(0)]^2} \approx \frac{2T_1}{T} + \frac{1}{\lambda T} \quad \text{for } T \gg T_1$$

derived by Peschel (1961) for a Poisson pulse process (0.42) with $E_n = 1$ for all n, $\Gamma(t) = A\exp(-t/T_1)$ for $t \geqslant 0$ and $\Gamma(t) = 0$ for $t < 0$. In (3.5') λ is the average number of Poisson points t_n per unit time. Since $b(\tau) = C\exp(-|\tau|/T_1)$ for the considered process (see Sec. 7, equation (2.48)), the first term on the right-hand side of (3.5') coincides with the result valid for the Gaussian process with the same correlation function (see (3.24)), but now it is supplemented by one more positive term, which greatly exceeds the first term at $\lambda \ll 1/T_1$. (This is, of course, quite natural, since for very rare pulses the averaging time required for reliable determination of the variance of $X(t)$ must be very long.) Refer also to the the results of Alekseev (1970), according to which, for centered processes of form $|Y(t)|$, or $Y(t) + c|Y(t)|$, where $\langle Y(t)\rangle = 0$, $\langle Y(t + \tau)Y(t)\rangle = C\exp(-|\tau|/T_0)$, reliable determination of the third-order moment (which is evidently different from zero) requires a much longer averaging time than that necessary for determining the second moment with the same accuracy. These results agree well with the empirical finding by Sreenivasan et al. (1978) of the fact that at fixed averaging time T the accuracy of the odd-order moments is usually lower than that of the neighboring even-order moments. Finally, we note that Alekseev's results for the log-normal process $X(t) = \exp\{Y(t)\} - \langle\exp\{Y(t)\}\rangle$ (where $Y(t)$ has the same correlation function as above) prove to be strikingly different from the results for Gaussian processes: in this case for $T \gg T_0$

$$\frac{\sigma^2(b^*_T(0))}{[b(0)]^2} = \frac{\sigma^2(M^{(2)}_T)}{[\mu^{(2)}]^2} \approx \frac{84T_1}{T}, \qquad \frac{\sigma^2(M^{(3)}_T)}{[\mu^{(3)}]^2} \approx \frac{5240T_1}{T},$$

$$(3.6') \qquad \frac{\sigma^2(M^{(4)}_T)}{[\mu^{(4)}]^2} \approx \frac{2\times10^6\, T_1}{T}$$

where $T_1 \approx 0.77T_0$ is the correlation time of the process $X(t)$.

[18]Examples of conditions ensuring that the estimators $B^*_T(\tau)$, $b^*_T(\tau) = B^*_T(\tau) - M^{*2}_T$, and $R^*_T(\tau) = b^*_T(\tau)/b^*_T(0)$ of the functions $B(\tau)$, $b(\tau)$, and $R(\tau)$ of the discrete argument τ have asymptotically normal

probability distributions can be found, e.g., in the books by Hannan (1970), Sec. IV.4, Anderson (1971), Secs. 8.4.2 and 8.4.5, and Rosenblatt (1985), Sec. III.2, and in the papers by Hannan (1976), Hannan and Heyde (1972), and Rosenblatt (1986); some examples where the corresponding limit distribution is more "exotic" are also considered by Rosenblatt (1985), Sec. III.5. Curiously, the asymptotic normality of the distributions of the estimators $R_T^*(\tau)$ can be proved under weaker conditions than those necessary to prove the asymptotic normality of the distributions of $B_T^*(\tau)$; see Anderson and Walker (1964), Anderson (1971), and Hannan and Heyde (1972). Some general conditions for the almost sure convergence of $B_T^*(\tau)$ to $B(\tau)$, as $T \to \infty$ (i.e. for the applicability of the strong law of large numbers to the function $Y_\tau(t) = X(t + \tau)X(t)$) are given by Doob (1953), Secs. X.7 and XI.7, and Hannan (1970), Sec. IV.3. Later Hannan (1974) showed also that if we replace the factor $(T - \tau)^{-1}$ on the left-hand side of (3.26a) by T^{-1} (i.e. consider, instead of $B_T^*(\tau)$, the estimators $B_T^{**}(\tau)$ which are most often used nowadays; see pp. 241–243 in Vol. I), then it is possible to prove, under very general conditions, that $B_T^{**}(\tau)$ converges uniformly to $B(\tau)$ with probability one, i.e.

$$(3.7')\qquad \lim_{T \to \infty}\ \sup_{0 \leqslant \tau \leqslant \infty} |B_T^{**}(\tau) - B(\tau)| = 0$$

almost surely. Some estimates of the rate of almost sure convergence of the left-hand side of (3.7') to zero are indicated by An,.Chen, and Hannan (1982).

[19]To obtain (3.29), we must only apply (3.2') to the variables $X_1 = X(t + s + \tau)$, $X_2 = X(t + s)$, $X_3 = X(t + \tau)$, and $X_4 = X(t)$. Note also that (3.29) can be rewritten in the form

$$(3.8')\qquad \begin{aligned} B^{(4)}(t,s) - [B(\tau)]^2 &= b^2(s) + b(s + \tau)b(s - \tau) \\ &\quad + m^2[2b(s) + b(s + \tau) + b(s - \tau)]. \end{aligned}$$

Using now the known inequality

$$\int_0^T b(s + \tau)b(s - \tau)dt \leqslant \left[\int_0^T b^2(s + \tau)d\tau \int_0^T b^2(s - \tau)d\tau\right]^{1/2}$$

(or a similar inequality for discrete τ) and the fact that the condition (3.20) implies the validity of Slutsky's condition (3.10) or (3.10a), it is easy to show that in the Gaussian case condition (3.28) (or a similar condition for discrete τ) is equivalent to condition (3.20).

The exact formula for the mean square error $\Delta_T^2(\tau)$ of the estimate $B_T^*(\tau)$ of the function $B(\tau)$ can easily be obtained from (3.11) and (3.11a) by replacing τ with s in these equations, T with $T-\tau$, and substituting the function $B^{(4)}(\tau,s) - B^2(\tau)$ for $b(s)$. The formula for the mean square error of the estimate $B_{TN}^*(\tau)$ can be obtained in a similar way. Proceeding from the exact formula for $\Delta_T^2(\tau)$, it is not hard to derive the exact formulae for the mean square errors (or, if desired, for the biases and variances separately) of the two estimates $b_T^*(\tau)$ and $\tilde{b}_T^*(\tau)$ of the function $b(\tau)$ and the approximate formulae for the mean square errors (or for the biases and variances) of the two estimates

$R_T^*(\tau)$ and $\tilde{R}_T^*(\tau)$ of $R(\tau)$. Similarly exact (in cases where estimates of $B(\tau)$ or $b(\tau)$ are considered) or approximate (in the case of estimates of $R(\tau)$) formulae can also be derived for covariances between the values of the estimates at two time points τ and τ_1 and for the second joint moments of the differences $B_T^*(\tau) - B(\tau)$, $b_T^*(\tau) - b(\tau)$, etc. at two values of τ.

If the function $X(t)$ is Gaussian, then all the above-mentioned mean square errors, biases, variances, covariances, and second joint moments of the estimates of $B(\tau)$, $b(\tau)$, and $R(\tau)$ are expressed only in terms of $b(\tau)$ and m. Moreover, in the case of the estimates of $b(\tau)$ or $R(\tau)$ these statistical characteristics of the estimates depend only on the estimated function $b(\tau)$ or, respectively, $R(\tau)$. (With reference to estimates of $R(\tau)$ the indicated result is valid even for some classes of random functions $X(t)$ more general than the class of Gaussian functions.) The general formulae for the mean square errors, variances, covariances, etc. of the estimates are often very cumbersome but they may be simplified considerably (and derived under wider conditions) if T is large enough (i.e. if only the asymptotic expressions are considered). The formulae for the covariances show, in particular, that there is a fairly close correlation between the values of the studied estimates of $B(\tau)$, $b(\tau)$, and $R(\tau)$ at comparatively close values of τ, i.e. the values of the estimates at neighboring points are strongly interdependent.

For information on all these results see, e.g., the pioneering papers by Slutsky (1929a,b) and Bartlett (1946) and the books on stochastic processes and time series analysis by Bartlett (1978), Chap. IX; Hannan (1960), Sec. II.3; Kendall and Stuart (1968), Chap. 48; Anderson (1971), Chap. 8; and Priestley (1981), Sec. 5.3. See also related papers by P. Robinson (1977) and Brillinger (1979) on asymptotic estimates for variances and covariances of the estimators

$b_T^{**}(\tau) = [(T - \tau)/T]\tilde{b}_T^*(\tau)$ of $b(\tau)$ (cf. pp. 241–243 in Vol. I) and $R_T^{**}(\tau) = b_T^{**}(\tau)/b_T^{**}(0)$ of $R(\tau)$ and on confidence intervals for $b_T^{**}(\tau)$; the recent time series paper by O. Anderson (1979a), De Gooijer (1981), and Anderson and De Gooijer (1982), where rather tedious exact formulae for biases, variances, and covariances of $b_T^{**}(\tau)$ are given, and numerous engineering books and papers on statistical estimation (e.g., the books by Livshits and Pugachev, 1963; Pugachev, 1965; Vilenkin, 1967; Jenkins and Watts, 1968; Sveshnikov, 1968; Ball, 1968; Bendat and Piersol, 1986; also the papers by Kutin, 1957; Volgin and Karimov, 1967a,b; Veselova and Girbanov, 1968, 1970a; Itskovish and Tsaturova, 1969; German, Morozov, and Svalov, 1974; and many others). These sources include, in particular, some examples of calculations of the above-mentioned mean square errors and other statistical characteristics of the estimates for several particular forms of correlation functions $b(\tau)$ or $R(\tau)$. Cf. also the following Notes 27 and 29 containing some additional references.

Explicit equations for the mean square error of the estimates $B_T^*(\tau)$, $b_T^*(\tau)$ (or $\breve{b}_T^*(\tau)$), and $R_T^*(\tau)$ (or $\breve{R}_T^*(\tau)$) can naturally also be obtained for a number of special models of non-Gaussian random processes permitting evaluation of the fourth moment $B^{(4)}(\tau,s)$. See in this connection, e.g., the papers by Peschel (1961); Geranin (1972); Vishnyakova, Geranin et al. (1975); Vishnyakova, Gaevoi et al. (1976); Vishnyakova, Gartstron et al. (1978); and Vishnyakova and Geranin (1978), which were mentioned in Note 17.

[20]The sufficiency of condition (3.20) for the convergence of time averages of a product $x(t)x(t + \tau_1) \ldots x(t + \tau_{n-1})$ to the mean value $\langle X(0)X(\tau_1) \ldots X(\tau_{n-1})\rangle$ can be proved with the aid of the Isserlis formula (3.1') from Note 11 similarly to the corresponding proofs for the cases where $n = 2$ or n is arbitrary but $\tau_1 = \ldots = \tau_{n-1} = 0$ (cf. Notes 19 and 14). The general ergodic theorem for Gaussian stationary random functions was proved independently and almost simultaneously by Maruyama (1949), Grenander (1950), Sec. 5.10, and (in somewhat different terms and under the assumption that the time is discrete) Fomin (1949, 1950). The proof of this theorem can also be found, e.g., in the books by Rozanov (1967), Sec. IV.6; and Cramér and Leadbetter (1967), Sec. 7.11. Some generalizations of the indicated theorem, which relate to more general classes of stationary random functions than the class of Gaussian functions and are formulated in terms of the higher moments of $X(t)$ or of its multidimensional characteristic functions, were given by Parzen (1958) and Leonov (1960) or (1964), Chap. 3. However, the validity of the ergodicity conditions from these works is difficult to verify in practice.

[21]Let us use (0.24) and (0.25) from the Introduction and the fact that the conditional probability density $p(y/x)$ of the probability distribution for the variable Y, given that $X = x$ (some fixed value), is equal to the ratio $p(x,y)/p_1(x)$. If a vector (X,Y) has a two-dimensional normal distribution and $\langle X \rangle = \langle Y \rangle = 0$, we then obtain that

$$p(y/x) = \frac{1}{\sqrt{2\pi\sigma_2^2(1 - r^2)}} \exp\left\{-\frac{[y - (\sigma_2/\sigma_1)rx]^2}{2\sigma_2^2(1 - r^2)}\right\},$$

i.e. the conditional probability distribution of Y, given $X = x$, is a one-dimensional normal distribution with a mean value $\sigma_2 rx/\sigma_1$ and a variance $\sigma_2^2(1 - r^2)$. Hence, the conditional mean value of $g(X(t + \tau))X(t)$, given that $X(t + \tau) = x(t + \tau)$, is equal to $g(x(t + \tau))x(t + \tau)R(\tau)$. The averaging of the last expression over all the possible values of $x(t + \tau)$ gives (3.30) (where the stationarity of $X(t)$ is also taken into account).

The fact that the conditional mean values of $X(t + \tau)$, given $X(t) = x$, are equal to $xR(\tau)$, may be used for approximate determination of $R(\tau)$ based on experimental data on the conditional distribution of $X(t + \tau)$, given the value of $X(t)$.   See in this respect, e.g., the Russian books by Gribanov, Veselova, and Andreev (1971), Mirsky (1972), and Zhovinsky and Arkhovsky (1974); and the papers by Egorov (1968), Kiriyanov (1971), and Mirsky (1978), where some modifications of the method are also described and additional references can be found.

The relation (3.30) was apparently first given by Bussgang (1952). Later Barrett and Lampard (1955) (see also Deutsch, 1962, Sec. 6.4.1) and Nuttall (1958) showed that in fact this relation is valid under considerably more general conditions than the condition that the random function $X(t)$ is a Gaussian one.   Another substantial generalization of the elementary relation (3.30) was indicated independently by Furutsu (1963), Novikov (1964), and Donsker (1964) (see also Rytov, Kravtsov, and Tatarsky, 1978, Sec. I.7).   The formula given by these authors allows $\langle G[X(t)]X(t)\rangle$, where $G[X(t)]$ is a functional depending on the values of the stationary Gaussian process $X(t)$ at various points t, to be expressed in the form of an integral containing the functional derivative of $G[X(t)]$ in the integrand.   (The functional derivatives of a functional have already been mentioned in Note 3 to Chap. 1.)

[22]The relay correlator was apparently first proposed, as a practical device for computing correlation functions, about 1960 (cf. Jespers, Chu, and Fettweis, 1962).   At present the relay method of correlation analysis, its performance, applications, and some of its modifications, are considered in many Russian books and papers intended for engineers (e.g., Khomyakov, 1968; Ball, 1968; Veselova and Gribanov, 1969; Khavkin and Grinberg, 1970; Gribanov, Veselova, and Andreev, 1971; Mirsky, 1972; Kosyakin and Filaretov, 1972; and Zhovinsky and Arkhovsky, 1974) and also in a number of works in English (e.g., Veltman and van den Bos, 1964; Huzii, 1962, 1964; Iwase, 1973; Andrews, 1980; Hertz, 1982; Cacopardi, 1983; Egau, 1984; Jacovitti, Neri, and Cusani, 1984; Jacovitti and Cusani, 1985; Koh and Powers, 1985; and Jordan, 1986); see also Note 26 below.   Some of the above-mentioned authors (e.g., Khavkin and Grinberg, Kosyakin and Filaretov, Iwase, and Jacovitti et al.) also calculated the mean square error of the relay estimate of the correlation function.   They found, quite surprisingly, that at a fixed averaging time T this error is sometimes even smaller than the mean square error of the conventional estimate $b_T^*(\tau)$ (i.e. the relay estimate of $b(\tau)$ is in some cases even more efficient than the conventional one).

[23]The problem of determination of the mean values $\langle g(X(t + \tau)g(X(t))\rangle = b_g(\tau)$ and $\langle g_1(X(t + \tau))g_2(X(t))\rangle = b_{g_1 g_2}(\tau)$ for a Gaussian

stationary function $X(t)$ is considered, e.g., in the books by Middleton (1960), Deutsch (1962), Tikhonov (1982), and Levin (1974), and in the papers by Price (1958), Baum (1969), and Cambanis and Masry (1978). Note that the last-mentioned authors especially studied the problem of the reconstruction of $b(\tau)$ by a given function $b_g(\tau)$. The substantially more general mean values of the form $\langle G_1[X(t)]G_2[X(t)]\rangle$, where G_1 and G_2 are two non-linear functionals of a Gaussian random process $X(t)$, are treated by Bochkov and Dubkov (1974).

[24]Let the random vector (X,Y) have a normal probability density (0.24), where $m_1 = m_2 = 0$ and let $q = P\{X > 0, Y > 0\} = P\{X/\sigma_1 > 0, Y/\sigma_2 > 0\}$. The vector $(X/\sigma_1, Y/\sigma_2)$ obviously also has a probability density of the form (0.24) but with $m_1 = m_2 = 0$, $\sigma_1 = \sigma_2 = 1$.

Passing from Cartesian coordinates (x,y) to polar coordinates (ρ,ϕ), we obtain

$$q = \frac{1}{2\pi(1 - r^2)^{1/2}} \int_0^\infty \int_0^\infty \exp\left\{- \frac{x^2 - 2rxy + y^2}{2(1 - r^2)} \right\}dx\,dy$$

$$= \frac{1}{2\pi(1 - r^2)^{1/2}} \int_0^{\pi/2} \int_0^\infty \exp\left\{- \rho^2\left[\frac{\cos^2\phi - 2r\cos\phi \sin\phi + \sin^2\phi}{2(1 - r^2)}\right]\right\}\rho\,d\rho\,d\phi$$

$$(3.9') \qquad = \frac{(1 - r^2)^{1/2}}{2\pi} \int_0^{\pi/2} \frac{d\phi}{1 - r\sin2\phi} = \frac{1}{2\pi}\left[\frac{\pi}{2} + \mathrm{tg}^{-1}\frac{r}{(1 - r^2)^{1/2}}\right]$$

$$= \frac{1}{2\pi}\left[\frac{\pi}{2} + \sin^{-1}r\right] = \frac{1}{2\pi}\cos^{-1}(- r).$$

Thus $P\{X(t + \tau) > 0, X(t) > 0\} = q(\tau) = \cos^{-1}[- R(\tau)]/2\pi$ and hence $P\{X(t + \tau)X(t) > 0\} = 2q(\tau) = \cos^{-1}[- R(\tau)]/\pi$, $P\{X(t + \tau)X(t) < 0\} = 1 - 2q(\tau)$ and, finally, $\langle \mathrm{sgn}(X(t + \tau))\mathrm{sgn}(X(t))\rangle = P\{X(t + \tau)X(t) > 0\} - P\{X(t + \tau)X(t) < 0\} = 4q(\tau) - 1 = 2\sin^{-1}[R(\tau)]/\pi$.

Formula (3.9') was given by both T. J. Stieltjes and W. F. Sheppard as far back as the end of the last century (see Kedem, 1980, Chap. 4; and Lawson and Uhlenbeck, 1950, p. 58). In the form (3.32) it was rediscovered by Van Vleck (1943); see also Van Vleck and Middleton (1966).

[25]Polarity coincidence correlators for Gaussian random functions (i.e. correlators which use (3.32) for determination of $R(\tau)$) are considered, e.g., by Gershman and Feinberg (1955); Lomnicki and Zaremba (1955); Veltman and Kwakernaak (1961); Wolff, Thomas, and Williams (1962); Huzii (1962); Korn (1966), Chap. 6; McNeil (1967); Ball (1968); Veselova and Gribanov (1970b); Gribanov, Veselova, and Andreev (1971), Sec. 2.2; Mirsky (1972), Secs. 4.5 and 4.6; Kurochkin (1972), Secs. 4.2 and 5.1; Hurt (1973); Zhovinsky and Arkhovsky (1974); Otnes and Enochson (1978), Sec. 7.3; Kedem (1980), Chaps. 4 and 7; Max (1981), Sec. 10.1; Gabriel (1982); Egau (1984); and Jordan (1986). These sources also contain many

additional references to related papers. (In some works, in particular
in those by Lomnicki and Zaremba and by Kedem, the function bi[$X(t)$]
= 1 for $X(t) > 0$ and = 0 for $X(t) \leqslant 0$ is used instead of sgn[$X(t)$], but this
produces only minor modification of (3.32).) The problem of accurate
evaluation of the mean square error of such a correlator is quite
difficult and has not yet been solved satisfactorily.

[26]The possible generalizations of the relay and polarity coincidence
methods of correlation analysis to some classes of non-Gaussian
stationary function $X(t)$ are discussed in many works mentioned in
Notes 22 and 25; see also Barrett and Lampard (1955); J. Brown (1957);
Jespers, Chu, and Fettweis (1962); Deutsch (1962); McFadden (1965);
Bogner (1965); Kopilovich (1966); Berndt (1968); Chang and Moore
(1970); and Jordan (1986). Note that the relay method and the method
of a conditional mean value (see Note 21) are more widely applicable,
in their conventional form, to non-Gaussian random functions $X(t)$
than the polarity coincidence method. However, there are some
modifications of both the relay method and the polarity coincidence
method that can be applied to rather general non-Gaussian stationary
functions $X(t)$. These modifications are based on some special curious
relations involving the sign function. In particular, it is not hard to
show that *if* X(t) *is an arbitrary stationary sequence satisfying the relation*
P{|X(t) $\geqslant$ A} = 0 *while* E(t) *is a random sequence independent of the
sequence* X(t) *and such that* E(t) *at any* t *is uniformly distributed between* −
A *and* A, *then*

$$(3.10') \quad \langle X(t)\mathrm{sgn}[X(t + \tau) + E(t + \tau)]\rangle = \frac{2}{A}\,\langle X(t)X(t + \tau)\rangle = \frac{2}{A}B(\tau)$$

(see Bogner, 1965). Therefore, the time averaging of the function
$x(t)\mathrm{sgn}[x(t + \tau) + e(t + \tau)]$, where $x(t)$ and $e(t)$ are realizations of $X(t)$
and $E(t)$, can be used for approximate determination of $B(\tau)$ (the
modified relay correlation method). Note that the boundedness
condition P{$|X(t)| \geqslant A$} = 0 is clearly of no substantial importance in
practical situations if A is selected large enough (cf. Knowles and
Tsui, 1967). Similarly, *if* P{|X(t)| $\geqslant$ A} = 0 *and both* E(t) *and* E(t + τ)
are independent of each other and of the sequence X(t) *and are
uniformly distributed between* − A *and* A, *then*

$$(3.10a') \quad \langle \mathrm{sgn}[X(t + \tau) + E(t + \tau)]\mathrm{sgn}[X(t) + E(t)]\rangle = \frac{1}{A^2}\langle X(t + \tau)X(t)\rangle$$

$$= \frac{1}{A^2}B(\tau)$$

(see, e.g., Veltman and Kwakernaak, 1961; Berndt, 1968; and
Theorem 7.2 in Kedem, 1980; the generalization of (3.10a') to
higher-order moments is given in Jespers, Chu, and Fettweis, 1962).
Equation (3.10a') implies that the time averaging of sgn[$x(t + \tau)$ +
$e(t + \tau)$]sgn[$x(t) + e(t)$] can also be applied to estimation of $B(\tau)$ (the
modified polarity coincidence correlation method; see, e.g., Veltman
and Kwakernaak, 1961; Jespers, Chu, and Fettweis, 1962; Veltman

and van den Bos, 1964; Berndt, 1968; and Max, 1981). Knowles and Tsui (1967) calculated the mean square errors of the correlation function estimates obtained by the modified relay correlator and by the modified polarity coincidence correlator (which estimates $B(\tau)$ by the modified polarity coincidence method) for the case where $E(t)$ is a uniformly distributed white noise (i.e. a sequence of independent uniformly distributed random variables). They compared these errors with the mean square error of a conventional estimate $B_T^*(\tau)$ and found that the error of a modified polarity coincidence estimate of $B(\tau)$ is greater than the errors of two other estimates, and the error of a modified relay estimate is intermediate between the errors of an estimate $B_T^*(\tau)$ and of a modified polarity coincidence estimate. See, in this connection, also the error estimates by Berndt (1968) and the related more general results by Chang and Moore (1970).

[27]Some early system of analog correlators are described in the books by Solodovnikov (1960) and Barber (1961); see also Korn (1966), Chap. 5, Papoulis (1984), pp. 253–254, and Jordan (1986); the German book by Lange (1959); the French book by Max (1981); and the Russian books by Ball (1968), Gribanov, Veselova, and Andreev (1971), Kurochkin (1972), Mirsky (1972), and Zhovinsky and Archovsky (1974).

Let us also mention the parametric method of correlation analysis which is used sometimes in practice. Then it is assumed that the unknown correlation function belongs to some parametric family of functions depending on several numerical parameters and determined uniquely by the values of these parameters. Further, the observed values of $x(t)$ are used for the approximate estimation of the parameters which determine the correlation function. The advantage of this approach is that it is easy to ensure here that the obtained correlation function $b(\tau)$ will necessarily be smooth enough and will fall off rapidly with $|\tau|$; moreover, it is also possible to guarantee that $b(\tau)$ will have a form convenient for solving the specific statistical problems of interest to the investigator (e.g., will be rational in $\exp(i\omega)$). For information on "parametric correlometers" see, e.g., Gorbatsevich (1971) or Volkov, Motov, and Prokhorov (1974); cf. also Sec. 19, pp. 288-304 in Vol. I, where closely related parametric methods of spectral analysis are discussed.

It should be emphasized that at present the widespread use of digital computers has greatly diminished the practical applicability of analog correlators. Digital methods of correlation analysis are most widely adopted now, and this gives rise to an extensive literature on the possible improvements in the speed and accuracy of the calculation of the sums on the right-hand sides of (3.26) and (3.26b). Some elementary practical recommendations for simplifying such calculations were given by Popenko and Tikhonov (1972), W. Kendall (1974), and Blankinship (1974). Much more profound changes in the conventional numerical methods of correlation analysis were proposed by Ahmed and Natarajan (1974), Lopresti and Suri (1974), and Larsen (1976) who made wide use of binary arithmetic in their "fast algorithms" (see also Anishin and Tivkov, 1978). Another approach to the same problem permitting one to considerably increase the speed of computations at large values of T is given in Sec. 19 (see pp. 284-287 in Vol. I).

[28]It follows easily from (0.11) that if $\langle [B_T^*(\tau) - B(\tau)]^2 \rangle = \Delta_T^2(\tau)$ and

$\langle [B_T^{**}(\tau) - B(\tau)]^2 \rangle = \Delta_{1T}^2(\tau)$, then $\Delta_{1T}^2(\tau) = (1 - \tau/T)^2 \Delta_T^2(\tau) +$

$\tau^2 B^2(\tau)/T^2$. This implies obviously that if $\Delta_T^2(\tau) \to 0$, as $T \to \infty$, then also $\Delta_{1T}^2(\tau) \to 0$, as $T \to \infty$.

In connection with the bias of the estimate $B_T^{**}(\tau)$ it should be remarked that in practice interest is often focused on the centered correlation function $b(\tau) = B(\tau) - m^2$ or the normalized correlation function $R(\tau) = b(\tau)/b(0)$. (In particular, it is precisely the estimates of the function $R(\tau)$ that are depicted in Figs. 37 and 38.) However, if $\langle X(t) \rangle = m$ is unknown, then both the estimates $b_T^*(\tau) = B_T^*(\tau) - m_T^{*2}$ and

$\tilde{b}_T^*(\tau)$ (see p. 232 in Vol. I) of the function $b(\tau)$ are biased, while an unbiased estimate of $b(\tau)$ cannot be constructed at all. As for the function $R(\tau)$, its estimate $R_T^*(\tau) = b_T^*(\tau)/b_T^*(0)$, where $b_T^*(\tau) = B_T^*(\tau) - m^2$ or $B_T^*(\tau) - m_T^{*2}$, is biased even when m is known exactly; the same is true

for the estimate $\tilde{R}_T^*(\tau) = \tilde{b}_T^*(\tau)\tilde{b}_T^*(0)$ of $R(\tau)$. Therefore the estimate $B_T^*(\tau)$ of the function $B(\tau)$ has no obvious advantages over $B_T^{**}(\tau)$ when the aim is either the estimation of the function $R(\tau)$ or the estimation of the function $b(\tau)$ at an unknown m.

M. H. Quenouille's approximate method for reduction of the bias in the estimate of $R(\tau)$ consists in replacing the estimate $R_T^*(\tau)$ by $2R_T^*(\tau) - [R_{T/2}^{(1)*}(\tau) + R_{T/2}^{(2)*}(\tau)]/2$, where $R_{T/2}^{(1)*}(\tau)$ and $R_{T/2}^{(2)*}(\tau)$ are estimates of the same kind as $R_T^*(\tau)$, but constructed from the first and second halves of the observed series $x(t)$ (see, e.g., Kendall and Stuart, 1968, Sec. 48.4). Such a replacement greatly reduces the bias, but it increases the variance of the estimate. Section 48.4 of Kendall and Stuart's book also contains some references to papers on the bias of the estimate $R_T^*(\tau)$ for some classes of stationary random sequences with an unknown mean value; see also O. Anderson

(1979b). The general formula for the bias of the estimate $\tilde{b}_T^*(\tau)$ in the case of discrete time was given by O. Anderson (1979a); some examples of bias calculations for estimates $b_T^*(\tau)$ of the function $b(\tau)$ can be found in German, Morozov, and Svalov (1974).

Note also that the contribution of the bias to the total mean-square

error in estimates $b_T^*(\tau)$ (or $\tilde{b}_T^*(\tau)$) and $R_T^*(\tau)$ (or $\tilde{R}_T^*(\tau)$) is usually relatively small, because the bias (the systematic error) decreases in general as T^{-1}, as $T \to \infty$, while the standard deviation (the root-mean-square random error) decreases as $T^{-1/2}$ (cf. the corresponding discussion for the case where $\tau = 0$ in Note 13).

[29]The assumption that for Gaussian stationary random functions $X(t)$ with $\langle X(t) \rangle = 0$ the inequality $\Delta_{1T}^2(\tau) \leqslant \Delta_T^2(\tau)$ is always valid, was made by Parzen (1961). Considerable space in Schaerf's paper (1974) is given to the verification of this assumption. Some of Schaerf's results are also presented in the book by Jenkins and Watts (1968), Sec. 5.3, while Priestley (1981, pp. 323–324) set forth some convincing arguments favoring the estimate $B_T^{**}(\tau)$ as compared with $B_T^*(\tau)$. Finally, one typical example of error computations for estimates of $b(\tau)$, related to $B_T^*(\tau)$ and $B_T^{**}(\tau)$, in the case where $\langle X(t) \rangle$ is unknown, can be found in the paper by German, Morozov, and Svalov (1974).

Parzen proceeded from the fact that $\Delta_{1T}^2(0) = \Delta_T^2(0)$ (since $B_T^{**}(0) = B_T^*(0)$) while, as can easily be checked, $d\Delta_{1T}^2(\tau)/d\tau < d\Delta_T^2(\tau)/d\tau$ for $\tau = 0$, so that necessarily $\Delta_{1T}^2(t) < \Delta_T^2(\tau)$ for small values of τ. (Note, in this connection, that $\Delta_{1T}^2(\tau)$ always decreases, while $\Delta_T^2(\tau)$ may even increase with τ in the vicinity of the point $\tau = 0$.) Moreover, he also took into account that $\Delta_{1T}^2(T) = b^2(T) \ll b^2(0) + b^2(T) = \Delta_T^2(T)$; therefore he concluded that apparently $\Delta_{1T}^2(\tau) \leqslant \Delta_T^2(\tau)$ for all T and τ. Schaerf calculated the values of the functions $\Delta_{1T}^2(\tau)$ and $\Delta_T^2(\tau)$ for Gaussian processes $X(t)$ with $\langle X(t) \rangle = 0$, having the correlation function of the form $b(\tau) = C\exp(-\alpha|\tau|)$, or $b(\tau) = C\cos\omega\tau$, or $b(\tau) = C(1 - \alpha|\tau|)$ for $|\tau| \leqslant \alpha^{-1}$ and $= 0$ for $|\tau| > \alpha^{-1}$, or, finally, $b(\tau) = b(0) = $ constant. Moreover, he also performed similar calculations for arbitrary Gaussian stationary sequences $X(t)$ with $\langle X(t) \rangle = 0$ under the assumption that $T \leqslant 4$. In all these cases the inequality $\Delta_{1T}^2(\tau) \leqslant \Delta_T^2(\tau)$ was fulfilled for all τ, and in some cases the difference between $\Delta_{1T}^2(\tau)$ and $\Delta_T^2(\tau)$ proved to be rather great. Priestley remarked that $\langle B_T^{**}(\tau) - B(\tau) \rangle = |\tau|B(\tau)/T$ and hence the bias of $B_T^{**}(\tau)$ at large enough values of T is in general small for all τ (since $|B(\tau)|$ is usually very small at large values of τ, i.e. if $|\tau|/T$ is not small). Besides, it is possible to show that the variance of $B_T^{**}(\tau)$ is usually of the order of $1/T$, whereas the variance of $B_T^*(\tau)$ is of the order $1/(T - |\tau|)$. Therefore, the variance of $B_T^*(\tau)$ increases substantially as $|\tau|$ approaches T and this increase of the variance produces an erratic behavior of the "tail" of the function $B_T^*(\tau)$ and forces one to prefer the biased estimate $B_T^{**}(\tau)$.

German, Morozov, and Svalov (1974) performed calculations of the mean square errors $\Delta_T^2(\tau)$ and $\Delta_{1T}^2(\tau)$ of the estimates $b_T^*(\tau)$ and $b_T^{**}(\tau)$ (similar to $B_T^*(\tau)$ and $B_T^{**}(\tau)$) of the function $b(\tau)$ for a Gaussian stationary sequence with a correlation function $b(\tau) = Ca^{|T|}$, $|a| < 1$, and an unknown mean value $\langle X(t) \rangle$ (which was estimated by the time average M_T^*). It was found in this work that $\Delta_{1T}^2(\tau) > \Delta_T^2(\tau)$ for some small values of τ, but the difference $\Delta_T^2(\tau) -$

$\Delta^2_{1T}(\tau)$ at such τ is rather small (as it must be since $b_T^*(\tau)$ and $b_T^{**}(\tau)$ differ only slightly at small values of τ), while with a further increase of τ it changes sign and can take relatively large negative values. Therefore they concluded that also in the case of an unknown $\langle X(t)\rangle$ an estimate of the correlation function which is similar to $B_T^{**}(\tau)$ should be preferred to one similar to $B_T^*(\tau)$.

[30]Indeed, let $x_T(0) = x(t)$ for $0 \leqslant t \leqslant T$ and $x_T(t) = 0$ for $t < 0$ or $t > T$. We assume the time to be continuous for definiteness; then (3.33a) can be written as

$$B_T^{**}(\tau_1 - \tau_2) = \frac{1}{T}\int_{-\infty}^{\infty} x_T(t+\tau_1-\tau_2)x_T(t)dt$$
$$= \frac{1}{T}\int_{-\infty}^{\infty} x_T(t+\tau_1)x_T(t+\tau_2)dt \ .$$

It readily follows that

$$\sum_{j,k=1}^{n} B_T^{**}(\tau_j - \tau_k)c_j c_k = \frac{1}{T}\int_{-\infty}^{\infty}\left[\sum_{j=1}^{n} c_j x(t + \tau_j)\right]^2 dt \geqslant 0.$$

[31]It is not expedient to use the estimate $B_T^{**}(\tau)$ for comparatively large values of $|\tau|/T$ since the corresponding *relative error* $|B_T^{**}(\tau) - B(\tau)|/B(\tau)$ can be rather large in this case. However, the absolute error $|B_T^{**}(\tau) - B(\tau)|$ will necessarily be quite small at all τ if T is large enough by virtue of equation (3.7'). (Note that an equation of such a form cannot be proved for the estimate $B_T^*(\tau)$.)

[32]The *Parzen window* shown in Fig. 39(c) is given by the equation $a_T(\tau) = a(\tau/k_T)$, where $a(s) = 1 - 6s^2 + 6|s|^3$ for $0 \leqslant |s| \leqslant 1/2$, $a(s) = 2(1- |s|)^3$ for $1/2 \leqslant |s| \leqslant 1$, and $a(s) = 0$ for $|s| > 1$ (see, e.g., Parzen, 1961). The positive definiteness of such a function $a(s) = a_P(s)$ can be derived from the easily verifiable fact that $a_P(s)$ coincides, with an accuracy to the choice of length units on the coordinate axis, with the convolution of the function $a_B(s) = \max(1$

$- |s|, 0)$ with itself, i.e. with the function $\int_{-\infty}^{\infty} a_B(s - s_1)a_B(s_1)ds_1$. (The

function $a_B(s)$ specifies the *Bartlett window* of Fig. 39(b).) The function $a_B(s)$, in turn, coincides, with the same degree of accuracy, with the convolution of the two "rectangular functions" $a_r(s)$ shown in Fig. 39(a); proceeding from this it is easy to prove that the function $a_B(s)$ is also positive definite. Additional information on this subject can also be found in Notes 50-53 (see pp. 93-94 below) and in Sec. 18 in Vol. I, where some additional examples of lag windows and references to the extensive literature on such windows can be found.

[33]As a few typical examples we can mention the books by Granger

and Hatanaka (1964) and Fishman (1969) on spectral methods in economics, by Båth (1974) on applications of spectral analysis to seismology and other branches of geophysics, and by Krylov (1966) and Konyaev (1981) on some oceanographic applications.

[34]The following books and collections of articles are devoted (entirely or at least mainly) to statistical spectral analysis: Blackman and Tukey (1959), Harris (1967), Jenkins and Watts (1968), Konyaev (1973), Gribanov and Mal'kov (1974, 1978), Koopmans (1974), Childers (1978), Zhurbenko (1982), Haykin and Cadzow (1982), Haykin (1983), Durrani (1983), Grenier (1986), and many others. A chapter (or several chapters) on spectral estimation can also be found, e.g., in texts by Grenander and Rosenblatt (1956), Hannan (1960, 1970), Korn (1966), Vilenkin (1967), Robinson (1967), Kendall and Stuart (1968), Anderson (1971), Mirsky (1972), Otnes and Enochson (1972, 1978), Brillinger (1975), Bloomfield (1976), Bartlett (1978), Robinson and Silvia (1978), Priestley (1981), Papoulis (1984), Rosenblatt (1985), and Bendat and Piersol (1986). Among the many published surveys of the field, only the following three typical recent examples will be mentioned: Alekseev and Yaglom (1980; 74 references), Parzen (1981; 24 references), and Kay and Marple (1981; 278 references). As for the immense number of spectral analysis research papers, many of them will be indicated in the following notes to this chapter, but many more will not be mentioned at all.

[35]It was shown in Note 30 that in the case of continuous time

$$B_T^{**}(\tau) = \frac{1}{T}\int_{-\infty}^{\infty} x_T(t + \tau)x_T(t)dt,$$

where $x_T(t) = x(t)$ for $0 \leqslant t \leqslant T$ and $x_T(t) = 0$ for $t < 0$ or $t > T$. Quite a similar formula can, of course, also be obtained for discrete time, the only difference being that the integral from $-\infty$ to ∞ must now be replaced by the sum over all the integral values of t. Let us substitute now this expression for $B_T^{**}(\tau)$ into the right-hand side of (3.35) (or (3.35a)) and note that the integration (or summation) limits in (3.35) (or (3.35a)) can be replaced by $-\infty$ and ∞, since $B_T^{**}(\tau) = 0$ for $|\tau| \geqslant T$. Then, using the obvious identity $\exp(-i\omega\tau) = \exp[i\omega(t+\tau)]\exp(-i\omega\tau)$ we easily get the result (3.36) (or (3.36a)). This result is, in fact, a simple corollary of the "convolution theorem" for Fourier's transforms, since the above formula for $B_T^{**}(\tau)$ implies that $TB_T^{**}(-\tau)$ is a convolution of the functions $x_T(-t)$ and $x_T(t)$.

[36]See Schuster (1898, 1900, 1906). Schuster's definition of the periodogram differs slightly from the one used in this book, but the difference is quite insignificant. Schuster used the periodogram to search the "hidden periodicities" in fluctuating time series (see p. 307 in Vol. I).

[37]Averaging both sides of the equation of the form (3.35) that

expresses the periodogram $I_T(\omega)$ via the estimator $B_T^{**}(\tau)$ and then replacing $\langle B_T^{**}(\tau)\rangle$ by $(T - \tau)B(\tau)/T$ and $B(\tau)$ by its spectral representation (2.67), we obtain

$$\langle I_T(\omega)\rangle = \frac{1}{2\pi T}\int_{-T}^{T}\int_{-\infty}^{\infty}(T - |\tau|)e^{-i(\omega-\omega')\tau}\,f(\omega')d\omega'd\tau.$$

Now integrating with respect to τ, we get (3.39). The same result can be derived from (3.37) if we apply to its right-hand side equation (1.56) for the mean value of a product of two integrals involving $X(t)$, then replace $B(t - s)$ by its spectral representation, and finally, integrate with respect to t and s. (Note that in this second derivation of (3.39) it is convenient to replace the integral from 0 to T on the right-hand side of (3.37) by the integral from $-T/2$ to $T/2$. The permissibility of such a replacement, when the mean value of $I_T(\omega)$ is evaluated, is clearly justified by the stationarity of $X(t)$.) See also Note 55 on p. 95, where a third method for deriving (3.39) is applied to a somewhat more general problem. Equation (3.39a) relating to the case of discrete time is proved quite similarly.

To prove equation (3.40) for the case of a Gaussian $X(t)$ we take advantage of the fact that in calculating the moments of $I_T(\omega)$ the periodogram can be replaced by $J_1^2(\omega) + J_2^2(\omega)$, where

$$J_1(\omega) = (2\pi T)^{-1/2}\int_{-T/2}^{T/2}\cos\omega t\, X(t)dt \quad \text{and} \quad J_2(\omega) = (2\pi T)^{-1/2}\int_{-T/2}^{T/2}\sin\omega t\, X(t)dt$$

are Gaussian random variables having mean values zero. Let us apply (3.2') (see p. 70) to the calculation of the moments $\langle J_i^2(\omega_1)J_j^2(\omega_2)\rangle$, where $i = 1, 2$ and $j = 1, 2$, and use the fact that $\langle J_1(\omega_1)J_2(\omega_2)\rangle = \langle J_2(\omega_1)J_1(\omega_2)\rangle = 0$ (because according to (1.56) both these mean values are represented by integrals over the square $-T/2 \leqslant t, s \leqslant T/2$ of some odd functions of two variables t and s). Then we obtain

$$b_I(\omega_1\omega_2) = 2[\langle J_1(\omega_1)J_1(\omega_2)\rangle]^2 + 2[\langle J_2(\omega_1)J_2(\omega_2)\rangle]^2$$

$$= [\langle J_1(\omega_1)J_1(\omega_2) + J_2(\omega_1)J_2(\omega_2)\rangle]^2$$

$$+ [\langle J_1(\omega_1)J_1(\omega_2) - J_2(\omega_1)J_2(\omega_2)\rangle]^2$$

$$= \left[\frac{1}{2\pi T}\int_{-T/2}^{T/2}\int_{-T/2}^{T/2}\cos(\omega_1 t - \omega_2 t')B(t - t')dtdt'\right]^2$$

$$+ \left[\frac{1}{2\pi T}\int_{-T/2}^{T/2}\int_{-T/2}^{T/2}\cos(\omega_1 t + \omega_2 t')B(t - t')dtdt'\right]^2.$$

Replacing now the correlation function $B(t - t')$ on the right-hand side of the last equation by its spectral representation and then integrating with respect to t and t', we arrive at (3.40). For discrete time, the same arguments lead to an equation for $b_I(\omega_1,\omega_2)$ which differs from (3.40) only by replacement of the limits of integration by $-\pi$ and π, and of the function $g_T(x) = (2/\pi T)^{1/2}\sin(Tx/2)/x$ in the integrand by the functions $g_T^{(1)}(x) = (1/2\pi T)^{1/2}\sin(Tx/2)/\sin(x/2)$.

The integral from $-\infty$ to ∞ of the square of the function $g_T(x)$ is equal to unity, and the function $[g_T(x)]^2$ is even and tends to the δ-function $\delta(x)$ as $T \to \infty$. The graph of the function $g_T(x)$ at large T has a sharp peak at $x = 0$ (where the value $g_T(0) = (T/2\pi)^{1/2}$ is reached) and falls off rapidly with $|x|$. At the points $x = \pm\pi/T$ the function $g_T(x)$ vanishes, and with a further increase in $|x|$ it fluctuates about zero, taking relatively small (compared with the main maximum) values of both signs (cf. a similar graph in Fig. 27). The function $g_T^{(1)}(x)$, arising when we deal with sequences $X(t)$ (i.e. with the case of discrete time), behaves similarly, with the only difference that the integral of $[g_T^{(1)}(x)]^2$ from $-\pi$ to π is equal to unity, while the function $g_T^{(1)}(x)$ is periodic with period 2π. These facts easily imply relations (3.41) (plus their analog for discrete time) and (3.42). Moreover, the shape of the functions $g_T(x)$ (and $g_T^{(1)}(x)$) appearing in the expression for $b_I(\omega_1,\omega_2)$ shows that the

correlation between the variables of $I_T(\omega_1)$ and $I_T(\omega_2)$ at large, but finite values of T practically vanishes already at a distance $|\omega_2 - \omega_1|$ of the order of several segments of length $1/T$. The same conclusion can be deduced from the following simplified asymptotic equation for $b_I(\omega_1,\omega_2)$, which is valid at large T:

$$(3.11') \quad b_I(\omega_1,\omega_2) \approx \left\{ \left[\frac{\sin[T(\omega_2-\omega_1)/2]}{T(\omega_2-\omega_1)/2} \right]^2 + \left[\frac{\sin[T(\omega_2+\omega_1)/2]}{T(\omega_2+\omega_1)/2} \right]^2 \right\} f(\omega_1)f(\omega_2).$$

(In the case of discrete time, i.e. for stationary sequences, $(\omega_2 - \omega_1)/2$ and $(\omega_2 + \omega_1)/2$ in the denominators of the right-hand side of (3.11') must only be replaced by $\sin[(\omega_2 - \omega_1)/2]$ and $\sin[(\omega_2 + \omega_1)/2]$. However, for very small values of $|\omega_2 - \omega_1|$ or $|\omega_2 + \omega_1|$, which is the most interesting case, this replacement is of little consequence.) The heuristic derivation of (3.11') is given in the book by Jenkins and Watts (1968), Chaps. 6 and 9. The same result can be derived more rigorously (for discrete time) from the equations proved in Hannan (1960), Sec. III.1, or in Hannan (1970), Sec. V.2, or in Brillinger (1975), Sec. 5.2, or in Priestley (1981), Sec. 6.2.2. Equation (3.11') describes the asymptotic (for large T) behavior of the function $b_I(\omega_1,\omega_2)$ in a much simpler way than (3.40). Note, in particular, that (3.11') implies that the correlation between $I_T(\omega_1)$ and $I_T(\omega_2)$ at large T is especially small if both $T(\omega_2 - \omega_1)/2\pi$ and $T(\omega_2+ \omega_1)/2\pi$ are integers (e.g., when $\omega_1 = 2\pi n_1/T$, $\omega_2 = 2\pi n_2/T$, where n_1 and n_2 are integers).

It follows from (3.38) that the periodogram $I_T(\omega)$ at $\omega = 0$ (or, for discrete time, at $\omega = 0$ and $\omega = \pm\pi$) is equal to the square of a normally distributed (Gaussian) random variable with mean value zero, while at other values of ω it is equal to the sum of the squares of two Gaussian random variables with mean values zero. Moreover, it is easy to show that at large values of T these two Gaussian random variables are practically uncorrelated (and, hence, independent) and have equal variances. These last results can also be formulated as follows: *if T is* *large enough, then at $\omega = 0$ (and for discrete time at $\omega = 0$ and $\omega = \pm\pi$) the probability distribution of the appropriately normalized periodogram $I_T(\omega)$ is very close to a χ_1^2 distribution, and at the other values of ω, to a χ_2^2 distribution.* (In probability theory χ_ν^2 always means the "chi-square distribution with ν degrees of freedom", which is the probability distribution of the sum of squares of ν independent Gaussian variables with mean value zero and unit variance; see, e.g., any of the books on statistics given in Note 1 to this chapter, or Priestley, 1981, Sec. 2.11, or Papoulis 1984, p. 187.) The properties of chi-square distributions imply, in particular, that the variables $I_T(\omega_1)$ and $I_T(\omega_2)$, where $\omega_2 \neq \pm \omega_1$, become, as $T \to \infty$, not only asymptotically uncorrelated, but also asymptotically independent (since for chi-square distributions, as well as for normal distributions, the lack of correlation implies independence).

So far we have only been discussing the case of Gaussian (normal) functions $X(t)$. However, in the above reasoning we actually used not the normality of $X(t)$, but only the normality of the integrals (or sums, if time is discrete) $J_1(\omega)$ and $J_2(\omega)$. Recall now that according to the central limit theorem for stationary random functions (see Note 9 on p. 67) the probability distributions of the random variables $J_1(\omega)$ and $J_2(\omega)$ will tend, as $T \to \infty$, to normal probability distributions under very wide conditions and even the two-dimensional distribution of $J_1(\omega)$ and $J_2(\omega)$ will tend to two-dimensional normal distribution. Thus it is clear that the statistical properties of the periodogram $I_T(\omega)$ considered on pp. 249–250 in Vol. I and in this Note must be valid not only for Gaussian functions $X(t)$, but also for wide classes of non-Gaussian stationary functions $X(t)$. Indeed, the books by Jenkins and Watts (1968), Chap. 9; Hannan (1960), Sec. III.1, and (1970), Chap. V; Anderson (1971), Chap. 8; Brillinger (1975), Chap. 5; Priestley (1981), Sec. 6.2.2; and Rosenblatt (1985), Sec. V.1, contain the generalizations of the indicated results to several different wide classes of non-Gaussian stationary functions $X(t)$ (usually, only for discrete time). The statistical properties of the periodogram $I_T(\omega)$ at large values of T (also under the assumption that the time is discrete) are studied in the special papers by Walker (1965) and Olshen (1967) and are considered in almost all the books cited above.

[38]Note that if we assume the periodogram $i_T(\omega)$ (or the spectral density $f(\omega)$) to be a function of some nonlinear function $T_0(\omega)$ of ω (e.g., of the period $T_0 = 2\pi/\omega$), rather than of the frequency ω itself, then to preserve the area under the graph of the function (which must be equal to $B(0) = \langle|X(t)|^2\rangle$), we must multiply the ordinates by $|d\omega/dT_0|$ (i.e., by $2\pi/T_0^2$, if T_0 is the period); cf. Chiu (1967). The local maxima for the new ordinates will no longer coincide precisely with the local maxima for the initial function $i_T(\omega)$ or $f(\omega)$, though usually the shifts in the positions of maxima will not be very significant. Nevertheless, this argument shows that with a continuous spectrum the exact positions of the spectral peaks are of a conventional nature (since they depend on the choice of the independent variable) and have no strict physical

meaning. (This remark is, of course, unrelated to the frequencies of the
genuine periodic components of $X(t)$, i.e. to the discrete spectrum.)

A detailed graph of the dependence of the periodogram $i_T(\omega)$ for
the Beveridge time series on the frequency, which takes into account
many more values of the argument, can be found, e.g., in the books
by Kendall and Stuart (1968), Fig. 49.1, and by Anderson (1971), Fig.
A.1.3. This graph was found to be even appreciably more erratic
than the curve in Fig. 40. In the same two books, and also in those
by Jenkins and Watts (1968), Brillinger (1975), and some others, as
well as in many papers by various authors, one can find a great
number of additional examples of empirical periodograms $i_T(\omega)$ (see,
e.g., the next Note). All these periodograms appear to be extremely
irregular functions of frequency.

[39]This graph is taken from Polyak (1975). Many other examples of
empirical periodograms $i_T(\omega)$ for meteorological and hydrological time
series can be found in the books by Polyak (1975, 1979) and in a number
of papers published in "Trudy GGO" ("Works of Main Geophysical
Observatory", Leningrad); see, e.g., the references in Yaglom (1981).
All these periodograms have a form similar to that in Fig. 40(b).

[40]The method for obtaining a consistent estimate of $f(\omega)$ by
partitioning the full record of $x(t)$ into n pieces (where $1 << n << T$),
computing the periodograms for each of these pieces, and forming
the arithmetic mean of all the periodograms is due to Bartlett
(1948). Later this method was improved considerably by Welch
(1967) (these improvements are discussed in Sec. 19; see, in
particular, pp. 275–276 and 286–288 in Vol. I and Notes 57 and 67)
and now it is rather widely used in practice (cf., e.g., Nuttall and
Carter, 1980, 1982; Carter and Nuttall, 1980a,b and 1983; Kay and
Marple, 1981; Zhurbenko, 1979, 1980, 1982; and Dahlhaus, 1985a).

[41]The arguments given here were first outlined by Daniell (1946) in a
short written contribution to the discussion at the "Symposium on
Autocorrelations in Time Series". A little later Tukey (1949) presented
them at greater length and developed a practical method for computing
the estimates of spectral density.   See in this respect also Bartlett
(1950, 1978).

When the writing of the book was already finished, the author
became acquainted with an amazing early paper by Einstein
(1914) (cf. also Note 13a to Chap. 2). In this paper Einstein
introduces the new statistical characteristic of fluctuating time
series $x(t)$ ("the statistical value of fluctuating observations")
which he calls the *intensity* $I(\omega)$ of $x(t)$. (The original notations
of Einstein are slightly changed here to make them more similar
to notations adopted in this book.) Einstein's definition of
intensity $I(\omega)$ implies that it coincides with the periodogram $i_T(\omega)$
(where T is assumed to be very large) averaged over the frequency
interval of small width $\Delta\omega$ satisfying the condition that $T\Delta\omega$ is large

enough.   Then Einstein proves that the intensity $I(\omega)$ is the Fourier transform of the correlation function $B(\tau)$ of $x(t)$. (In this proof the average of $i_T(\omega)$ over a narrow frequency band is in fact replaced by the mean value of $I_T(\omega)$.) Hence we see that Einstein already knew in 1914 the definitions of the correlation function and of the spectral density of a stationary time series; moreover, he also knew the relation between these two quantities (which is the essence of the Wiener-Khinchin theorem) and proposed (without a formal proof) a correct method for the determination of the spectral density from a single lengthy realization of the series which was rediscovered (by P.J. Daniell) only 32 years later.

[42]See, e.g., the engineering thesis by Goldstein (1962) and the mathematical works by Hinich (1967), McNeil (1967), Brillinger (1968), and Kedem (1980), Chap. 8, where asymptotic statistical properties of the corresponding spectral estimates are studied and some applications of the method are described. The more special problem of estimating the second spectral moment (which is a functional of $f(\omega)$) from zeros of a Gaussian process is discussed by Lindgren (1974), where it is shown, in particular, that in some cases the estimate based on the observation of $\mathrm{sgn}\,x(t)$ appears to be more efficient than the one obtained conventionally from $x(t)$. The general problem of estimating the spectral density of a Gaussian sequence $X(t)$, when only a function $g\{x(t)\} = y(t)$ of $x(t)$ (belonging to some rather wide functional class) has been observed, is considered by Rodemich (1966).

[43]The terms "spectral window" and "lag window" are due to Tukey; see Blackman and Tukey (1959).
We will make one more general remark concerning the spectral estimators of the form

$$\Phi_T^{(A)}(\omega) = \int_{-\pi}^{\pi} A_T(\omega - \omega')I_T(\omega')d\omega' = (2\pi)^{-1} \sum_{\tau=-T+1}^{T-1} e^{-i\omega\tau} a_T(\tau)B_T^{**}(\tau)$$

(the time is now assumed to be discrete). If $X(t)$ is a physical quantity having a definite dimension (e.g., the dimension of length, or of velocity, or of electric voltage, etc.), then both $B(\tau)$ and its Fourier transform $f(\omega)$ will be quadratic in this dimension. Therefore it seems natural to seek an estimate of $f(\omega)$ as a quadratic form in the observations, i.e. to consider the estimator

$$\Psi_T^{(a)}(\omega) = \sum_{t=1}^{T} \sum_{s=1}^{T} a_T(t,s;\omega)X(t)X(s).$$

The class of estimators $\Psi_T^{(a)}(\omega)$ includes all the estimators $\Phi_T^{(A)}(\omega)$ (which are obtained if we assume that

$$a_T(t,s;\omega) = \frac{\exp(-i\omega(t-s))}{2\pi T(T-|t-s|)}\, a_T(|t-s|)),$$

but it is, of course, much wider than the class of estimators $\Phi_T^{(A)}(\omega)$. It was, however, proved by Grenander and Rosenblatt (1956a), Sec. 4.2, that at large T little is lost by restricting oneself to estimators $\Phi_T^{(A)}(\omega)$, since under wide conditions *for every asymptotically unbiased estimator* $\Psi_T^{(a)}(\omega)$ *of the spectral density* $f(\omega)$ *there exists an estimator* $\Phi_T^{(A)}(\omega)$ *with the same bias and a mean square error that is asymptotically not larger than that of the estimator* $\Psi_T^{(a)}(\omega)$.

The stated result shows that it is unnecessary to consider any quadratic estimators different from $\Phi_T^{(A)}(\omega)$ (at least, for discrete time). However, several types of nonquadratic estimators are also widely used in practice now; some of them are considered in Sec. 19 of this book.

[44]If we do not claim perfect mathematical rigor, then it is very easy to obtain the important equation (3.49) by using equation (1.56) for the mean value of a product of two integrals involving $X(t)$, the approximate equation (3.11') for $b_I(\omega_1,\omega_2)$, and the facts that $f(\omega)$ is an even function of ω and that

$$\frac{2\sin^2(Tx/2)}{\pi T x^2} \to \delta(x) \text{ as } T \to \infty.$$

In fact,

$$\sigma^2\left\{\Phi_T^{(A)}(\omega)\right\} = \left\langle\left\{\int A_T(\omega - \omega')[I_T(\omega') - \langle I_T(\omega')\rangle]d\omega'\right\}^2\right\rangle$$

$$= \iint A_T(\omega - \omega')A_T(\omega - \omega'')b_I(\omega',\omega'')d\omega'd\omega''$$

$$\approx \iint A_T(\omega - \omega')A_T(\omega - \omega'')\left\{\frac{4\sin^2[T(\omega' - \omega'')/2]}{T^2(\omega' - \omega'')^2}\right.$$

$$+ \left.\frac{4\sin^2[T(\omega' + \omega'')/2]}{T^2(\omega' + \omega'')^2}\right\}f(\omega')f(\omega'')d\omega'd\omega''$$

$$\approx \frac{2\pi}{T}\iint A_T(\omega - \omega')A_T(\omega - \omega'')\left\{\delta(\omega' - \omega'')\right.$$

$$+ \left.\delta(\omega' + \omega'')\right\}f(\omega')f(\omega'')d\omega'd\omega''$$

$$= \frac{2\pi}{T}\int A_T(\omega - \omega')[A_T(\omega - \omega') + A_T(\omega + \omega')]f^2(\omega')d\omega'$$

(see, e.g., Jenkins and Watts, 1968, Chap. 6). For information on a more rigorous derivation of this result, its extension to certain wide classes of non-Gaussian stationary functions $X(t)$, and discussion of some related questions, see, e.g., the books by Grenander and Rosenblatt (1956a), Chap. 4; Hannan (1960), Chap. III, and (1970), Chap. V; Jenkins and Watts (1968), Chap. 9; Anderson (1971), Chap. 9; Koopmans (1974), Sec. 8.3; Brillinger (1975), Chap. 5; Priestley (1981), Chap. 6 and Rosenblatt (1985), Sec. V.2; and the papers by Grenander and Rosenblatt (1953, 1956b), Parzen (1957a,b, 1958b, 1961), Jenkins (1961, 1965), Priestley (1962, 1965a), Brillinger (1969), Alekseev (1971a, 1973a), Khachaturova (1972), and Bentkus and Rutkauskas (1973). (In almost all these sources attention is centered on the case of discrete time.) The validity of (3.49) not only for Gaussian, but also for wide classes of non-Gaussian stationary functions $X(t)$ is, of course, a consequence of the applicability, under wide conditions, of the central limit theorem to the functions $X(t)\cos\omega t$ and $X(t)\sin\omega t$ (cf. Notes 37 and 9 to this chapter).

Neave (1970a,b) derived some modified asymptotic expressions for $\sigma^2\{\Phi_T^{(A)}(\omega)\}$ which are more accurate than (3.49) in some cases of not-too-large values of T. Moreover, he also later evaluated the exact variance $\sigma^2\{\Phi_T^{(A)}(\omega)\}$ and compared the exact results for some forms of $B(\tau)$ with the asymptotic expressions; see Neave (1971).

Equations (3.48) and (3.49) permit one to prove easily that under fairly general conditions the estimate $\varphi_T^{(A)}(\omega)$ is consistent, i.e. $\Phi_T^{(A)}(\omega) \rightarrow f(\omega)$ (in the mean, and also in probability), as $T \rightarrow \infty$. The same equations and some of its refinements can be used for estimating the rate of convergence of $\Phi_T^{(A)}(\omega)$ to $f(\omega)$ for several more restricted (but nevertheless rather wide) classes of stationary functions $X(t)$. These rate estimates can then be used to prove that in many cases (including the vast majority of instances enountered in practice) $\Phi_T^{(A)}(\omega) \rightarrow f(\omega)$ not only in the mean, but also almost surely (i.e. with probability one) as $T \rightarrow \infty$, and that the convergence of $\Phi^{(A)}(\omega)$ to $f(\omega)$ is quite often uniform, i.e. takes place for all the values of ω with the same rate of convergence (see, e.g., Alekseev, 1974b, and Gaposhkin, 1980).

[45]This kind of terminology was originated by Parzen; cf. Parzen (1963).

[46]The situation of interest here is very similar to those related to Heisenberg's famous "uncertainty principle" in quantum mechanics, which shows that for a microscopic particle the "accuracy of the coordinate measurement" and the "accuracy of the velocity measurement" are antagonistic (cf. also Note 24 to Chap. 2 on pp. 41–45). Grenander (1951, 1958) showed that the antagonism of bias reduction to variance reduction can also be formulated mathematically as a specific "uncertainty principle". To derive this principle he proposed a strict mathematical definition of "overall width" (or "bandwidth", as it is usually called) of a spectral window and proved that the bandwidth B_A of a window $A_T(\omega)$ satisfied the condition

$$(3.12')\quad B_A \times \sigma^2\{\Phi_T^{(A)}(\omega)\} = \text{const.}$$

This formula is often called *"Grenander's uncertainty principle"*; it shows that *the bandwidth and the estimate variance are antagonistic.* It is clear that the increase in bandwidth implies an increase in the estimate bias, and vice versa; hence (3.12') also gives the mathematical formulation of the fact that *the variance reduction implies a bias increase, and vice versa.* Relating bandwidth to the "lack of resolvability" and the variance to the "lack of reliability", Grenander (1958) also formulated his uncertainty principle as follows: *"reliability and resolvability are antagonistic"*. In the same paper he also showed that (3.12′) remains valid if his original definition of the bandwidth B_A is replaced by another reasonable definition of this quantity. More detailed discussion of various definitions of the window bandwidth and of the uncertainty principle can be found in Priestley (1981), Sec. 7.3.2; cf. also Nuttall and Carter (1982), p. 1117.

[47]The result (3.54a) is due to Parzen; see, e.g., Parzen (1957a, 1958b, 1961) and also Hannan (1960), Sec. III.2, and (1970), Sec. V.4; Anderson (1971), Sec. 9.3.2; Brillinger (1975), Chaps. 3 and 5; Priestley (1981), Sec. 6.2.4; and Rosenblatt (1985), Sec. V.2.

[48]It is easy to verify that the minimal value of $C(T) = C_1 k_T^{-2q} + C_2 k_T T^{-1}$, where C_1 and C_2 are independent of T, is reached when $k_T = (C_1/C_2)^{1/(2q+1)} T^{1/(2q+1)}$ and is equal to $2 C_1^{1/(2q+1)} C_2^{2q/(2q+1)} T^{-2q/(2q+1)}$. (This is an obvious consequence of the fact that $dC(T)/dT$ vanishes at the above-mentioned value of k_T.) When (3.54a) is valid $\Delta^2\{\Phi_T^{(A)}(\omega)\} = \Delta_T^2$ is close to a function of the form $C(T)$ for large values of T. Moreover, since $T^{-2q/(2q+1)}$ decreases with an increase of q, the stated result implies that, if $f(\omega)$ is a very smooth function of ω (i.e. can be differentiated very many times), it is advantageous to utilize estimators $\Phi_T^{(A)}(\omega)$ with a large value of q (see Alekseev, 1973b; cf. also Vol. I, pp. 271–272). However, if $f(\omega)$ can be differentiated not more than r times, where r is a fixed integer, then too large an increase in q will give no gain. Namely, it can be shown that *if* f(ω) *has only* r *finite derivatives and its* r-*th derivative* f$^{(r)}(\omega)$ *is a function of the class* Lipα, *where* $0 < \alpha \leqslant 1$, (i.e. $|f^{(r)}(\omega') - f^{(r)}(\omega'')| \leqslant K |(\omega'-\omega'')|^{\alpha}$, $0 < K < \infty$, *for any* ω' *and* ω''), *then* Δ_T^2 *will asymptotically fall off like* $T^{-(2r+2\alpha)/(2r+2\alpha+1)}$ *with an increase of* T *for any* q $\geqslant$ r + α (see Alekseev, 1974a; cf. also Alekseev and Yaglom, 1980). This result apparently holds for both discrete and continuous time (though it has been proved only for discrete time) and under rather wide conditions imposed on $X(t)$ and guaranteeing the validity of the used asymptotic formulae for the estimator bias and variance.

Note now that the following important result can also be proved: *for any estimator* $\Psi_T^*(\omega)$ *of a spectral density* f(ω) *depending on the values of* X(t) *at* t = 1, ..., T *or* 0 $\leqslant$ t $\leqslant$ T *a Gaussian stationary function* X(t) *exists that has the following properties:* (i) *the spectral density of* X(t) *has an* r-*th derivative* f$^{(r)}(\omega)$ *belonging to the class* Lipα, *and* (ii) *the corresponding mean square error* $\Delta^2\{\Psi_T^*(\omega)\} = \langle\{\Psi_T^*(\omega) - f(\omega)\}^2\rangle$ *falls off, as* T $\rightarrow \infty$, *as* $T^{-(2r+2\alpha)/(2r+2\alpha+1)}$ *or even slower* (see Samarov, 1977, and Farrell, 1979; cf. also Alekseev and Yaglom, 1980, Rosenblatt, 1985, Sec. V.6, and Bentkus, 1985a, b). We see that no spectral estimator can exist that for any spectral density of given smoothness yields a rate of decrease with T of the mean square error which is superior to the rate attainable for the best smoothed periodogram estimator $\Phi_T^{(A)}(\omega)$.

[49]See Blackman and Tukey (1959), p. 11; Tukey (1961), Parzen (1967b),

Jenkins and Watts (1968), Chap. 7.   The book by Blackman and Tukey (1959), Chap. 7 in Jenkins and Watts (1968), and Chap. 7 in Priestley (1981), are entirely devoted to the practical aspects of empirical spectral analysis; cf. also Hannan (1960), Sec. III.3, and (1970), Sec. V.7, and Bendat and Piersol (1986), Secs. 8.5 and 11.5.

[50]It has already been mentioned in Note 32 (see p. 83) that by an appropriate change of scales the Bartlett (i.e. triangular) lag window $a_B(\tau)$ can be made to coincide with the convolution of two identical rectangular lag windows $a_r(\tau)$, while the Parzen lag window $a_P(\tau)$ can be made to coincide with the convolution of two identical Bartlett windows $a_B(\tau)$.   This implies that if the time is continuous, the spectral window $A_B(\omega)$ corresponding to the lag window $a_B(\tau)$ is of the form $k_1[A_r(k_2\omega)]^2$, where $A_r(\omega)$ is the Fourier transform of $a_r(\tau)$ and k_1 and k_2 are constants, while the spectral window $A_P(\omega)$ corresponding to the lag window $a_P(\tau)$ is of the form $k_1'[A_B(k_2'\omega)]^2$. The explicit equations for the three spectral windows corresponding to these three lag windows in the case of discrete time can easily be obtained from the general formula given in the footnote on p. 263 in Vol. I; see also, e.g., Hannan (1960), Sec. III.2; Anderson (1971), Sec. 9.2.3; or Priestley (1981), Sec. 6.2.3.
Note that smoothing of a function of ω by a convolution with an appropriate kernel (or window) $A(\omega)$ is often used in many branches of advanced calculus; therefore, many such kernels bear special names.   Thus, the kernel $A_r(\omega)$ corresponding to the rectangular lag window is often called the *Dirichlet kernel*, while the kernel $A_B(\omega)$ is the *de la Vallé-Poussin–Jackson kernel* (see, e.g., Akhiezer, 1956, or Brillinger, 1975, Sec. 3.3).

[51]See, e.g, Blackman and Tukey (1959), where the properties of both Tukey's windows are studied in detail and it is also noted in passing that the first of these windows is called "Hanning" after the prominent Austrian meteorologist and climatologist Julius von Hann (1839–1921), who had applied, at the beginning of this century, the related smoothing to series of meteorological observations, while the second window is called "Hamming" after the contemporary applied mathematician R. W. Hamming. The formulae for Fourier transforms $A(\omega)$ of the lag window generators (3.55) and for spectral windows $A_T(\omega)$ corresponding to lag windows $a_T(\tau) = a(\tau/k_T)$, $\tau = 0,\pm1,\pm2, ...,$ can be found, e.g., in Blackman and Tukey (1959); Hannan (1960), Sec. III.2; Anderson (1971), Sec. 9.2.3; Brillinger (1975), Sec. 3.3; and Priestley (1981), Sec. 6.2. The minimal value of the spectral window generator $A(\omega)$ in the case of the "Hanning" lag window is negative, but in absolute value it is close to only 2% of the main peak of $A(\omega)$ at $\omega = 0$; for the "Hamming" window the minimum of $A(\omega)$ is also negative, but now its height is about 1/3 of the corresponding height for the "Hanning" window.
The window (3.55) can, of course, be rewritten as $a(\tau) = (1 - 4a) - 4a\cos^2(\pi\tau/2)$, $|\tau| \leqslant 1$. The generalization of this window as $a(\tau) = c +$

$(1 - c)\cos^n(\pi\tau/2)$ was suggested by Cook and Bernfield (1967), while Webster (1978) considered a generalized Tukey window of the form $a(\tau) = c\,\cos^n(\pi\tau/2) + (1 - c)\cos^{n+2}(\pi\tau/2)$.   For other, more practical, generalizations of Tukey's windows see, e.g., F. Harris (1978) and Nuttall (1981).

[52]Many examples of lag windows and lag window generators can be found, e.g., in Parzen (1958b, 1961), Palmer (1969), F. Harris (1978), Papoulis (1973a, 1984), Nuttall (1981) and Palenskis et al. (1985); see also Mirsky (1972), Sec. 5.4; Brillinger (1975), Sec. 3.3; Max (1981), Chap. 14; and Priestley (1981), Sec. 6.2.   Here we indicate only the following three additional examples of positive definite lag window generators: a Gaussian window $a(\tau) = \exp(-\alpha\tau^2)$ mentioned by Daniels (1962) and also considered by Max (1981) and Palenskis et al. (1985), a window of the form $a(\tau) = \exp(-\alpha\tau^2)\max(1- |\tau|, 0)$ suggested by Nidekker (1968), and Hext's window (1966) which is a convolution of the Bartlett and Parzen windows and has a Fourier transform

$$A(\omega) = \frac{10}{33\pi} \left\{ \frac{\sin\omega/6}{\omega/6} \right\}^6.$$

[53]The utilization of a triangular spectral window was recommended, in particular, by Cooley, Lewis, and Welch (1970), and a parabolic window was introduced by Priestley (1962, 1965a) and, independently, by Bartlett (1963).   Therefore, this last window is called the *Bartlett–Priestley window* in Priestley's book (1981). Priestley studied the problem on determining the form of a non-negative spectral window which minimizes the asymptotic expression for the mean square error of the spectral density estimate. Assuming that a true spectral density $f(\omega)$ is twice differentiable, he found that the parabolic window is optimal (cf. Priestley, 1981, pp. 444, 445, 569-571, and also Bloomfield, 1976, pp. 202-203). Alekseev (1971b) considered the same problem under the condition that $f(\omega)$ has r derivatives and its r-th derivative $f^{(r)}(\omega)$ is a function of class Lipα, where $0 < \alpha \leqslant 1$. He proved that in this case the optimal non-negative spectral window generator $A(\omega)$ has the form

$$(3.13') \qquad A(\omega) = \begin{cases} \dfrac{\beta + 1}{2\beta}\,(1 - |\omega|^\beta) & \text{for } |\omega| \leqslant 1, \\[2mm] 0 & \text{for } |\omega| > 1, \end{cases}$$

where $\beta = \min(r+\alpha,\ 2)$.   Later the same result was proved by Zhurbenko (1980, 1982) under somewhat more general conditions imposed on the stationary function $X(t)$; this is why the spectral estimate corresponding to the window $(3.13')$ was called the Zhurbenko estimate by Shiryaev (1980).

[54]The utilization of spectral windows with alternating signs for bias reduction was suggested at the beginning of the 1960s (see,

e.g., Daniels, 1962; Bartlett, 1963, 1967), but then it attracted little attention.   Later Parzen (1972) recommended, at a meeting of turbulence experts, a special "bias reducing lag window" $a(\tau)$, which corresponds to an alternating spectral window $A(\omega)$ satisfying the condition $\int_{-\infty}^{\infty} \omega^2 A(\omega) d\omega = 0$.   Special discrete analogues of alternating spectral windows are also considered in Polyak's book (1975, Secs. 1.5 and 5.8) intended for meteorologists.   A systematic investigation of the spectral estimates corresponding to even spectral window generators $A(\omega)$ satisfying the conditions

$$(3.14') \qquad \int_{-\infty}^{\infty} \omega^{2k} A(\omega) d\omega = 0, \qquad k = 1, ..., m - 1; \qquad \int_{-\infty}^{\infty} \omega^{2m} A(\omega) d\omega \neq 0$$

(so-called *spectral windows of 2m-th order)* was carried out (for discrete time only) by Alekseev (1973b, 1974a,b, 1980, 1981, 1985); see also Alekseev and Yaglom (1980).   It was noted in Note 53 that the optimal form of a nonnegative spectral window is a parabolic one for any twice differentiable spectral density $f(\omega)$; for this window $q = 2$ in (3.54a).   Moreover, it was also stated in Note 48 that for sufficiently smooth spectral densities $f(\omega)$ alternating spectral windows with greater values of q lead to a considerably smaller asymptotic mean square error of an estimate, than any nonnegative spectral window.   Alekseev (1973b) suggested three different sequences of alternating spectral windows $A^{(l)}(\omega)$ (where l = 2, 4, ... is the window order); a few terms of one of them are reproduced in Alekseev and Yaglom (1980).   He also performed some numerical experiments with simulated realizations (of large lengths T) of stationary sequences $X(t)$ having spectral densities of several given forms.   (All the selected forms of $f(\omega)$ were analytical, i.e. they had derivatives of all orders.) The experiments described in Alekseev and Yaglom (1980) and Alekseev (1980, 1981) dealt with rather long realizations (T of the order of $10^5 - 10^6$); they showed that a considerable gain in accuracy can be achieved by using an alternating spectral window of a rather high order l (e.g., $l \geqslant 10$). Subsequent experiments (described in Alekseev, 1985) involved much shorter realizations of length $T = 2048$.   In this case also some gain in accuracy was achieved by increasing the window order above $l = 2$ (at least if $f(\omega)$ had a sharp peak at some frequency ω_0), but windows of rather high orders prove not to be advantageous.

[55]Formulae (3.57) and (3.58)–(3.58a) can be derived quite similarly to either of the two derivations of (3.39) and (3.39a) outlined in Note 37. We can also make use of the spectral representation (2.204) or (2.61) of $X(t)$ to obtain the result

$$(3.15') \qquad d_T^{(h)}(\omega) = \int D_T^{(h)}(\omega - \omega') dZ(\omega'),$$

where

$$(3.16') \qquad d_T^{(h)}(\omega) = \sum_{t=1}^{T} h_T(t)X(t)e^{-i\omega t} \qquad \text{or} \qquad d_T^{(h)}(\omega) = \int_0^T h_T(t)X(t)dt,$$

the limits of integration are $-\pi$ and π or $-\infty$ and ∞, and

$$(3.17') \qquad D_T^{(h)}(\omega) = \sum_{t=1}^{T} d_T(t)e^{-i\omega t} \qquad \text{or} \qquad D_T^{(h)}(\omega) = \int_0^T h_T(t)e^{-i\omega t}dt,$$

depending on whether the time t is discrete or continuous. Since $I_T^{(h)}(\omega) = (2\pi T)^{-1}|d_T^{(h)}(\omega)|^2$, (3.57) and (3.58)-(3.58a) follow easily from (3.15'), (3.17'), and (2.81).

Not tapering (i.e. the utilization of the conventional periodogram $I_T(\omega)$) corresponds to $K_T^{(h)}(\omega) = (2\pi T)^{-1}\sin^2(T\omega/2)\sin^{-2}(\omega/2)$ or $K_T^{(h)}(\omega) = 2(\pi T)^{-1}\sin^2(T\omega/2)\omega^{-2}$. These kernels have a sharp peak at $\omega = 0$, vanish at $\omega = \pm 2\pi/T, \pm 4\pi/T, ...$, and have many secondary maxima on both sides of the main peak, i.e. besides the main lobe at $-2\pi/T \leqslant \omega \leqslant 2\pi/T$ they have many rippling sidelobes, whose heights die off rather slowly (approximately as ω^{-2}; cf. Fig. 28(b)). It is possible to show that such a behavior of $K_T^{(h)}(\omega)$ is related to the existence of sharp discontinuities of the rectangular data window at the points $t = 0$ and $t = T$ (since $d_T(t) = 1$ for $0 < t \leqslant T$ but $d_T(t) = 0$ for $t \leqslant 0$ to $t > T$). For a continuous *triangular data window*, which decreases linearly on both sides of the middle point $t = T/2$ to zero values at the edges, the kernel $K_T^{(h)}(\omega)$ dies off as ω^{-4}, i.e. much faster than for a rectangular window (cf. (2.124); the time t is now assumed to be continuous for simplicity and definiteness). The smoother *Parzen data window* (see Fig. 39(c) in Vol. I and p. 83) corresponds to sidelobe heights dying off as ω^{-8}. Many other examples of data windows with even better sidelobe behavior can also be found in the literature; see, e.g., Babič and Temes (1976), Thomson (1977, 1982), F. Harris (1978), Geckinli and Yavuz (1978), Prabhu and Reddy (1980), and Nuttall (1981).

By virtue of the well-known Parseval theorem for Fourier series and its analogue for Fourier integrals, the normalizing condition (3.56) implies that $\int K_T^{(h)}(\omega)d\omega = 1$ (where the integration limits are, as usual, $-\pi$ and π for discrete time and $-\infty$ and ∞ for continuous time). Moreover, according to the "uncertainty principle" for Fourier transforms (see p. 116 in Vol. I and Note 24 to Chap. 2 on p. 44) the width of $D_T^{(h)}(\omega)$ (and hence also that of $K_T^{(h)}(\omega) = (2\pi T)^{-1}|D_T^{(h)}(\omega)|^2$) must tend to zero as the width of the window $d_T(t)$ tends to infinity. This implies that $K_T^{(h)}(\omega) \to \delta(\omega)$ as $T \to \infty$ if the width of the data window $h_T(t)$ increases infinitely with T.

The utilization of data tapering in spectral analysis was suggested, in particular, by Tukey (1967), Bingham, Godfrey, and Tukey (1967), and Welch (1967). See in this respect also the notes by Nuttall and Carter (1980), Carter and Nuttall (1980a, b), and Brillinger (1981), the paper by Alekseev (1985), and the books by Koopmans (1974), Brillinger (1975), Bloomfield (1976), Otnes and Enochson (1978), Priestley (1981), Rosenblatt (1985), and Bendat and Piersol (1986).

[56]Let us apply to the calculation of the covariance function

$b_{I(h)}(\omega_1,\omega_2) = \langle(I_T^{(h)}(\omega_1) - \langle(I_T^{(h)}(\omega_1)\rangle)(I_T^{(h)}(\omega_2) - \langle I_T^{(h)}(\omega_2)\rangle)\rangle$ the method used in Note 37 to derive equation (3.40) for $b_I(\omega_1,\omega_2)$. Assuming that $X(t)$ is a Gaussian stationary process and that $h_T(t) = h_T(T - t)$ for $0 \leqslant t \leqslant T/2$, we obtain for $b_{I(h)}(\omega_1,\omega_2)$ an equation similar to (3.40), the only difference being that the function $g_T(x) = (2/\pi T)^{1/2}[\sin(Tx/2)]/x$ appearing several times in (3.40) is now replaced by the function

$$(3.18') \quad g_T^{(h)}(x) = \left(\frac{2}{\pi T}\right)^{1/2}\int_{-T/2}^{T/2} e^{ixt}h_T^{(1)}(t)dt = \left(\frac{2}{\pi T}\right)^{1/2}\int_{-T/2}^{T/2}\cos xt\, h_T^{(1)}(t)dt,$$

where $h_T^{(1)}(t) = h(t - T/2)$. (For discrete time the integrals in the middle and on the right-hand side of (3.18') must, of course, be replaced by the corresponding sums.) Since $[g_T^{(h)}(x)]^2 = K_T^{(h)}(\omega)$ (see the previous Note), $[g_T^{(h)}(x)]^2 \to \delta(x)$ as $T \to \infty$ and the width of the window $h_T(t)$ increases infinitely. Since the behavior of $g_T^{(h)}(x)$ at large T is similar to that of $g_T(x)$, the formula for $b_{I(h)}(\omega_1,\omega_2)$ implies that the limiting relations (3.41) and (3.42) (with the usual modification of (3.41) when time is discrete) are valid not only for an ordinary periodogram $I_T(\omega)$ but also for a modified periodogram $I_T^{(h)}(\omega)$.

The asymptotic formula (3.11') for $b_I(\omega_1,\omega_2)$ at large values of T can also be generalized to the correlation function $b_{I(h)}(\omega_1,\omega_2)$ as

$$(3.19') \quad b_{I(h)}(\omega_1,\omega_2) \approx \{L_T^{(h)}(\omega_1 - \omega_2) + L_T^{(h)}(\omega_1 + \omega_2)\}f(\omega_1)f(\omega_2),$$

where $L_T^{(h)}(\omega) = \left|T^{-1}\int_0^T h_T^2(t)\exp(-i\omega t)dt\right|^2$ for continuous time and $L_T^{(h)}(\omega)$

$= \left|T^{-1}\sum_{t=1}^T h_T^2(t)\exp(-i\omega t)\right|$ for discrete time (see Brillinger, 1975, Sec. 5.2).

It is easy to check that $L_T^{(h)}(\omega) \approx \left[(2\pi/T^2) \int_0^T h_T^4(t)dt\right]\delta(\omega)$ when T is

very large and the width of the data window $h_T(h)$ is very large too. For discrete time the integral in the asymptotic expression for $L_T^{(h)}(\omega)$ must, of course, be replaced by the corresponding sum. In the important case where $h_T(t) = h(t/T)$ (i.e. (3.59) and (3.60) hold) we obtain

$$(3.20') \quad L_T^{(h)}(\omega) \approx \left[\frac{2\pi}{T} \int_0^1 h^4(\tau)d\tau\right]\delta(\omega)$$

for both continuous and discrete time; only this case will be considered below.

Let now $\Phi_T^{(A,h)}(\omega)$ be a smoothed modified periodogram obtained by using $I_T^{(h)}(\omega)$ in place of $I_T(\omega)$ on the right-hand side of (3.43a). Applying to $\sigma^2\{\Phi_T^{(A,h)}(\omega)\}$ the method used in Note 44 on p. 90 to derive an equation for $\sigma^2\{\Phi_T^{(A)}(\omega)\}$ and taking into account (3.19') and (3.20'), we obtain the following result:

$$(3.21') \quad \sigma^2\{\Phi_T^{(A,h)}(\omega)\} \approx \frac{2\pi}{T} \int_0^1 h^4(\tau)d\tau \int A_T(\omega - \omega')\{A_T(\omega - \omega') + A_T(\omega + \omega')\}f^2(\omega')d\omega'$$

(for more rigorous derivation of (3.21') see, e.g., Brillinger, 1975, Sec.

5.6; cf. also Koopmans, 1974, Sec. 9.2). We see that the variance of the "tapered" periodogram estimator $\Phi_T^{(A,h)}(\omega)$ at large values of T differs from that of the conventional ("untapered") estimator $\Phi_T^{(A)}(\omega)$ corresponding to the same spectral window $A_T(\omega)$ by the factor $\int_0^1 h^4(\tau)d\tau$, where the function $h(\tau)$ satisfied the condition $\int_0^1 h^2(\tau)d\tau = 1$.

Since

$$(3.22') \qquad \{\int_0^1 h^2(\tau)d\tau\}^2 \leqslant \int_0^1 h^4(\tau)d\tau$$

by the Buniakovsky–Schwarz inequality (see, e.g., Note 12 on p. 70) and the equality in (3.22') holds only when $h(\tau)$ = constant, this factor cannot be less than one and it is equal to one only when there is no tapering. (To obtain (3.22') we must put $a = 0$, $b = 1$, $f(\tau) = h^2(\tau)$, $g(\tau) = 1$ in the inequality from Note 12. The inequality becomes an equality only when $f(\tau)$ and $g(\tau)$ are proportional, i.e., in our case, when $h(\tau)$ = constant.) We see that tapering always increases the variance of a smoothed periodogram estimator. However, the increase in variance is often rather slight from the point of view of applications (see, e.g., the numerical examples below in this note).

The argument concerning the usefulness of data tapering in cases where periodogram is smoothed by convolution with the Daniell window (see pp. 274–275 in Vol. I) is due to Priestley (1981), p. 560.

Tukey (1967) and Bingham, Godfrey, and Tukey (1967) recommended that only a small portion (of length about $0.1T$) of the whole observed realization should be tapered at each of its ends. In other words, they considered such taper functions $h(\tau)$ that $h(\tau) = C$ = constant over the central portion of the segment $0 \leqslant \tau \leqslant 1$. Specifically, Tukey (1967) and Bingham et al. (1967) introduced the *cosine taper*:

$$(3.23') \qquad h(\tau) = \begin{cases} (C/2)\{1 - \cos(\pi\tau/a)\} & \text{for } 0 \leqslant \tau \leqslant a, \\ C & \text{for } a < \tau < 1 - a, \\ (C/2)\{1 - \cos(\pi(1-\tau)/a)\} & \text{for } 1 - a \leqslant \tau \leqslant 1, \\ 0 & \text{otherwise,} \end{cases}$$

where the normalizing constant C is determined by (3.60). They supposed that $a = 0.1$, but in fact a can be selected arbitrarily within the limits $0 \leqslant a \leqslant 0.5$ (for $a = 0.5$ we obtain the *Hanning taper*). A similar useful taper is the *trapezoidal taper*; in this case also $h(\tau) = C$ = constant for $a < \tau < 1 - a$, where $0 \leqslant a \leqslant 0.5$, but now $h(\tau)$ changes linearly from 1 to zero at the end portions $0 \leqslant \tau \leqslant a$ and $1 - a \leqslant \tau \leqslant 1$. Zhurbenko (1978) used a simple *binomial taper* (not of the form (3.59)) and also a related, but a more general, *polynomial* taper which he attributed to Kolmogorov (see Zhurbenko 1979, 1980, 1982); similar taper was also proposed by Bentkus (1977). Thomson (1977) described the application of the very useful *prolate spheroidal taper* to a specific practical problem related to a spectrum having enormous variability (see also

Thomson, 1982).   Many other examples of tapers (data windows) can be found in the literature cited in the previous Note.

Brillinger (1975), p. 151, calculated that $\int_0^1 h^4(\tau)d\tau = 1.116$ for a cosine taper (3.23') with $a = 0.1$. Thus, such tapering increases the variance of the spectral estimate by less than 12 percent. For a trapezoidal taper extended over the first and last 10 percent we obtained similarly that $\int_0^1 h^4(\tau)d\tau = 1.118$, while $\int_0^1 h^4(\tau)d\tau = 1.350$ for a trapezoidal taper with $a = 0.25$ (cf. Koopmans, 1974, p. 302). We see that in all these cases the increase in the spectral estimate variance due to tapering is relatively small.

[57]Let us assume for definiteness that the time t is discrete. We have already indicated (see Note 40) that the method for constructing a consistent spectral estimate by partitioning a given realization of length T into n nonoverlapping pieces of length $T_1 = T/n$, computing the periodogram for each piece, and averaging it over all the n available pieces is due to Bartlett (1948). A modern version of the method which utilizes overlapped pieces and modified periodograms was proposed by Welch (1967). According to Welch, if $x(1), ..., x(T)$ is an observed realization, where $T = (P - 1)L + T_1$, $L < T_1 \ll T$ and P, L, and T_1 are integers, we may consider P overlapping pieces of length T_1: $x(1), ..., x(T_1)$; $x(L_1 + 1), ..., x(L_1 + T_1)$; $x(2L_1 + 1), ..., x(2L_1 + T_1)$; ...; $x((P - 1)L + 1), ..., x((P - 1)L + T_1) = x(T)$. Then we compute for all the pieces the modified periodograms $i_{T_1}^{(h,1)}(\omega)$, $i_{T_1}^{(h,2)}(\omega)$, ..., $i_{T_1}^{(h,P)}(\omega)$ corresponding to a given data window $h_{T_1}(t)$, and use the arithmetic mean $\sum_{k=1}^{P} i_{T_1}^{(h,k)}(\omega)$ as an estimate of $f(\omega)$. Welch's method is rather widely used at present (see, e.g., Carter and Nuttall 1980a, and also Carter and Nuttall, 1980b; Nuttall and Carter, 1980, 1982, and pp. 286–288 in Vol. I , where a further development of the method is described). The statistical properties of the corresponding estimator were studied by Zhurbenko (1979, 1980, 1982, Chap. 4) and Dahlhaus (1985a); see also Rosenblatt (1985), Sec. V.5.

[58]Recall that we have already mentioned in Note 53 the optimality property of the parabolic spectral window and of the somewhat more general Alekseev's window (3.13'). We have also indicated in Note 54 that if the spectral density $f(\omega)$ is smooth enough, the spectral window giving the minimal mean-square error must necessarily alternate signs (but the precise shape of the optimal window is in this case unknown). See also related works by Watts (1964), Palmer (1968), Neave (1972), Papoulis (1973a), Eberhard (1973), Harris (1978), Blomqvist (1979), and V. Levin (1986).

[59]The probability distribution for a spectral estimator $\Phi_T^{(A)}(\omega)$ is studied in many works. As far back as 1949 Tukey (see also Blackman and Tukey, 1959) recommended that this probability distribution (after an appropriate normalization of $\Phi_T^{(A)}(\omega)$) should be approximated by a χ_ν^2 distribution (introduced in Note 37) and proposed an equation determining the corresponding *equivalent degrees of freedom* ν; see Jenkins and Watts (1968), Sec. 6.4.2; Koopmans (1974), Sec. 8.3; Brillinger (1975), Sec. 5.5; and Priestley (1981), Sec. 6.2.4. Hannan (1970), Secs. V.4 and V.5, tried to justify Tukey's recommendation mathematically, but nevertheless the accuracy of the proposed approximation remains unclear even now. Brillinger (1975), Theorem 5.5.3, found that under some reasonably general conditions the asymptotic probability distribution for $\Phi_T^{(A)}(\omega)$ coincides with the distribution of some linear combination of independent χ_2^2-distributed random variables; this result is, however, too complicated to be useful in practice. Other approximations to the probability distribution for the spectral estimator $\Phi_T^{(A)}(\omega)$ have been proposed, in particular, by Freiberger and Grenander (1959) and Grenander, Pollak, and Slepian (1959). Note also that the probability distribution of $\Phi_T^{(A)}(\omega)$ tends, under fairly general conditions, to the normal distribution, as $T \to \infty$ and $k_T \to \infty$, but $k_T/T \to 0$. Different proofs of this important result can be found, e.g., in the books by Hannan (1970), Sec. V.5; Anderson (1971), Sec. 9.4; Brillinger (1975), Sec. 5.6; and Rosenblatt (1985), Sec. V.3; see also Priestley (1981), p. 469. More refined results of the same type, which supplement the proof of the asymptotic normality by the more precise expansion for the asymptotic distribution and some theorems on the probabilities of large deviations are given by Bentkus and Rudzkis (1982); see also Dahlhaus (1985b).

The approximate knowledge of the probability distributions of spectral estimates $\Phi_T^{(A)}(\omega)$ can easily be used for constructing the approximate confidence intervals for a spectral density; see, e.g., any of the books mentioned in this Note and also Bendat and Piersol (1985), Sec. 8.5.

[60]See, e.g., Korn (1966), Chap. 7, and Max (1981), Chap. 19, where some additional references can be found. There is also a vast number of reports and manuals issued by various companies describing commercial analogue spectrum analyzers of many types.

[61]Some examples of error calculations for spectrum analyzer readings can be found in the books mentioned in the preceding Note. The analogy between the smoothed periodogram estimates $\varphi_T^{(A)}(\omega)$ and the time-averaged squared output signal of a narrow band-pass filter has already been revealed by Grenander and Rosenblatt (1956), Sec. 4.5, and Blackman and Tukey (1959), Sec. 10; see also Priestley (1981), pp. 496–499. A detailed study of errors in the readings of some particular models of analogue spectrum analyzers was carried out, e.g., by Priestley and Gibson (1965) and

Alekseev (1965, 1967).

[62]The fast progress in computer technology made it possible in the mid-1960s to efficiently perform periodogram smoothing over a rectangular Daniell's window within an acceptable time in cases where $T \leqslant 1000$; see Jones (1965). Jones' conclusion was based on the use of the conventional numerical technique for periodogram computation. However, this approach was never utilized later on since the appearance of Cooley and Tukey's paper (1965) made this conventional technique quite obsolete (cf. pp. 282–284 in Vol. I).

[63]The method was proposed by Tukey (1949) and it became very popular after the publication of Blackman and Tukey's book (1959) (originally published in 1958 in the Bell System Technical Journal).

[64]The first attempt to trace the history of FFT was made by Cooley, Lewis, and Welch (1967a). However, it was later discovered that in fact the FFT algorithm had already been known to famous mathematician C. F. Gauss about 1805, and that some related methods of numerical Fourier analysis were also worked out by a number of scientists in the nineteenth century but were then forgotten; see Heideman, Johnson, and Burrus (1984). The important paper by Cooley and Tukey (1965) is also reprinted (together with some other papers on FFT) in a collection edited by Oppenheim (1969). A synopsis of the FFT can be found, e.g., in the books by Jenkins and Watts (1968), Hannan (1970), Anderson (1971), Koopmans (1974), Brillinger (1975), Bloomfield (1976), Priestley (1981), Rosenblatt (1985), and Bendat and Piersol (1986). The various versions of FFT are treated at length, together with some related fast algorithms and many specific applications, by Gentleman and Sande (1966), Cooley, Lewis, and Welch (1967b,c), Bergland (1969), Brigham (1974), Oppenheim and Schafer (1975), Rabiner and Gold (1975), Roshal (1976), Lifermann (1977), Otnes and Enochson (1978), Huang (1981), Chap. 4 (written by S. Zohar), Nussbaumer (1982), Elliott and Rao (1982), and Perera and Rayner (1986). In particular, the last four sources contain also the description of the more recent fast algorithms by Rader (1968) and Winograd (1976, 1978) for Fourier transform computations, as well as the algorithm by Agarwal and Cooley (1977) for fast convolution computations.

[65]See, e.g., Kholmyansky (1972) and Alekseev (1980, 1981). Various aspects of spectral estimation via FFT are studied, e.g., by Bingham, Godfrey, and Tukey (1967), Welch (1967), Cooley, Lewis, and Welch (1970), Parzen (1972), Zhurbenko (1982), and in the references cited in the previous Note.

[66]See, e.g., Oppenheim and Schafer (1975), Sec. 11.6.2; Rabiner and Gold (1975), Chap. 6; Otnes and Enochson (1978), Chap. 7; Priestley (1981), pp. 577–579; Bendat and Piersol (1986), Sec. 11.4; and also

Stockham (1966), Borgioli (1968), and Rader (1970). An example of application of this method to the evaluation of correlation functions for some micrometeorological turbulent fluctuations is given by Sreenivasan, Chambers, and Antonia (1978).

The applicability of FFT to computation of correlation function estimates $B_T^{**}(\tau)$ is based on the well-known convolution theorem of the Fourier transform theory. According to this theorem the Fourier transform of a convolution of any two numerical functions is equal to the product of the Fourier transforms of two factors. Since FFT provides for very easy computation of a Fourier transform, it is advisable, in many cases where a convolution of two functions is of interest, to compute first the Fourier transforms of both the functions, then multiply the two Fourier transforms obtained (this is also a very simple task for modern computers) and, finally, evaluate the inverse Fourier transform of this product. Thus, calculation of a convolution can be reduced to three Fourier transforms and one function product evaluation; if we compute all three Fourier transforms by the FFT method, then for series of, say, 1000 numbers or more it is often much faster to evaluate the convolution in this way than compute it directly. (This is precisely the result noted by Stockham, 1966; see, also, e.g., Cooley, Lewis, and Welch, 1967b, or Bergland, 1969). As indicated in Note 30 on p. 81, the function $B_T^{**}(\tau)$ is the convolution of $x_T(-t)$ and $x_T(t)$; this explains the applicability of FFT to $B_T^{**}(\tau)$ calculation.

Of course, instead of FFT we can also apply some other fast algorithm for Fourier transform computation (e.g., the Winograd algorithm; see Note 64). We can also base the computation of $B_T^{**}(\tau)$ on some fast convolution algorithm, e.g., that introduced by Agarwal and Cooley (1977); see Reddy and Reddy (1980), Nussbaumer (1982), and Elliott and Rao (1982).

[67]The generalized method of spectral estimation is due to Nuttall and Carter (1980, 1982); see also Carter and Nuttall (1980a,b; 1983). Recall now that the mean value of all the modified periodograms $i_{T_1}^{(h,k)}(\omega)$, $k = 1, ..., P$, and of their arithmetic mean $f_{T_1}^{(P,T_1)}(\omega)$ (i.e. the mean value of the corresponding estimators) is given by (3.57) and (3.58) or (3.58a) (with $K_{T_1}^{(h)}(\omega)$ replaced by $K_{T_1}^{(h)}(\omega)$) depending on whether time t is discrete or continuous. On pp. 286–288 of Vol. I it was assumed that time is discrete; now, however, it is convenient to assume for simplicity that time is continuous. (In this case the statement on p. 288 about the effective spectral window $A_e(\omega)$ will be exact. The case of discrete time implies some minor additional complications related to the contents of the footnote on p. 263 of Vol. I; it is unnecessary to study them here in detail.) Since $B_T^{(P,T_1)}(\tau)$ is a Fourier transform of $f_T^{(P,T_1)}(\omega)$, it is clear

that $\langle B_{T_1}^{(P,T_1)}(\tau)\rangle = B(\tau)k_{T_1}^{(h)}(\tau)$, where $k_{T_1}^{(h)}(\tau)$ is the Fourier transform

of $K_{T_1}^{(h)}(\omega)$. Therefore $\langle \tilde{B}_T(\tau)\rangle = B(\tau)k_{T_1}^{(h)}(\tau)a_{T_1}(\tau)$ and this implies that

(3.24') $\quad \langle \tilde{B}_T(\tau)\rangle = \int A_e(\omega - \omega')f(\omega')d\omega'$,

where $A_e(\omega)$ is a convolution of the Fourier transforms of $k_{T_1}^{(h)}(\tau)$ and

$a_{T_1}(\tau)$.

Nuttall and Carter (1982) and Carter and Nuttall (1983) showed that in the case where the data window $h_{T_1}(t)$ is a rectangular one while the

effective lag window $a_e(\tau)$ has a form similar to that sketched in Figs. 39(c) and 42, the lag window $a_{T_1}(\tau)$ need not necessarily be a

monotonically decreasing function of lag τ for positive lags. Moreover, they also computed some examples of effective spectral windows $A_e(\omega)$ corresponding to simple and widely used forms of the data window $h_{T_1}(\tau)$ and the lag window $a_{T_1}(\tau)$ and some examples of lag windows

$a_{T_1}(\tau)$ corresponding to the rectangular data window and some

convenient forms of the effective spectral window $A_e(\omega)$. Finally, they obtained some expressions for the variance of the second-stage spectral

estimate $\tilde{f}_T(\omega)$.

[68]See, e.g., the recent survey by Kay and Marple (1981) and the collections of papers edited by Childers (1978), Haykin (1983), and Haykin and Cadzow (1982) which are mostly devoted to parametric methods of spectral analysis. Of course, only a small portion of the material from these sources is used in this book.

[69]The autoregressive spectral estimates were introduced independently by Parzen (1964, 1969, 1972) and Akaike (1969b). However, their wide utilization in practice is related, first of all, to the work of Burg (1967); see Note 72 below.

[70]Let $X(t)$, $t = 0,\pm1, ...$, be a Gaussian stationary random sequence with a spectral density $f(\omega)$. Then it is easy to see that the best predictor $X^*(t + 1)$ of the future values $X(t + 1)$ in terms of the present value $X(t)$ and all the past values $X(t - 1)$, $X(t - 2)$, ... is a linear function of $X(t - s)$, $s = 0,1,2, ...$, (see, e.g., Yaglom, 1962, pp. 99-100). According to the well-known theorem of Kolmogorov (1941a) on the best linear predictor (see also, e.g., Doob, 1953, Sec. XII.4; Yaglom, 1962, Sec. 36; Rozanov, 1967, Sec. II.5; Anderson, 1971, Sec. 7.6; Hannan, 1971, Sec. III.3; Koopmans, 1974, Sec. 7.4; Gihman and Skorohod, 1974, Sec. IV.9; Priestley, 1981, Sec. 10.1; or Rosenblatt, 1985, Sec. II.3) the mean square

error of the best linear predictor $X^*(t + 1)$ is equal to

$$(3.25') \quad \langle [X^*(t + 1) - X(t + 1)]^2 \rangle = 2\pi\exp\left\{\frac{1}{2\pi}\int_{-\pi}^{\pi}\ln f(\omega)d\omega\right\},$$

where ln denotes the natural (base e) logarithm. (Of course, the logarithm base in the definition of the entropy H can, in fact, be selected quite arbitrarily since the base change leads only to the multiplication of H by a constant.) Thus, the sequence $X(t)$ from a given class corresponding to the maximal value of the entropy H is the *most unpredictable* one. For an arbitrary (non-Gaussian) sequence $X(t)$ the variable $X^*(t + 1)$ is not the best possible, but only the best linear, predictor; therefore, entropy maximization is equivalent here to the selection of a spectral density $f(\omega)$ corresponding to the linearly most unpredictable sequence from a given class.

For more profound discussion of the probabilistic meaning of the entropy maximization condition see, e.g., Ables (1974), Smith (1981), Papoulis (1981, 1984), and Jaynes (1982). Note also that the existence and uniqueness, under rather general conditions, of the correlation (i.e. positive definite) function $B(\tau)$ on $-\infty < \tau < \infty$, which coincides with a given positive definite function on $-m \leqslant \tau \leqslant m$ and corresponds to the maximal entropy, was proved (for the case of continuous time) by Chover (1961).

[71]The maximization of $H = \int_{-\pi}^{\pi}\ln f^*(\omega)d\omega$ under the condition that $\int_{-\pi}^{\pi}e^{ik\omega}f^*(\omega)d\omega = B^*(k)$, $k = 0,\pm1,...,\pm m$, is equivalent to the requirement that

$$\delta(H + L) = 0,$$

where $L = -\sum_{k=-m}^{m}\lambda_K\int_{-\pi}^{\pi}e^{ik\omega}f^*(\omega)d\omega$ and λ_k, $k = 0,\pm1, ..., \pm m$, are the Lagrange multipliers satisfying the conditions $\lambda_k = \bar{\lambda}_{-k}$, since the functional L must be real. (Of course, we could also include only the terms with $k = 0, 1, ..., m$ in the definition of L, replace $\exp(ik\omega)$ by $\cos k\omega$, and assume λ_k to be real; the final result then would be the same.) Moreover,

$$\delta(H + L) = \int_{-\pi}^{\pi}\left[\frac{1}{f^*(\omega)} - \sum_{k=-m}^{m}\lambda_k e^{ik\omega}\right]\delta f^*(\omega)d\omega,$$

where $\delta f^*(\omega)$ is an arbitrary function. Therefore $\delta(H + L) = 0$ if and only if

$$(3.26') \quad f^*(\omega) = \frac{1}{\sum_{k=-m}^{m}\lambda_k e^{ik\omega}}, \quad \lambda_{-k} = \bar{\lambda}_k.$$

Since $f^*(\omega)$ must be nonnegative and integrable, the denominator on the right-hand side of (3.26') must have the representation

$$\sum_{k=-m}^{m} \lambda_k e^{ik\omega} = \left| \sum_{j=0}^{m} c_m e^{-ijm} \right|^2,$$

i.e. (3.26') must coincide with (3.64).

The above elementary proof is due to Edward and Fitelson (1973). (Burg, 1967, first gives the result, but without its explicit derivation.) Later Grandell, Hamrud, and Toll (1980) and Akaike (see also Priestley, 1981, pp. 604–606, and Papoulis, 1984, p. 488) showed that the result can easily be derived on the basis of theory of the best linear prediction (cf. the previous Note).   If the values $B(0)$, $B(1)$, ..., $B(m)$ are known, then we can find the best predictor $X_m^*(t + 1)$ of the form $X_m^*(t + 1) = \alpha_0 X(t) + \alpha_1 X(t - 1) + ... + \alpha_{m-1} X(t - m + 1)$ and compute the corresponding mean square error $\langle [X(t + 1) - X_m^*(t + 1)]^2 \rangle$.   (The determination of α_0, ..., α_{m-1} requires only that a simple system of m linear equations be solved; see, e.g., Yaglom 1962, Sec. 20.) It is clear that if $X^*(t + 1)$ is the best linear predictor in terms of all the past, then we always have $\langle [X(t + 1) - X^*(t + 1)]^2 \rangle \leqslant \langle [X(t + 1) - X_m^*(t + 1)]^2 \rangle$. Therefore, if we wish to maximize $\langle [X(t + 1) - X^*(t + 1)]^2 \rangle$ for a class of sequences with given values of $B(0)$, $B(1)$, ..., $B(m)$, we must choose a sequence $X(t)$ such that the condition $X^*(t + 1) = X_m^*(t + 1)$ is satisfied.   It is easy to see that this condition is satisfied for an m-th order autoregressive sequence determined by (2.13) and that such an autoregressive sequence with given values of $B(0)$, ..., $B(m)$ necessarily exists.   Hence the spectral density $f(\omega)$ satisfying (3.65) and maximizing H is an autoregressive spectral density of the form (3.64), where c_0, ..., c_m must be determined from $m + 1$ conditions (3.65).

Another simple proof was recently given by Choi and Cover (1984). These authors considered the class of random sequences $X(t)$, $t = 1, 2, ...,$ satisfying conditions: $\langle X(k)X(k + \tau) \rangle = B(\tau)$ for $\tau = 0, 1, ..., m$ and any integer k, where $B(0)$, ..., $B(m)$ are given numbers. Let H_n be the entropy of the n-dimensional probability distribution of the vector $\{X(1), ..., X(n)\}$ and $H_0 = \lim_{n \to \infty} H_n/n$. Choi and Cover proved that within the considered class of random sequences H_0 is maximal for the (unique) mth order autoregressive sequence (their proof is based on some known information theoretic inequalities). It is easy to see that this general theorem by Choi and Cover implies Burg's result as a special case.

[72]The maximum entropy spectral estimates were introduced by Burg (1967, 1968, 1975), and since that time they have gained great popularity and have been used extensively in practice. At present, a vast literature is devoted to these estimates, which include, in particular, many hundreds of specialized papers describing the applications of maximum entropy spectral analysis to a great many specific problems from various applied fields. (The papers by Gersch, 1970, on encephalograms, by R. Currie, 1973, on solar activity, by Bain, 1976, on temperature variations in central England, by Petersen and Larsen, 1978, on Earth temperature variations during the last 700,000

years, and by Kesler and Haykin, 1978, and Haykin, B. Currie, and Kesler, 1982, on noise and confused echoes in radar systems are typical examples of such applied works.) See in this connection the books by Priestley (1981), pp. 604–607, and Papoulis (1984), Chap. 14, and also the surveys and collection of papers by Ulrych and Bishop (1975), Childers (1978), Kay and Marple (1981), Parzen (1981), Smith (1981), Papoulis (1981), Haykin and Cadzow (1982), and Haykin (1983), where many additional references can be found.

[73]In the case of an autoregressive sequence $X(t)$ the properties of the estimates a_1^*, ..., a_m^*, $(c^2)^*$ were investigated, under very general conditions, by Mann and Wald (1943); see also Hannan (1960), Sec. II.2, and (1970), Sec. IV.2, Anderson (1971), Sec. 5.4 and 5.6, or Rosenblatt (1985) Sec. IV.1.

[74]The standard procedure to computing the Yule–Walker estimates a_1^*, ..., a_m^*, $(c^2)^*$ requires the order of m^3 elementary arithmetical operations (multiplications and additions), but a computationally more efficient procedure can also be developed on the basis of the fact that the correlation matrix $B_T^{**} = \|B_T^{**}(k,j)\| = \|B_T^{**}(k-j)\|$ is

both symmetric (i.e. $B_T^{**}(k,j) = B_T^{**}(j,k)$) and Toeplitz (i.e. $B_T^{**}(k,j) = B_T^{**}(k+1,j+1)$). The procedure was developed first by Levinson (1947) in application to a similar system of equations of another origin and was later also used by Durbin (1960); therefore, it is commonly referred to as the Levinson (or Levinson–Durbin) algorithm. This algorithm is of a recursive type, i.e., it begins with the computation of the parameters $a_1^{(1)}$ and $c_{(1)}^2$ for a first-order autoregressive model ($m = 1$) fitting the following observations:

$$a_1^{(1)} = -B_T^{**}(1)/B_T^{**}(0), \quad c_{(1)}^2 = (1 - [a_1^{(1)}]^2)B_T^{**}(0),$$

and after this the parameters $a_1^{(j)}$, ..., $a_j^{(j)}$, $c_{(j)}^2$ of an appropriate j-th order model are computed successively for $j = 2, 3, ..., m$:

$$a_j^{(j)} = -[B_T^{**}(j) + \sum_{k=1}^{j-1} a_k^{(j-1)} B_T^{**}(j-k)]/c_{(j-1)}^2 ,$$

$$a_k^{(j)} = a_k^{(j-1)} + a_j^{(j)}a_{j-k}^{(j-1)}, \quad k = 1, ..., j-1, \quad c_{(j)}^2 = (1 - [a_j^{(j)}]^2)c_{(j-1)}^2 .$$

It is easy to show that this algorithm permits one to compute the estimates $a_1^{(m)} = a_1^*$, ..., $a_m^{(m)} = a_m^*$, $c_{(m)}^2 = (c^2)^*$ of the m-th order model (and also the estimates of the parameters for all lower order models as a byproduct) using the order of m^2 arithmetical operations; see, e.g., Papoulis (1984), pp. 432–435, Haykin (1983), Kay and Marple (1981), and Cadzow (1982).

Recently Cadzow (1982) suggested the use of more than m Yule–Walker equations in estimating the coefficients a_1, ..., a_m. He stated that the estimates can be improved by considering not only equations (2.16) with $k = 1, ..., m$, but also a number of equations with k exceeding m and then applying some reasonable method (e.g., the least square fit) which permits finding uniquely the values of the coefficients $a_1, ..., a_m$ from the overdetermined system (containing more equations than there are unknowns). However, this approach was apparently not used in real applications till now.

Chan and Langford (1982) suggested the estimation of the coefficients $a_1, ..., a_m$ via the high-order Yule–Walker equations which do not include the value of $B(0)$. Such an estimation method is advantageous in the presence of uncorrelated noise (i.e., observation errors) which affects the variance estimate $B_T^{**}(0)$ but does not change $B_T^{**}(\tau)$ for $\tau > 0$. However, the method has a serious drawback, also discussed by Chan and Langford.

[75]Burg's recursive algorithm allows one to compute estimates for the parameters of autoregressive models of all orders k from $k = 1$ up to a fixed order m. It was first outlined in Burg's paper of 1968 and was later worked out in detail by Andersen (1974); see also Papoulis' book (1984), pp. 496–498, and the papers by Andersen (1978), Marple (1980), and Swingler (1979, 1984) where some further simplifications and modifications of this algorithm are presented. The simplified version of Burg's algorithm is based on the fact that the coefficients $- a_1, ..., - a_m$ on the right-hand side of (3.64) coincide with the coefficients of the best linear predictor $X_m^*(t)$ of $X(t)$ in terms of m past values $X(t - 1), ..., X(t - m)$, and c^2 is equal to the mean square error $\langle [X(t) - X_m^*(t)]^2 \rangle$ of this predictor. Therefore, the values of $a_1, ..., a_m$ that minimize the expression

$$S_m^T = \sum_{t=m+1}^{T} [x(t) + \sum_{j=1}^{m} a_j x(t - j)]^2$$

are reasonable estimates of the corresponding coefficients in (3.64) and the minimal value of S_m^T divided by $T - m$ can be used as an estimate of c^2. The most popular and widely used form of Burg's algorithm is slightly more complicated since it uses, instead of a sum of the squared errors for the usual ("forward") predictors of $x(m + 1), ..., x(T)$, a sum of the squared errors for both the forward and backward (i.e. in terms of $x(t + 1), ..., x(t + m)$) predictors of the observations $x(t)$. This form has some important advantages; however, we will not pursue here its detailed examination and will only refer to the book by Haykin (1983) and the survey by Kay and Marple (1981).
Some other very useful sequential (adaptive) algorithms for the computation of autoregressive parameter estimates were also proposed by several authors. These algorithms permit one to compute very easily the estimates in terms of the observation $x(1), ..., x(T)$ when the estimates in terms of $x(1), ..., x(T - 1)$ have already been computed (see, e.g., Kay and Marple, 1981, Sec. II.E). Many various numerical algorithms for maximum entropy (i.e.

autoregressive) spectral analysis and for some of its generalizations are studied at length in the book by Haykin (1983) and the survey by Kay and Marple (1981), where numerous additional references related to this subject can also be found; cf. also Dickinson (1978), Scott and Nikias (1981, 1982), and Cadzow (1982).

[76]It is easy to show that the solutions $\overset{\circ}{a}{}_1^{(j)}$, ..., $\overset{\circ}{a}{}_j^{(j)}$, $\overset{\circ}{c}{}^2_{(j)}$ of the exact

Yule–Walker equations (2.16), $k = 1, 2, ..., j$, and (2.16a) coincide with the coefficients in the formula $X_j^*(t) = -\overset{\circ}{a}{}_1^{(j)}X(t-1) - ... - \overset{\circ}{a}{}_j^{(j)}X(t-j)$ describing the best linear predictor of $X(t)$ in terms of $X(t-1), ...,X(t-j)$ and with the corresponding mean square error $\overset{\circ}{c}{}^2_{(j)} = \langle[X(t) - X_j^*(t)]^2\rangle$. Therefore, the sequence $\overset{\circ}{c}{}^2_{(1)}, \overset{\circ}{c}{}^2_{(2)}, ..., \overset{\circ}{c}{}^2_{(j)}, ...$ is always strictly nonincreasing and, as $j \to \infty$, it tends to a constant (equal to the mean square error for the best linear predictor $X^*(t)$ of $X(t)$ in terms of all the past). Moreover, if $X(t)$ is an m-th order autoregressive sequence, the best linear predictor $X^*(t)$ coincides with the best linear predictor $X_m^*(t)$ in terms of $X(t-1), ..., X(t-m)$ (cf. Notes 71 and 75) and hence in this

case $\overset{\circ}{c}{}^2_{(1)} > \overset{\circ}{c}{}^2_{(2)} > ... > \overset{\circ}{c}{}^2_{(m)} = \overset{\circ}{c}{}^2_{(m+1)} = \overset{\circ}{c}{}^2_{(m+2)} = ...$. Since $c^2_{(j)}, j = 1, 2,$

..., are reasonable estimates of $\overset{\circ}{c}{}^2_{(j)}, j = 1,2, ...,$ it is advisable to take the value of j for which the "leveling off" of the sequence $c^2_{(j)}, j = 1, 2, ...,$ occurs as the estimate of the parameter m on the right-hand side of (3.64).

The above-described method of the choice of the order of autoregression was proposed by Whittle (1963a), p. 35; see also Koopmans (1974), p. 328, and Priestley (1981), pp. 370–371.

[77]The final prediction error criterion was proposed by Akaike (1969a, 1970) on the basis of some consideration related to the evaluation of linear predictor errors. In this respect see the discussion of various criteria for the autoregression order determination in the books by Priestley (1981), Sec. 5.4.5 and 7.8, and Haykin (1983), Sec. 2.10, and in the surveys by Kay and Marple (1981), Sec. II.E, Parzen (1981), Poskitt and Tremayne (1984) and DeGooijer et al. (1985); cf. also Lütkepohl (1985) and Bhansali (1986a,b,c).

[78]See Akaike (1973, 1974, 1977) and Bhansali (1986c). To prove the asymptotic relation (3.69) one must only note that

$$\ln[FPE(m)] = \ln\left[\frac{1 + m/T}{1 - m/T}\, c^2_{(m)}\right] = \ln\left(1 + \frac{m}{T}\right) - \ln\left(1 - \frac{m}{T}\right)$$
$$+ \ln\, c^2_{(m)} = \ln\, c^2_{(m)} + \frac{2m}{T} \quad \text{for large } T.$$

In cases where the form (3.67a) of $FPE(m)$ is appropriate the term $2/T$ is added to the right-hand side; it does not depend on m and therefore plays no role when only that value of m is of interest for which the criterion is the minimum.

Akaike's information criterion (AIC) is based on some rather general information-theory concepts, and it can be applied to many different situations related to Gaussian random functions (or vectors). Later on, Akaike (1978, 1979) developed a new order determination criterion called BIC (a Bayesian modification of the AIC), which leads, in general, to lower autoregression orders than those derived from the AIC criterion (see Priestley, 1981, pp. 375-376).

[79]See Parzen (1974, 1977) and Bhansali (1986a,b). The CAT criterion is based on consideration of the integrated relative mean square error of the autoregressive spectral estimate

$$I = \int_{-\pi}^{\pi} \left\langle \left[\frac{\varphi_T^{AR}(\omega) - f(\omega)}{f(\omega)} \right]^2 \right\rangle d\omega,$$

and its main idea is to determine m as that value of the autoregression order which minimizes I. By means of some manipulations and simplifications Parzen shows that the above procedure can be approximately reduced to the search of the minimum value of the quantity (3.70). More accurate derivation of CAT criterion was given by Bhansali (1986a,b). In his papers and in the paper by Tong (1979), some modifications of CAT are also proposed and the relation between various criteria for determining the model order is studied.

[80]The results obtained with the use of various criteria for selecting the autoregression order are discussed, e.g., by Priestley, Sec. 5.4.5 (where some other procedures for determining the order are also described); Kay and Marple (1981), Sec. II.E; Cadzow (1982), Sec. IX; and Poskitt and Tremayne (1984); in these sources many additional references can also be found. A number of criteria are enumerated by DeGooijer et al. (1985), Bhansali (1986a,b,c), Pukkila and Krishnaiah (1986a), and Lütkepohl (1985). (The last author applied them to multidimensional stationary processes; see Sec. 20.) The comparison of the orders selected by various criteria when applied to different types of simulated time series and varying data length T is presented by, e.g., Beamish and Priestley (1981), Poskitt and Tremayne (1984), Bhansali (1986a). For the inconsistency of the estimates for the autoregression order determined by the AIC and FPE criteria (which are asymptotically equivalent to each other), see, e.g., Shibata (1976) and Kashyap (1980a). The criteria leading, under appropriate general conditions, to consistent order estimates (in cases where $X(t)$ is a true autoregressive sequence) were proposed by, among others, Akaike (1977), Schwarz (1978), Rissanen (1978, 1986), Hannan and Quinn (1979), Fine and Hwang (1979), Kashyap (1980b), and Bhansali (1986b,c). (These new criteria are often of the form $F(m) = \ln c^2_{(m)} + [mC(T)]/T$, where $C(T) = \ln T$ according to Schwarz and Rissanen, $C(T) = c \ln\ln T$ according to Hannan and Quinn, while Kashyap formulated general conditions for $C(T)$.) The suggestion of selecting the value of m between $T/3$ and $T/2$ for short observation records (i.e. in the cases where T is small) is due to T.J. Ulrych and M. Ooe; see Haykin (1983), p. 82.

[81]Equation

$$\lim_{m\to\infty} \lim_{T\to\infty} \frac{T}{m} \left\langle \left[\Phi^{AR}_{T,m}(\omega) - f(\omega) \right]^2 \right\rangle = \begin{cases} 2f^2(\omega) \text{ for } \omega \neq 0,\pm\,\pi, \\ 4f^2(\omega) \text{ for } \omega = 0 \text{ or } \omega = \pm\,\pi, \end{cases}$$

which is quite similar to (3.71), was first proved by Parzen's student Kromer (1969) under the assumption that a random sequence $X(t)$ is a Gaussian one and its spectral density $f(\omega)$ is everywhere positive and sufficiently smooth. Kromer also proved that the autoregressive (maximum entropy) estimator $\Phi^{AR}_{T,m}(\omega)$ is asymptotically normal (i.e. the distribution of $(T/m)[\Phi^{AR}_{T,m}(\omega) - f(\omega)]$ tends to a normal distribution as $T \to \infty$ and then $m \to \infty$). Later the same results were proved by Berk (1974) and Pisarenko (1977) under the wider condition that $X(t)$ is a linear stationary sequence, i.e., it can be represented as a one-sided moving average sequence of infinite order (2.6), where $E(t)$, $t = 0,\pm1,\pm2, ...,$ are independent and identically distributed random variables.

Pisarenko proved his results assuming, as Kromer did, that at first $T \to \infty$ and then $m \to \infty$. Berk proved that (3.71) holds if $m_T \to \infty$, $m_T^3/T \to 0$ as $T \to \infty$, but the rate of increase of m_T with T is not too low either. It would be interesting to determine the lowest admissible rate of increase m_T with T since this would be equivalent to the determination of the asymptotic mean square error for the best possible autoregressive spectral estimator; however, this last problem still remains unsolved. Alekseev and Yaglom (1980) indicated one rough estimate from above of this lowest rate and noted also that the accuracy of the best autoregressive spectral estimator will coincide with the accuracy of the best smoothed periodogram estimator if, and only if, it is possible to take $m_T \sim T^{1/(2r+2\alpha+1)}$ in (3.71) in all cases where the true spectral density $f(\omega)$ has a r-th derivative $f^{(r)}(\omega)$ belonging to the class Lipα (cf. Note 48). Whether this latter rate of increase of m_T with T is in fact admissible or not is yet unknown.

The confidence intervals for autoregressive spectral estimators were studied by Newton and Pagano (1984) and Koslov and Jones (1985).

[82]See, e.g., Burg (1967), Akaike (1969b), Jones (1974), Radoski, Fougere, and Zawalick (1975), Pisarenko (1977), Kozhevnikova (1979), Theodoridis and Cooper (1981), Beamish and Priestley (1981), and Priestley (1981), Sec. 6.9. The results of some comparisons between two types of autoregression estimates using either Yule–Walker equations or Burg's method for parameter estimation are discussed, e.g., by Ulrych and Bishop (1975), Radoski et al. (1975), and Beamish and Priestley (1981). These results show that at small and moderate values of T the difference between these two types of estimates $\varphi^{AR}_T(\omega)$ can be substantial, but at large T it is usually very small.

[83]See Beamish and Priestley (1981) or Priestley (1981), pp. 611–612. Note that the moving average sequence considered can also be

represented as an autoregressive sequence of infinite order of the form $\sum_{k=0}^{\infty}(-0.95)^k X(t-k) = E(t)$, i.e. with coefficients which decay very slowly. Therefore, the autoregression order m determined by means of any of the criteria considered on pp. 293–295 of Vol. I, is in this case, always rather high (it depends, of course, on the given realization $x(t)$ and on its length) and the resulting autoregressive spectral estimate is very poor.

The numerical examples from Beamish and Priestley (1981) show also that in cases of simulated autoregressive sequences of low order the estimates $\varphi_T^{AR}(\omega)$ are usually smoother and more accurate than the estimates $\varphi_T^{(A)}(\omega)$. In many practical situations the observation series can be approximated rather accurately by an autoregressive sequence of comparatively low order. This explains why autoregressive spectral estimates often look better in applications than the smoothed periodogram estimates.

[84]Typical examples of applications of autoregressive spectral analysis to situations involving only short observation series are presented, in particular, by Kesler and Haykin (1978) and Konyaev (1981).

[85]The resolvability of the autoregressive (maximum entropy) spectral estimates $\varphi_T^{AR}(\omega)$ is discussed and compared with the resolvability of the conventional smoothed periodograms $\varphi_T^{(A)}(\omega)$ in many articles and books; see, e.g., Ulrych and Bishop (1975), Radoski, Fougere, and Zawalick (1975), Pisarenko (1977), Kesler and Haykin (1978), Theodoridis and Cooper (1981), Beamish and Priestley (1981), Kay and Marple (1981), and Haykin (1983). It is worth mentioning the strange fact that many of the resolvability studies are based on investigations of applications of autoregressive spectral analysis to time series composed of sinusoids in white noise, i.e. to an inappropriate model of sequences $X(t)$ having a mixed spectrum which includes both continuous and discrete parts. The results of all the indicated works show that autoregressive estimates $\varphi_T^{AR}(\omega)$ have, as a rule, better resolvability than the smoothed periodograms $\varphi_T^{(A)}(\omega)$. Moreover, these works also produce an impression that autoregressive estimates using Burg's method to assess unknown autoregressive parameters often have better resolvability than the estimates $\varphi_T^{AR}(\omega)$ with parameters computed via Yule–Walker equations.

Note also that several special types of new spectral estimates were especially proposed because of their very high resolvability; see, e.g., Capon (1969), Pisarenko (1972), and Byrne and Fitzgerald (1984). The discussion of works by Capon and Pisarenko can be found in the book by Haykin (1983) and the survey by Kay and Marple (1981); cf. also Tufts and Kumaresan (1982).

[86]In particular, Bloomfield (1973) proposed approximation of $\log f(\omega)$ by a finite Fourier series, i.e. utilization of a parametric model of the form

$$(3.27') \qquad f(\omega) = \frac{c^2}{2\pi} \exp\left\{ c_1\cos\omega + c_2\cos2\omega + \ldots + c_m\cos m\omega \right\},$$

where $c_1, \ldots, c_m, c^2$ are adjustable parameters. Bloomfield described a comparatively simple method of estimating the unknown parameters of the model from the observations $x(1), \ldots, x(T)$ that can be used to obtain a parametric spectral estimate of the form (3.27'). However, such estimates are apparently not yet used in any applied problem.

[87]ARMA spectral estimates were used by, among others, Tretter and Steiglitz (1967), Akaike (1974), and Privalsky (1977, 1985). These estimates are also studied at length by Haykin (1983), Kay and Marple (1981), and Haykin and Cadzow (1982); see also the papers cited in Note 92.

[88]Recall, however, that the class of moving average stationary sequences coincides with the class of sequences whose correlation functions $B(\tau)$, $\tau = 0, \pm 1, \ldots$, have only a finite number of non-zero values (see Vol. I p. 77 and Note 1 to Chap. 2 on p. 27 of this volume). It follows that the use of the conventional Blackman–Tukey spectral estimate of the form (3.46a), where the lag window $a_T(\tau)$ has a truncation point k_T (i.e. $a_T(\tau) = 0$ for $|\tau| \geqslant k_T$), means in fact that the true spectral density $f(\omega)$ is approximated by the spectral density of a moving average sequence of order $k_T - 1$. Such a sequence is determined by k_T numerical parameters, e.g., by k_T coefficients b_k in a relation of the form (2.1) which defines a moving average sequence of order k_T or by k_T nonzero values $B(0)$, $B(1)$, $\ldots$, $B(k_T - 1)$ of the correlation function $B(\tau)$. The Blackman–Tukey procedure of spectral estimation consists in utilization of the estimates $B^{(a)}(\tau) = B_T^{**}(\tau)a_T(\tau)$ (or $= B_T^{*}(\tau)a_T(\tau)$) according to the original proposal by Blackman and Tukey, 1959) for the parameters $B(0)$, $\ldots$, $B(k_T - 1)$ and subsequent computation of the spectral density via the correlation function. Therefore the Blackman–Tukey spectral estimates can also be considered as special parametric estimates, namely, moving average spectral estimates. Nevertheless, it is traditional not to include the Blackman–Tukey procedure into the set of parametric spectral estimation procedures.

[89]See, e.g., the books by Box and Jenkins (1970), Chap. 7; Hannan (1970), Chap. 5; Priestley (1981), Sec. 5.4; Rosenblatt (1985), Chap. IV; Azencott and Dacunha-Castelle (1986), Chaps XI and XII; Dzhaparidze (1986), Chaps. II and III; and the survey by Dzhaparidze and Yaglom (1983), containing 182 references.

[90]Note that the matrix of the system which is composed of the first m equations (2.18) is Toeplitz but asymmetric; therefore, the Levinson algorithm (see Note 74) cannot be directly applied to this system. Nevertheless, the algorithm which permits one to solve this system using the order of m^2 elementary arithmetical operations also exists; see, e.g., Zohar (1979). Moreover, Cadzow (1982) suggested not

confining oneself to m equations (2.18) with $k = n + 1, ..., n + m$, but also considering a number of equations with $k > n + m$ and then determining the values of $a_1, ..., a_m$ from the obtained overdetermined system of more than m equations by the least squares method; see also Choi (1986) and Martinelli, Orlandi, and Burraccano (1986).

[91]See Wilson (1969), whose results are also presented by Box and Jenkins (1970), p. 303, and Dzhaparidze and Yaglom (1983). The last two sources also contain a study of some other, more accurate consistent estimates of the parameters $b_1, ..., b_n$.

[92]The estimate (3.77) of the spectral density $f(\omega)$ was proposed by Kaveh (1979) and Kinkel et al. (1979); see also Kay (1980) and Cadzow (1982). Another convenient form of the spectral estimate $\varphi_T^{ARMA}(\omega)$ due to Cadzow and expressed in terms of the estimates a_j^* and $B_T^{**}(\tau)$ is also presented by Kay (1980), who notes that both the forms of the estimate $\varphi_T^{ARMA}(\omega)$ considered may sometimes result in an estimated spectral density having not only positive but also some meaningless negative values. (This is impossible, of course, if the ARMA estimate is obtained by replacing the parameters $a_1, ..., a_m, b_0, ..., b_n$ on the right-hand side of (3.72) by some of their estimates.) It has already been explained on pp. 271–272 of Vol. I that spectral estimates which, under special circumstances, may yield negative values at some frequencies ω, are not necessarily unsatisfactory and practically useless (if only their negative values are replaced by zero in all the applications). Nevertheless, Kay (1980) proposes in this connection some alternative forms of ARMA spectral estimates which also require no computation of the estimates for $b_0, b_1, ..., b_n$, but always result in strictly nonnegative values of the estimated spectral density. See also recent papers by Friedlander (1982, 1983), Davis and Cowley (1982), Kaveh (1983), Gan, Eman, and Wu (1984), and Hannan and Kavalieri (1984) on ARMA spectral estimation containing effective numerical procedures and a number of additional relevant references.

[93]See, e.g., the literature referred to in Note 89.

[94]See, e.g., the discussion of the order determination problem for ARMA estimates in the books by Box and Jenkins (1970), Chap. 6, and Priestley (1981), Sec. 5.4.

[95]For the derivation of Akaike's information criterion (3.78), applicable to order determination of ARMA estimates, see, e.g., Akaike (1973, 1974), Haykin (1983), Sec. 3.6, or Priestley (1981), Sec. 5.4. The paper by Ozaki (1977) is especially devoted to the study of the AIC criterion. Hannan (1980) showed that order estimates obtained via AIC are not consistent in cases where a series $x(t)$ is generated by an autoregressive-moving average sequence $X(t)$, but the estimates of m and n derived from the minimization of $BIC(p,q) = \ln(c_{(m,n)}^2) + [2(m + n)\ln T]/T$ or $\phi(p,q) = \ln(c_{(m,n)}^2) + [m + n)c\ln\ln T]/T$ are much more satisfactory from the point of view of consistency. Order selection procedures for ARMA estimates are also studied in the papers by

Hannan and Rissanen (1982), Chan and Wood (1984), Poskitt and Tremayne (1984), Tsay and Tiao (1985), DeGooijer et al. (1985), Rissanen (1986), An and Chen (1986), and some others.

[96]According to (3.80), in the case of continuous time the inverse form of the Fourier transform (3.35) can be rewritten as

$$(3.28') \qquad B_T^{**}(\tau) = \int_{-\infty}^{\infty} e^{i\omega\tau} dF_T^{*}(\omega).$$

We thus see that $B_T^{**}(\tau)$ plays the role of a characteristic function corresponding to the probability distribution function $F_T^{*}(\omega)$ (cf. Note 3 to the Introduction on p. 2).

Let us now use the following well-known theorem of probability theory: *If a family* $B_T^{**}(\tau)$ *of the characteristic functions* (3.28') *converges, as* $T \to \infty$, *to a continuous limiting function* $B(\tau)$, *then this* $B(\tau)$ *is necessarily also a characteristic function of some distribution function* $F(\omega)$ *and* $F_T^{*}(\omega) \to F(\omega)$, *as* $T \to \infty$, *at all the points of continuity of* $F(\omega)$ (see, e.g., Cramér, 1946, Sec. 10.4, Parzen, 1960, Sec. 10.3, or Gnedenko, 1962, Sec. 38). It follows at once that if the estimator $B_T^{**}(\tau)$ converges, as $T \to \infty$, to the true correlation function $B(\tau)$ with probability one, then $F_T^{*}(\omega) \to F(\omega)$, as $T \to \infty$, at all the points of continuity of the function $F(\omega)$ with probability one. On the other hand, if $B_T^{**}(\tau) \to B(\tau)$, as $T \to \infty$, only in probability (i.e., in particular, if $B_T^{**}(\tau) \to B(\tau)$ in the mean), then it is not hard to prove that $F_T^{*}(\omega)$ also tends in probability to $F(\omega)$ at all the continuity points of $F(\omega)$. The case of discrete time is interpreted quite similarly. See, e.g., Doob (1953), Sec. X.7 and XI.7, and Hannan (1960), Sec. III.2.

[97]In particular, the estimated spectral distribution function can be used to construct a reasonable "goodness-of-fit test" to check the hypothesis that the observed series $x(t)$ is a realization of a particular specified stationary random function $X(t)$ (see, e.g., Grenander and Rosenblatt, 1956a,b, Bartlett, 1978, and Priestley, 1981, Sec. 6.2.6). A number of examples of estimates $F_T^{*}(\omega)$ for some meteorological and climatological series of observations are presented by Polyak (1975, 1979) and Petersen and Larsen (1978).

[98]See, e.g., Grenander and Rosenblatt (1956a,b), Ibragimov (1963), Malevich (1975), Bentkus (1972), and Priestley (1981), Secs. 6.2.5 and 6.2.6, where many additional references can be found.

[99]See, e.g., Hannan (1960), Sec. IV.1, and (1970), Sec. VII.6, Walker (1973), Pisarenko (1974), and especially Priestley (1981), Sec. 6.1 and Chap. 8, where many additional references are listed.

[100]On the rather interesting early history of investigations of hidden periodicities see, e.g., Strumpff (1937), Wold (1954) (or the first edition of this book of 1938), Serebrennikov and Pervozvansky (1965). More modern approaches are described in the references given in the previous Note.

Chapter 4

[1]There is no standard terminology for some of the subjects considered in this chapter. In particular, finite collections of one-dimensional random functions $\{X_1(t), ..., X_n(t)\} = \mathbf{X}(t)$ are also often called *multivariate* (or *multiple*) *random functions* in the literature, while the term *multidimensional random functions* is used by many authors for random functions $X(t_1, ..., t_n) = \mathbf{X}(t)$ which depend on several variables (and are called *random fields* in this book); see, e.g., Bartlett (1978) and Priestley (1981).

A random function $\mathbf{X}(t) = \{X_1(t), ..., X_n(t)\}$ can, of course, be considered as a random function $X(j, t) = X_j(t)$ on the set $T_{(n)} = (1, ..., n) \times T$ of all the pairs (j, t) where $1 \leqslant j \leqslant n$ and t is an element of T.

[2]In other words, to specify a function $\mathbf{X}(t) = \{X_1(t), ..., X_n(t)\}$ we must give the probability distributions of all the random variables $X_{j_1}(t_1)$, $X_{j_2}(t_2)$, ..., $X_{j_n}(t_n)$, where $\{(j_1,t_1), (j_2,t_2), ..., (j_n,t_n)\}$ is an arbitrary finite set of elements from $T_{(n)}$ (see Note 1).

[3]The random function $\mathbf{X}(t) = X(j,t)$, where t runs either through the set of all integers or through the set of all real numbers, is called *strict sense stationary* if for any n, $j_1, ..., j_n, t_1, ..., t_n$ and τ, the probability distribution of the n-dimensional random variable $X_{j_1,j_2,...,j_n}(t_1, t_2, ..., t_n) = \{X_{j_1}(t_1), X_{j_2}(t_2), ..., X_{j_n}(t_n)\}$ coincides with the probability distribution of the variable

$$X_{j_1,j_2, ..., j_n}(t_1 + \tau, t_2 + \tau, ..., t_n + \tau).$$

[4]Let us consider the following four stationary random functions $X_j(t) + X_k(t) = X^{(1)}(t)$, $X_j(t) - X_k(t) = X^{(2)}(t)$, $X_j(t) + iX_k(t) = X^{(3)}(t)$, and $X_j(t) - iX_k(t) = X^{(4)}(t)$. Denote by $B^{(l)}(\tau)$ and $F^{(l)}(\tau)$ the

correlation function and spectral distribution function of the random function $X^{(l)}(t)$, $l = 1, 2, 3, 4$. Then it is easy to check that

$$B_{jk}(\tau) = [B^{(1)}(\tau) - B^{(2)}(\tau) + iB^{(3)}(\tau) - iB^{(4)}(\tau)]/4.$$

Hence, the function $B_{jk}(\tau)$ can be represented as a Fourier–Stieltjes integral (4.6), where

$$F_{jk}(\omega) = [F^{(1)}(\omega) - F^{(2)}(\omega) + iF^{(3)}(\omega) - iF^{(4)}(\omega)]/4.$$

It is clear that the complex function $F_{jk}(\omega)$ satisfies the condition defining functions of bounded variation.

[5]To derive the equation expressing $F_{jk}(\omega)$ via $B_{jk}(\tau)$ one can use the representation of $F_{jk}(\omega)$ as the linear combination of the four functions $F^{(1)}(\omega)$, $F^{(2)}(\omega)$, $F^{(3)}(\omega)$, and $F^{(4)}(\omega)$ indicated in Note 4. Expressing the spectral distribution functions $F^{(l)}(\omega)$, $l = 1, 2, 3, 4$ via the corresponding correlation function $B^{(l)}(\tau)$ with the aid of a known equation relating to the one-dimensional case and then replacing $[B^{(1)}(\tau) - B^{(2)}(\tau) + iB^{(3)}(\tau) - iB^{(4)}(\tau)]/4$ by $B_{jk}(\tau)$, we obtain the desired result.

[6]Equation (4.10) can also be readily derived from the formula for $F_{jk}(\omega)$ given in Note 4.

[7]It was shown in Sec. 2 that the function $B(t,s)$ where t and s belong to an arbitrary set T, is the correlation function of a random function on T if and only if $B(t,s)$ is a positive definite kernel on T, i.e., it satisfies the condition (1.18) (see Vol. I, pp. 46-47 and Note 3 to Chap. 1 on p. 18 of this volume). When applied to the complex random function $X(j,t)$ of the argument (j,t) (see Note 1 to Chap. 4), the condition (1.18) evidently takes the following form: The inequality $\Sigma^N_{j,k=1} B_{l_j l_k}(t_j,t_k) c_j \bar{c}_k \geq 0$ must hold for any positive integer N, integers $l_1, ..., l_N$ between 1 and n, arbitrary integers (or real numbers) $t_1, ..., t_N$, and complex numbers $c_1, ..., c_N$. If $B_{jk}(t,s) = B_{jk}(t - s)$, where all the functions $B_{jk}(\tau)$ are representable in the form (4.6), then it is easy to see that

$$\sum_{j,k=1}^{N} B_{l_j l_k}(t_j,t_k) c_j \bar{c}_k = \int \sum_{j,k=1}^{N} e^{it_j\omega} c_j e^{-it_k\omega} \bar{c}_k dF_{l_j l_k}(\omega) = \int \sum_{p,q=1}^{n} C_p(\omega)\overline{C_q(\omega)} dF_{pq}(\omega)$$

where $C_p(\omega) = \sum_{l_j=p} \exp(it_j\omega)c_j$. According to condition (4.14) the integral on the right-hand side of the last equation is equal to the limit of a sequence of integral sums consisting only of nonnegative terms.   Therefore this integral is also nonnegative.   It follows from this that the matrix $B(t - s) = \|B_{jk}(t - s)\|$ is indeed a correlation matrix.

[8]See Cramér (1940).   Another proof of the same result was given by Kolmogorov (1941a).

[9]Let $X_j(t)$ and $X_k(t)$ be stationary and stationarily correlated random functions.   Consider again the stationary random functions $X^{(1)}(t)$, $X^{(2)}(t)$, $X^{(3)}(t)$, and $X^{(4)}(t)$ introduced in Note 4, and denote by $Z^{(1)}(\omega)$, $Z^{(2)}(\omega)$, $Z^{(3)}(\omega)$, and $Z^{(4)}(\omega)$ the functions $Z_j(\omega) + Z_k(\omega)$, $Z_j(\omega) - Z_k(\omega)$, $Z_j(\omega) + iZ_k(\omega)$, and $Z_j(\omega) - iZ_k(\omega)$ appearing in the spectral representations of the function $X^{(l)}(t)$, $l = 1, 2, 3$, and 4.   Let $[\omega_1, \omega_1 + \Delta\omega_1] = \Delta_1\omega$ and $[\omega_2, \omega_2 + \Delta\omega_2] = \Delta_2\omega$ be two disjoint intervals of the ω-axis.   Since the functions $X^{(l)}(t)$, $l = 1, 2, 3, 4$, are stationary,

$$\langle Z^{(1)}(\Delta_1\omega)\overline{Z^{(1)}(\Delta_2\omega)}\rangle = 0 \quad \text{for all four values of } l. \text{ Hence}$$

$$\langle Z^{(1)}(\Delta_1\omega)\overline{Z^{(1)}(\Delta_2\omega)}\rangle - \langle Z^{(2)}(\Delta_1\omega)\overline{Z^{(2)}(\Delta_2\omega)}\rangle + i\langle Z^{(3)}(\Delta_1\omega)\overline{Z^{(3)}(\Delta_2\omega)}\rangle -$$

$i\langle Z^{(4)}(\Delta_1\omega)\overline{Z^{(4)}(\Delta_2\omega)}\rangle = 0$, and it is easy to check that this relation is equivalent to (4.18).   Similarly, relation $\langle|Z^{(l)}(\Delta\omega)|^2\rangle = F^{(l)}(\Delta\omega)$ implies that $\langle|Z^{(1)}(\Delta\omega)|^2\rangle - \langle|Z^{(2)}(\Delta\omega)|^2\rangle + i\langle|Z^{(3)}(\Delta\omega)|^2\rangle - i\langle|Z^{(4)}(\Delta\omega)|^2\rangle = 4F_{jk}(\omega)$ (see Note 4).   The last result is obviously equivalent to (4.19).

[10]It has already been indicated in the footnote on p. 105 of Vol. I that in the applied literature the spectral density of a stationary random function is usually called its spectrum.   This explains the widespread use of abridged terms "cospectrum" and "quadrature spectrum".

Note also that the terms cospectrum and quadrature spectrum are sometimes applied to the double functions $2c_{jk}(\omega)$ and $2q_{jk}(\omega)$ determined only on the semiaxis $\omega \geqslant 0$.   Moreover, some authors apply the term *quadrature spectrum* to the function $-q_{jk}(\omega)$ or $-2q_{jk}(\omega)$ having the opposite sign.

[11]Here again there is no standard terminology.   In particular, some authors call the squared quantity $\gamma_{jk}^2(\omega) = |f_{jk}(\omega)|^2/f_{jj}(\omega)f_{kk}(\omega)$ the *coherence* (or the *coherency*) of $X_j(t)$ with $X_k(t)$ at frequency ω.   Sometimes the complex function $\gamma_{jk}^{(1)}(\omega) = f_{jk}(\omega)/[f_{jj}(\omega)f_{kk}(\omega)]^{1/2}$ is also called the coherence;   more often, however, the term *complex coherence* is used for $\gamma_{jk}^{(1)}(\omega)$ to avoid confusion. Now and then the term coherence is applied to the function $\gamma_{jk}^{(2)}(\omega) = c_{jk}(\omega)/[f_{jj}(\omega)f_{kk}(\omega)]^{1/2}$. The term *phase spectrum* is sometimes replaced by the shorter term, *phase*, and in some cases $\theta_{jk}(\omega)$ is replaced by $-\theta_{jk}(\omega)$.   Note also that $-d\theta_{jk}(\omega)/d\omega = -\theta'_{jk}(\omega)$ is often termed the *group delay* (or *envelope delay*; see, e.g., Priestley, 1981, p. 664).

[12]In fact, almost all the literature on relay and polarity coincidence correlation estimation given in Notes 22, 25, and 26 to Chap. 3 refers to the general estimation problem for cross-correlation functions of multidimensional stationary processes. The confidence intervals for the estimate (4.30) of a cross-correlation function are studied by Brillinger (1979). See also the literature on multidimensional random stationary functions given in the next Note.

[13]Very many books and papers cited above in connection with certain topics from the theory of one-dimensional stationary functions in fact also contain results referring to the more general multidimensional case. In particular, the books by Rozanov (1967), Hannan (1970), and Brillinger (1975) are almost entirely devoted to the study of multidimensional stationary functions $X(t)$, and the one-dimensional results are usually given in these books only as particular cases of the corresponding results pertaining to multidimensional functions. The statistical analysis of multidimensional stationary functions $X(t)$ is also the central subject of the book by Robinson (1967), which includes a number of computer programs for numerical estimation of some basic statistical characteristics of a multidimensional stationary function. Large sections on statistical methods relating to multidimensional stationary functions can also be found in the books by Jenkins and Watts (1968), Koopmans (1974), Priestley (1981), and some others. The literature on multidimensional parametric (autoregressive) spectral estimation is also very extensive; see, e.g., Whittle (1963b), Jones (1974, 1978), Strand (1977), Morf et al. (1978), Kay and Marple (1981), Marple and Nuttall (1983), Lütkepohl (1985), and Pukkila and Krishnaiah (1986b), where many additional references can be found.

[14]In Note 1 on p. 115 it was mentioned that random functions $X(t_1, ..., t_n) = X(t)$ of several variables are sometimes also called *multidimensional random functions*. There are also several other terms used in the literature to denote such random functions $X(t)$ (and their realizations $x(t)$), e.g., *spatial random functions* (or *spatial series, or spatial processes*), *multiparameter time series* (or *time series with multidimensional time*), and *random surfaces*.

[15]See the references in Note 12 to the Introduction (p. 10) and also, e.g., the books and papers by Pierson and Tick (1955) and Konyaev (1981) (applications to oceanography), Robinson (1977) and Haykin (1983) (seismology), Goodman (1968) (optics), Jain (1981) (image processing), W. Thompson (1958) and Dodds and Robson (1973) (description of the roughness of roads and runways), Berry and Marble (1968), King (1969), and Gaile and Willmott (1984) (geography), Nash and Jordan (1978) (geodesy), Harbaugh and Preston (1968), Matheron (1970), and Davis (1973) (geology), Matérn (1960) (forestry), and Whittle (1956, 1962) (agricultural science). Many examples of applications of random field theory to different scientific disciplines can also be found in the books by Hepple (1974), Bartlett (1975), Getis and Boots

(1978), Bennett (1979), Bartels and Ketellapper (1979), Ripley (1981), Vanmarcke (1983), and Rosenblatt (1985), which contain much material on the methods of statistical analysis of random fields data; see also surveys by Whittle (1963c), Unwin and Hepple (1974), and Larimore (1976).

[16]A random field $X(t)$ is called *strict sense homogeneous* (or *strictly homogeneous*) if the probability distributions of random vectors $X(t_1, ..., t_m) = \{X(t_1), ..., X(t_m)\}$ for any integer m and points $t_1, ..., t_m$ remain the same when the whole group of points $t_1, ..., t_m$ is arbitrarily translated, i.e. is replaced by the group $t_1 + s, ..., t_m + s$, where s is an arbitrary vector in $\mathbb{R}^n$ (or $\mathbb{Z}^n$). It is clear that if the first and second moments of such a field $X(t)$ exist, they will necessarily satisfy the condition (4.34) and (4.36).

Note also that homogeneous random fields are sometimes called *stationary random fields*.

[17]It can be shown that the set of all moving average fields on $\mathbb{Z}^n$ (i.e. homogeneous fields on $\mathbb{Z}^n$ which can be represented in the form (4.38)) coincides with the set of fields having spectral density $f(k)$ (the definition of the spectral density is given in Vol. I on p. 331). Cf., e.g., pp. 451-453 in Vol. I and also Bruckner (1971) where the general case of moving average fields on a discrete commutative group with a Haar measure is considered.

[18]Various nearest neighbor models (not all of them satisfy a difference equation of the form (4.41)) are considered, in particular, in Sec. 2 of Bartlett's book (1975) which contains many additional references.

[19]Several different proofs of Bochner's multidimensional theorem are known at present. In particular, a probabilistic proof of this theorem, which is based on the random field concept and is similar to the proof of Bochner's one-dimensional theorem outlined in Vol. I at the end of Sec. 11, was given by Blanc-Lapierre and Fortet (1947b). See also Note 23 below which contains the references to some sources where the proof of generalized Bochner's theorem can be found.

[20]See, e.g., Whittle (1963c), Brillinger (1970), and Vanmarcke (1983), Sec. 3.5.

[21]See, in particular, Whittle's paper (1954b), where a number of simple two-dimensional autoregressive models are considered and some conditions guaranteeing the existence of a unique homogeneous solution of the difference equation (4.40) are given. A number of exact or approximate calculations of the correlation function $B(\tau_1, \tau_2)$ for some autoregressive models of two-dimensional homogeneous fields are considered by Besag (1972). The values of

$R(1,0)$, $R(2,0)$, $R(2,1)$, and $R(1,1)$ where $R(\tau_1,\tau_2) = B(\tau_1,\tau_2)/B(0,0)$, corresponding to discrete parameter random fields with a spectral density of the form $f(k_1,k_2) = A[1 - 2\beta(\cos k_1 - \cos k_2)]^{-1}$ are tabulated in Bartlett's book (1975) for various values of β. Some further results concerning autoregressive (and more general ARMA) models of homogeneous fields are given by Tjøstheim (1978, 1981); see also Jain (1981).

[22]The derivation of (4.65) uses equations 6.565.4 and 8.411.1 from Gradshteyn and Ryzhik's book (1980). The model of a continuous random field in the plane satisfying the differential equation (4.63) was considered by Whittle (1954b), who evaluated the correlation function and the spectral density for this model. Later on, Heine (1955) calculated the correlation functions and spectral densities for two-dimensional homogeneous random fields satisfying differential equations of the form

$$[a(\partial^2/\partial t_1^2) + 2d(\partial^2/\partial t_1 \partial t_2) + b(\partial^2/\partial t_2^2) + 2g(\partial/\partial t_1) + 2h(\partial/\partial t_2) + e]X(t_1,t_2)$$

$$= cE(t_1,t_2)$$

for a number of values of the constants a, b, d, g, h, and e, while Whittle (1963c) examined the homogeneous solution $X(t_1, ..., t_n)$ of the partial differential equation

$$(4.1') \quad [\sum_{j=1}^{n}(\partial^2/\partial t_j^2) - a^2]^p X(t_1, ..., t_n) = cE(t_1, ..., t_n), \quad p > 0,$$

where $E(t_1, ..., t_n)$ is the n-dimensional "white noise" with spectral density $f_E(k_1, ..., k_n) = (2\pi)^{-n}$ (see also Yadrenko, 1983, Sec. I.1, Example 6). In the latter case the spectral density $f(k_1, ..., k_n)$ and the correlation function $B(\tau_1, ..., \tau_n)$ depend only on $k = (k_1^2 + ... + k_n^2)^{1/2}$ and $\tau = (\tau_1^2 + ... + \tau_n^2)^{1/2}$, respectively, and have the form

$$f(k) = \frac{c^2}{(2\pi)^n(k^2 + a^2)^{2p}} ,$$

$$(4.2') \quad B(\tau) = \frac{c^2}{2^{2p+(n/2)-1}\pi^{n/2}\Gamma(2p)a^{2p-(n/2)}} \tau^{2p-(n/2)}K_{2p-(n/2)}(a\tau).$$

[23]The necessary information about topological groups can be found, e.g., in the books by Weil (1940) and Rudin (1962); see also Nachbin (1962). The generalized Bochner theorem which gives the form of positive definite functions on commutative topological groups with Haar measure is due to Weil (1940) and Raikov (1945); it can also be found in Rudin's book (1962).

[24]The theorem on the spectral representation of homogeneous random fields on commutative topological groups is a simple corollary to the generalized Bochner theorem mentioned in the

preceding Note and to Karhunen's theorem on the generalized spectral representation given in Note 17 to Chap. 2. This representation is considered, in particular, by Kampé de Fériet (1948) and by Jajte (1967).

[25]Some results referring to *statistical analysis of multidimensional homogeneous random fields* can be found in Brillinger's papers (1970, 1974).

[26]Equation (4.75) can be simply deduced from the obvious relationship

$$\sigma^2(M^*_{T_1 T_2}) = \frac{1}{T_1 T_2} \int_0^{T_1} \int_0^{T_1} \int_0^{T_2} \int_0^{T_2} b(t_1 - t_1', t_2 - t_2') \, dt_1 dt_1' dt_2 dt_2'$$

and (4.75a) is obtained similarly (cf. derivation of (3.11) and (3.11a) outlined in Note 4 to Chapter 3; see also Vanmarcke, 1983, Sec. 6.1). A three-dimensional version of condition (4.74) is given without proof in Monin and Yaglom's book (1971), p. 252. Much earlier, Wiener (1939) showed that if $X(t)$ is any homogeneous field, then the limit (in the mean)

$$(4.3') \qquad \hat{X} = \lim_{T \to \infty} \frac{1}{|V_T|} \int_{V_T} X(t) dt$$

where $|V_T|$ is the volume of V_T, exists for a wide class of spatial domains V_T extending unboundedly as $T \to \infty$. This Wiener theorem is a multidimensional generalization of Neumann's quasi-ergodic theorem (1932a) and the related Khinchin's law of large numbers (1933, 1934); see Note 3 to Chap. 3. For further generalizations of the indicated Wiener result see, e.g., Pitt (1942), Dunford and Schwartz (1958), Chap. 8, Jajte (1967), and Tempelman (1986).

[27]The case where $f(k_1, k_2) = k_1^{\beta_1} k_2^{\beta_2} g(k_1, k_2)$, $-1 < \beta_1 < 1$, $-1 < \beta_2 < 1$, while $g(k_1, k_2)$ is continuous and different from zero in the neighborhood of the zero point $k_1 = 0$, $k_2 = 0$, was considered by Dang (1982), who showed that here $\sigma^2(M^*_{T_1 T_2}) \approx a T_1^{-1-\beta_1} T_2^{-1-\beta_2}$ for large values of T_1 and T_2 and also evaluated the coefficient a. Dang also obtained a similar result for a homogeneous field $X(t) = X(t_1, ..., t_n)$ in $\mathbb{R}^n$ under the condition that $f(k) = k_1^{\beta_1} ... k_n^{\beta_n} g(k)$, where $-1 < \beta_j < 1$, $j = 1, ..., n$, $g(k)$ is a continuous function and $g(0, ..., 0) \neq 0$; cf. also Note 8 to Chap. 3. The case where $n = 2$ but the averaging region is not necessarily a rectangle was also studied by Dang (1984).

A three-dimensional version of equation (4.77) was given without proof by Monin and Yaglom (1971), p. 252; its accurate derivation for homogeneous fields in $\mathbb{R}^n$ and on $\mathbb{Z}^N$ was presented by

Davidovich (1970) under the condition that the field $X(t) - \langle X(t) \rangle$ has a bounded and continuous spectral density $f(k)$, which differs from zero at the point $k_1 = 0, ..., k_n = 0$.

[28]The variance of the estimate (4.78) can be simply calculated with the aid of (4.75a); cf. S. Vilenkin (1979), Sec. 6.1. It should be noted that results similar to those discussed in Note 6 to Chap. 3 for stationary random processes are valid for homogeneous random fields too. Namely, it is possible to show that for a wide class of correlation functions $b(\tau)$ there exist values N_1 and N_2 for which the estimate (4.78) is more accurate than the integral estimate $m^*_{T_1 T_2}$ (see

Davidovich and Kartashov, 1968, or Vilenkin, 1979, Sec. 6.1; a similar result for the field in $\mathbb{R}^n$ is given by Skalsky, 1971). However, as in the case where $n = 1$, the difference between the root-mean-square errors of the integral estimate $m^*_{T_1 T_2}$ and the best

of the estimates $m^*_{T_1 T_2; N_1 N_2}$ is usually very small and the

determination of the optimal values N_1 and N_2 is possible only in those (rather rare) cases where the correlation function $b(\tau)$ is known exactly. Moreover, if $b(\tau)$ is known exactly, it is possible, in principle, to find also the best linear unbiased estimate (BLUE) of the mean value m, which is even more exact than the best one of the estimates $m^*_{T_1 T_2; N_1 N_2}$ (see, e.g., Vilenkin and Dubenko, 1971;

Vilenkin, 1979, Sec. 6.2). However, the explicit determination of BLUE for the mean values m is possible only in a few special cases and is, as a rule, a complicated problem, while the difference in the accuracy of BLUE and of the estimate $m^*_{T_1 T_2}$ is rather small in most cases.

[29]See, e.g., Ripley (1981), where many additional references can be found.

[30]Various formulations of the central limit theorem for random fields can be found, e.g., in the papers by Deo (1975), Bulinskii and Zhurbenko (1976a,b), Neaderhouser (1978), Nahapetian (1980), Khalilov (1982), Gorodetskii (1984); and Rosenblatt (1986); see also the books by Rosenblatt (1985), Secs. III.6 and IV.6, and by Leonenko and Ivanov (1986), Sec. 1.7, and the informative survey paper by Goldie and Morrow (1986).

[31]Note that, as in the case of stationary random functions of a single variable t (see Note 10 to Chap. 3 on p. 67), the proofs of almost sure convergence of the estimator $M^*_{T_1 T_2}$ (or of some similar estimator

differing by the choice on an averaging domain V and/or in the number of independent variables t_i) to m as V extends unboundedly (i.e. proofs of the strong law of large numbers for homogeneous random fields) naturally fall into two categories. On the one hand, there are results on the strong law of large numbers for strictly homogeneous random fields which generalize the classic Birkhoff–Khinchin ergodic theorem; see, e.g., Wiener (1939) (who investigated the asymptotic behavior of square averages M^*_{TT} as $T \to \infty$), Dunford and Schwartz (1956) and (1958), Sec. VIII.7 (where some more general results can be found), and the book by Tempelman (1986) containing further generalizations. On the other hand, there are proofs of various sufficient conditions for the validity of the strong law of large numbers for wide sense homogeneous random fields which generalize similar conditions for wide sense stationary random functions of a single variable t; see, e.g., Yurinskii (1974), Gaposhkin (1977b, 1981b), and Klesov (1981).

[32]The asymptotic formula describing the behavior of the variance $\sigma^2(B^{**}_{T_1 T_2})$ for large values of T_1 and T_2 in the case of a Gaussian field $X(t)$ can be found, e.g., in Priestley's book (1981), p. 722. The central limit theorem for homogeneous random fields is applied to investigations of the asymptotic behavior of the estimates $B^{**}_{T_1 T_2}(T_1, T_2)$ of the correlation function in the papers by Ivanov and Leonenko (1980, 1981a,b) and Rosenblatt (1986), in the book by Leonenko and Ivanov (1986), and in Rosenblatt's book (1985), Sec. IV.6.

[33]See, e.g., Pagano (1971), where the asymptotic properties of the two-dimensional periodogram are studied. Some more refined results referring to this asymptotic behavior can be found in the paper by Brillinger (1970).

[34]The smoothed periodogram estimates of the spectral density of a homogeneous random field are considered in more detail in the books by Rayner (1971), Priestley (1981), Sec. 7.2, Ripley (1981), Sec. 5.2, and Rosenblatt (1985), Secs. III.6, IV.6, and V.8, and in Brillinger's papers (1970, 1974b); see also McClellan (1982), Sec. III, and Rosenblatt (1986). (In particular, Rosenblatt gives conditions of the asymptotic normality of these estimates.) Specific examples of the spectral density estimation for two-dimensional fields can be found in the books by Rayner (1971) and Ripley (1981), the latter also containing a number of additional references to such examples.

[35]See, first of all, the survey by McClellan (1982) devoted mainly to the parametric spectral estimation for random fields. Additional information on this topic can be found, e.g., in the papers by Brillinger (1970, 1974b), Larimore (1977), Dickinson (1980), Woods (1980), Roucos and Childers (1980), Tjøstheim (1981), Lang and McClellan (1982), Malik and Lim (1982), McClellan and Lang (1983), Lim and Dowla (1983), and Nikias and Raghuveer (1985). Some

general results on the asymptotic distribution of estimators of parameters for finite parameter models of homogeneous random fields are given by Rosenblatt (1985), Sec. IV.6, and (1986).

[36]The term "isotropic random field" is often replaced in the literature by a longer (but a more precise) term, "homogeneous and isotropic random field". Note also that some authors use the term "isotropic random fields" to denote more general random fields in $\mathbb{R}^n$ whose mean value $\langle X(t) \rangle = m(t)$ depends on the distance $t = |t|$ of the point t from the coordinate origin O while the correlation function $B(t_1, t_2)$ satisfies the condition $B(t_1, t_2) = B(gt_1, gt_2)$, where g is an arbitrary rotation about the point O, and gt is the point obtained from t by the rotation g (see, e.g., Wong, 1971, Sec. 7.3, and Yadrenko, 1983, Sec. 1.6; cf. also Gaposhkin, 1984). We will not consider such fields, however, and will apply the term "isotropic" only to fields $X(t)$ which are both homogeneous and isotropic in the above-mentioned sense.
As usual, in the body of the book we restrict ourselves to the correlation theory of isotropic random fields, i.e. we only consider fields which are *wide sense isotropic*. To define the concept of a *strictly isotropic* random field $X(t)$ one should consider all the possible finite-dimensional probability distributions for random variables $X(t_1)$, $X(t_2)$, ..., $X(t_N)$. Then two different definitions of a strictly isotropic random field can be given. One can require that all such probability distributions be unaffected either by all translations and rotations of the group of points t_1, t_2, ..., t_N, or by all translations, rotations, and reflections of this group of points. If, however, only correlation theory is considered, then the two definitions become equivalent and both reduce to the definition of wide sense isotropy formulated in Vol. I on p. 348.

[37]For the use of the concept of an isotropic random field in turbulence theory see, e.g., Batchelor (1982), Hinze (1975), Chap. 3, and Monin and Yaglom (1975), Chaps. 6 and 7. Other applications of this concept have been discussed, in particular, by Longuet-Higgins (1975a,b) (description of ocean-wave behavior), and Kamash and Robson (1978) (road roughness modelling).

[38]The class of n-dimensional isotropic correlation functions $B(\tau)$ evidently coincides with the class of positive definite kernels in the n-dimensional Euclidean space $\mathbb{R}^n$, which depend only on the distance between their two arguments; cf. Sec. 2 in Vol. I. Equation (4.105) was first derived by Schoenberg (1938a) as an equation giving the general form of a positive definite kernel in $\mathbb{R}^n$ depending on the distance between pairs of points; cf. also Hartman and Wintner (1940), where the characteristic functions of n-dimensional spherically symmetric probability distributions are studied.

[39]See in this connection Example 7 on pp. 366-367 in Vol. I.

[40]It is easy to see that the function $\exp(-k^2\tau^2)$ belongs to the class D_∞ for any k; see Example 3 on p. 364 of Vol. I.   Hence, according to the property (iii) of the functional classes D_n (see p. 355 of Vol. I), all the functions of the form (4.107), where $\Phi(k)$ is a bounded nondecreasing function, also belong to D_∞ .   The proof of the converse theorem about the possibility of representing any function of the class D_∞ in the form (4.107) is due to Schoenberg (1938a), pp. 817-821; see also Gihman and Skorohod, 1974, Sec. IV.2.   This proof is based on the following readily verifiable relation: $\Lambda_n((2n)^{1/2}x) \to \exp(-x^2)$ as $n \to \infty$, where Λ_n is the function (4.109).

[41]Since the function $\Phi_1(k)$ is continuous at the point $k = 0$, while $|\Lambda_n(k\tau)| \leqslant 1$, for any arbitrarily small $\epsilon > 0$ there exists $\delta = \delta(\epsilon)$ such that

$$\left| \int_0^\delta \Lambda_n(k\tau)d\Phi_1(k) \right| \leqslant \int_0^\delta d\Phi_1(k) < \epsilon/2.$$

On the other hand, since $\int_0^\infty d\Phi_1(k) = B_0 < \infty$ and $\Lambda_n(x) \to 0$ as $x \to \infty$,

there exists a positive number $T_0 = T_0(\delta)$ such that $|\Lambda_n(k\tau)| < \epsilon/2B_0$ for $k \geqslant \delta$, $\tau \geqslant T_0$.   Hence, $|B_1(\tau)| < \epsilon/2 + B_0\epsilon/2B_0 = \epsilon$ for $\tau > T_0$ and therefore $B_1(\tau) \to 0$ as $\tau \to \infty$.

[42]The inequalities (4.112) and (4.113) were proved by Obukhov (1954a) (for $n = 2$) and Matérn (1960).   Tables of functions $\Lambda_n(x)$ (and the notation of Λ_n for these functions) can be found in the book by Jahnke, Emde, and Lösch (1960).   The tables show that $\min\Lambda_n(x)$ rapidly approaches zero as n increases; in particular, they imply that $R(\tau) > -0.06$ for n = 6, and $R(\tau) > -0.03$ for $n = 8$.

[43]See Schoenberg (1938a), p. 822; cf. also Yadrenko (1983), Sec. I.1.

[44]It can be shown that formula (4.107) implies, in fact, that any function $B(\tau)$ of class D_∞ is analytic within the sector $|\arg\tau| < \pi/4$ of the complex τ-plane (see Schoenberg, 1938a, p. 822).

[45]See, e.g., Lord (1954) and Tsuji (1955).   Formulae (4.124) are due to Kovasznay, Uberoi, and Corrsin (1949), and (4.123) to Lord (1954) (see also Kamash and Robson, 1977).   Clearly, (4.124) is an obvious consequence of (4.122), but the derivation of (4.123) is slightly more complicated, and we will briefly outline it here. Let us begin with the differentiation of both sides of the relation

$$f_1(k_1) = \frac{1}{2\pi} \int_{-\infty}^\infty e^{-ik_1\tau} B(\tau)d\tau$$

with respect to k_1. Then we obtain the equation

$$\frac{df_1(k_1)}{dk_1} = -\frac{i}{2\pi} \int_{-\infty}^{\infty} e^{-ik_1\tau} \tau\, B(\tau) d\tau$$

which implies that

$$(4.4')\qquad \tau B(\tau) = i \int_{-\infty}^{\infty} \frac{df_1(k_1)}{dk_1} e^{ik_1\tau} dk_1 = -2 \int_0^{\infty} \frac{df_1(k_1)}{dk_1} \sin k_1\tau\, dk_1.$$

Substituting (4.4') onto the right-hand side of the first equation (4.118) and using then equation 6.671.7 from Gradshteyn and Ryzhik (1980), we obtain the first equation (4.123). Some other forms of the equation expressing $f(k)$ in terms of $f_1(k_1)$ for an isotropic field in the plane can be found in the papers by Lord (1954), Tsuji (1955), and Kamash and Robson (1977).

Lord (1954) considered, in fact, not only n-dimensional and one-dimensional spectral densities $f(k) = f_n(k)$ and $f_1(k)$, but also all the intermediate q-dimensional spectral densities $f_q(k)$, where $q = 2$,

3, ..., $n - 1$, of an isotropic random field $X(t) = X(t_1, ..., t_n)$ in $\mathbb{R}^n$. (Note that Lord used entirely different terminology, since he studied the characteristic functions of multidimensional spherically symmetric probability distributions, but not the correlation functions and spectral densities of isotropic random fields.) The density $f_q(k)$ for all integral values of q from 1 to n is defined as the (q-dimensional) spectral density of an isotropic field $X_q(t_1, ..., t_q) = X(t_1, ..., t_q, 0, ..., 0)$ determined in a q-dimensional space $\mathbb{R}^q$. It is easy to show that

$$f_q(k) = 2\pi^{m/2}\, [\Gamma(\tfrac{m}{2})]^{-1} \int_{k_1}^{\infty} f_{q+m}(k)(k^2 - k_1^2)^{(m-2)/2} k\, dk$$

$$(4.5')$$

$$= 2\pi^{(n-q)/2}\, [\Gamma(\tfrac{n-q}{2})]^{-1} \int_{k_1}^{\infty} f(k)(k^2 - k_1^2)^{(n-q-2)/2} k\, dk.$$

Lord's equation (4.5') clearly generalizes (4.120). Moreover, by first using (4.114), the obvious formula is, secondly,

$$f_q(k) = \frac{1}{(2\pi)^{q/2}} \int_0^{\infty} \frac{J_{(q-2)/2}(k\tau)}{(k\tau)^{(q-2)/2}} \tau^{q-1} B(\tau) d\tau,$$

where $B(\tau)$ is the common correlation function of isotropic fields $X(t)$ and $X_q(t)$, and, thirdly, the well-known recursion relation

$$\left[-\frac{1}{z}\frac{d}{dz} \right] \frac{J_k(z)}{z^{k+1}} = \frac{J_{k+1}(z)}{z^{k+1}},$$

we can prove that

$$(4.6')\quad f_{q+2m}(k) = \left[-\frac{1}{2\pi k}\frac{d}{dk} \right]^m f_q(k),\quad f(k) = \left[-\frac{1}{2\pi k}\frac{d}{dk} \right]^{(n-1)/2} f_1(k),$$

if n is an odd integer.   Now the first equations (4.5') and (4.6') with $m = 1$ clearly imply that

$$(4.7')\qquad f_q(k) = -\frac{1}{\pi}\int_k^\infty \frac{df_{q-1}(k_1)}{dk_1}\,\frac{dk_1}{(k_1^2 - k^2)^{1/2}}\;.$$

Equations (4.6') and (4.7') generalize (4.124) and (4.123).

[46]See, e.g., Dykhovichny's paper (1983) where a special estimate $B^*(\tau)$ of the function $B(\tau)$ is studied under the assumption that T is the n-dimensional sphere $V_R(O)$ of radius R with center at a given point O of $\mathbb{R}^n$, while the isotropic field $X(t)$ is Gaussian and has the spectral density $f(k)$. The studied estimate $B^*(\tau)$, where $\tau < R$, is obtained by averaging first of all the values of $x(t_1)x(t_2)$, where t_1 is a variable point within the sphere $V_{R-\tau}(O)$ of radius $R - \tau$ with center O, over all t_2 in the sphere $S_\tau(t_1)$ of radius τ with center at the point t_1 and then averaging the resulting spherical average over all t_2 within the sphere $V_{R-\tau}(O)$. A similar estimate of $B(\tau)$ from the observations within a sphere was later investigated by Dykhovichny (1985) at greater length and was used for the estimation of the corresponding spectral distribution function $\Phi(k)$. Another estimate of the function $B(\tau)$ from the data at points within $T = V_R(O)$, which uses averaging of $x(t_1)x(t_2)$ over a narrower set of pairs (t_1,t_2), has been considered by Ivanov and Leonenko (1981b) and Leonenko and Ivanov (1986).

[47]Turbulence, which is approximately isotropic (i.e. such that the fields of its velocity components, pressure, temperature, etc. are isotropic), is usually generated in a wind tunnel downstream from a regular array of rods forming a grid. Under these conditions, the turbulent fluctuations at the point $\mathbf{x} = (x,y,z)$ at the time $t - \tau$ can be identified, to a good accuracy, with the same fluctuations at the time moment t at the point with coordinates $(x - U\tau, y, z)$, where U is the mean flow velocity, which is directed along the x-axis (this is the so-called *frozen turbulence hypothesis* due to G. I. Taylor). Therefore, the data on the fluctuations of some fluid mechanical parameter at a single point enables one to determine approximately the fluctuations of the same parameter along a line parallel to the mean flow direction. In practice, it is just fluctuations at a fixed point that are usually used for estimating spatial correlation functions and spectral densities of isotropic turbulence (see, e.g., Batchelor, 1982; Bradshaw, 1971; Hinze, 1975; or Monin and Yaglom, 1975).

[48]This result was communicated to the author by M. I. Yadrenko. It is clear that in proving it, it will suffice to restrict oneself to the particular case of the function (2.123), where $C = 1$, $T = 1$. Then, according to (4.112), the corresponding two-dimensional spectral density $f(k)$ of the form

$$(4.8') \qquad f(k) = (2\pi)^{-1} \int_0^1 \tau(1 - \tau) \, J_0(k\tau) \, d\tau.$$

By replacing the Bessel function $J_0(k\tau)$ on the right-hand side of (4.8') by its power series and then integrating all the terms of this series it is not hard to show that

$$(4.9') \qquad f(k) = (2\pi k^3)^{-1} \, [- kJ_0(k) + \int_0^k J_0(x)dx].$$

The second term in the brackets on the right-hand side of (4.9') is bounded, whereas the first term can take arbitrary large, in absolute value, negative values. Hence the function $f(k)$ can also take negative values.

[49]Formulae (2.139) and (4.132), where $\nu = 1/3$, provide convenient models for the correlation function and spectral density of the fluctuation of temperature (or of the contamination concentration, or of the refractive index) in a fully developed turbulence; see, e.g., Monin and Yaglom (1975) or Tatarskii (1971).

[50]The first proofs of the fact that the functions (2.132) for $0 < m \leqslant 2$ belong to the class $\mathfrak{D}_\infty$ were given by Wintner (1936) and Schoenberg (1938b). Wintner's proof is, in fact, based on consideration of the isotropic random field $X_m(t)$ in the space having an n-dimensional spectral density of the form $f_m(k) = \min\{1, k^{-n-m}\}$, $0 < m \leqslant 2$. (Wintner used different terminology, since he studied characteristic functions of spherically symmetric multidimensional probability distributions and not the n-dimensional isotropic correlation functions.) With the aid of (4.115) one can prove in this case that

$$B_m(\tau) = C\{1 - a^m \tau^m + o(\tau^m)\} \quad \text{for } \tau \to 0,$$

where $B_m(\tau)$ is the correlation function of the field $X_m(t)$, C and a are positive constants, and $o(\tau^m)$ is a term of a higher order of smallness, for $\tau \to 0$, than τ^m. Since $B_m(\tau)$ is an n-dimensional isotropic correlation function, $B_{mj}(\tau) = [C^{-1}B_m(\tau/aj^{1/m})]^j$ is also an n-dimensional isotropic correlation function for any integer j. It is easy to see that, for $j \to \infty$, the limiting relation $B_{mj}(\tau) \to \exp(-\tau^m)$ holds uniformly in each fixed interval $0 \leqslant \tau \leqslant \text{const}$. Since the limit of the sequence of n-dimensional isotropic correlation functions is also such a function, it follows from this that the function $\exp(-\tau^m)$ for $0 < m \leqslant 2$ belongs to the class $\mathfrak{D}_n$ for any n, i.e. it belongs to the class $\mathfrak{D}_\infty$.

V.M Zolotarev (private communication) also noted that the last result can easily be derived from the well-known formula

$$(4.10') \qquad \exp(-\tau^\alpha) = \int_0^\infty e^{-\tau x} g(x;\alpha,1)dx, \quad 0 < \alpha < 1,$$

where $g(x;\alpha,1)$ is the special probability density on the semiaxis $0 \leqslant x < \infty$ (the so-called *stable probability density* corresponding to parameters α, $\beta = 1$, $\gamma = 0$, $\lambda = 1$); see, e.g., Zolotarev, 1983, Sec. 2.6. It follows from (4.10') and from the result relating to Example 1 on p. 362 of Vol. I that the functions (2.132), where $0 < m \leqslant 1$, belong to $\mathfrak{D}_\infty$. Further, replacing the variable τ by $s = \tau^{1/2}$ in (4.10'), we get

$$(4.11')\quad \exp(-s^{2\alpha}) = \int_0^\infty e^{-s^2 x} g(x;\alpha,1)\,dx, \quad 0 < \alpha < 1.$$

From (4.11') and from the result relating to Example 3 on p. 364 of Vol. I it follows that all the function (2.132), where $0 < m \leqslant 2$, belong to the class $\mathfrak{D}_\infty$.

Formulae (4.10') and (4.11') allow us to express the n-dimensional spectral densities, which corresponds to the correlation functions (2.132), in terms of the stable probability density $g(x;\alpha,1)$.

[51]Example 5 is due to Matérn (1960). By virtue of (4.134) and (4.136) the n-dimensional spectral density $f(k)$ which corresponds to the correlation function (4.135), is given by the equation

$$
\begin{aligned}
f(k) &= \frac{2C}{\Gamma(\nu)} \int_0^\infty \frac{1}{2^n \pi^{n/2}} \left\{ \exp\left(- \frac{k^2}{4u^2} - \alpha^2 u^2\right) \right\} u^{2\nu-1-n}\,du \\[2mm]
(4.12')\qquad &= \frac{C}{2^n \pi^{n/2} \Gamma(\nu)} \int_0^\infty \left\{ \exp\left(- \frac{k^2}{4x} - \alpha^2 x\right) \right\} x^{\nu-1-n/2}\,dx \\[2mm]
&= \frac{1}{2^{n/2+\nu-1}\,\pi^{n/2}\Gamma(\nu)\alpha^{2\nu-n}} (\alpha k)^{\nu-n/2} K_{\nu-n/2}(\alpha k).
\end{aligned}
$$

(Substitution of the variable $x = u^2$ instead of u and also Eq. 3.471.9 from Gradshteyn and Ryzhik's book, 1980, are used in the derivation of (4.12').)

[52]This result, which has not been published earlier, was obtained by M.I. Fortus during her graduate studentship.

[53]See, e.g., Shkarofsky (1968) and Khalidov (1978).

[54]Representation of the form (4.142) for an isotropic random field in a plane was given by Yaglom (1961, 1963). Later on it was also considered by a number of other authors (see, e.g., Gihman and Skorohod, 1969, 1974; Hannan, 1969, 1970; H. Ogura, 1966; Khudyakov, 1974). The term "polar spectral representation" is due to Ogura (1966).

[55]Spectral representation of the form (4.146) for an isotropic random field in the n-dimensional space $\mathbb{R}^n$ was obtained by Yaglom (1961); see also Wong (1969, 1971) and Yadrenko (1983). The particular case of this representation where $n = 3$ was also

considered by Ogura (1966) and Khudyakov (1974). The equation expressing the random functions $Z^l_m(k)$ in terms of $X(t)$ can be found in Yadrenko's book (1983).

[56]Some references to the literature on isotropic turbulence are indicated in Note 47; see also Yaglom (1948) and Moyal (1952).

[57]Let U_g be the $(s \times s)$-matrix describing the linear transformation of the components of the quantity $\mathbf{X}$ under the rotation and/or reflection g of the space $\mathbb{R}^n$. It is clear that $U_{g_1} U_{g_2} = U_{g_1 g_2}$, $U_{g^{-1}} = [U_g]^{-1}$, so that the set of matrices U_g forms an *s-dimensional representation* of the group G of all rotations and reflections in $\mathbb{R}^n$. (The literature on the theory of group representation is, in fact, immense; here we refer only to the books by N. Vilenkin, 1968, and Elliott and Dawber, 1979.) The homogeneous random field $\mathbf{X}(t)$ of quantities $\mathbf{X}$ is said to be isotropic if, and only if, $U_g \mathbf{m} = \mathbf{m}$, $U_g B(g\tau)U_g^* = B(\tau)$ for all the elements g of the group G, where, as usual, $\mathbf{m} = \langle \mathbf{X}(t) \rangle$, $B(\tau) = \| \langle X_j(t + \tau)\overline{X_k(t)} \rangle \|$, and the star denotes Hermitian conjugation of a matrix. The general form of the correlation matrix of an isotropic vector field is given on pp. 372-376 of Vol. I; a more general case of polyvector fields was studied by Itô (1956), while Malyarenko (1985a) considered the further generalization and investigated the forms of correlation matrices $B(\tau)$ of the isotropic fields of quantities $\mathbf{X}$ which transform accordingly to an arbitrary representation U of the group G of all rotations and reflections in $\mathbb{R}^n$. In the particular case where an s-dimensional representation U is the identity one (i.e., $U_g = \mathcal{E}$ for all g, where $\mathcal{E}$ is the unit $(s \times s)$-matrix), we once again arrive at the case where $\mathbf{X}(t) = \{X_1(t), ..., X_s(t)\}$ is a collection of s isotropic random fields isotropically correlated with each other. Thus, the first multidimensional generalization of the concept of isotropic random field is in fact a particular case of the second. Note also that both of the above definitions of the multidimensional isotropic random field refer to *wide sense isotropy*. To define *strict sense isotropy* of a multidimensional random field $\mathbf{X}(t) = \{X_1(t), ..., X_s(t)\}$ in $\mathbb{R}^n$ we must consider all the multidimensional probability distributions for the groups of random variables $\{X_{i_1}(t_1), X_{i_2}(t_2), ..., X_{i_n}(t_n)\}$ and require that these distributions be unaffected by any translation, rotation, and reflection of the points $t_1, t_2, ..., t_n$ accompanied, in the case where the quantity $\mathbf{X}$ differs from the collection of s scalars, by the appropriate linear transformation of the components of $\mathbf{X}$ (see, e.g., Monin and Yaglom, 1975, pp. 35-37).

[58]We give here one more derivation of (4.155) based on an important general idea (see Robertson, 1940). Let us note that the

tensor function $B_{jl}(\tau)$ of the n-dimensional vector τ is invariant under all rotations and reflections of $\mathbb{R}^n$. Therefore, if **a** and **b** are arbitrary unit vectors, then the quadratic form

$$B(\tau,\mathbf{a},\mathbf{b}) = B_{jl}(\tau)a_j b_l$$

(where, as also below in this Note, summation from 1 to n over the twice repeated indices is assumed) is a scalar depending on the three vectors τ, **a**, **b** (the last two have unit length), and its dependence on **a** and **b** is linear. It is known from the theory of invariants under the group of rotations and reflections (see, e.g., Weyl, 1939) that the scalar $B(\tau,\mathbf{a},\mathbf{b})$ should be expressible in terms of the principal invariants of the above three vectors, namely, the length $\tau = (\tau_j\tau_j)^{1/2}$ and the scalar products $\tau a = \tau_j a_j$, $\tau b = \tau_j b_j$, and $\mathbf{ab} = a_j b_j$ (we recall that $a_j a_j = b_j b_j = 1$). Therefore the general form of the function $B(\tau,\mathbf{a},\mathbf{b})$ which depends linearly on **a** and **b** is given by

$$B(\tau,\mathbf{a},\mathbf{b}) = A_1(\tau)\tau_j a_j \tau_l b_l + A_2(\tau)a_j b_j$$

and, hence, the general form of the tensor $B_{jl}(\tau)$ is

$$(4.13') \qquad B_{jl}(\tau) = A_1(\tau)\tau_j\tau_l + A_2(\tau)\delta_{jl}.$$

In other words, the tensor function of the vector τ which is invariant under rotations and reflections should be a linear combination of the constant invariant tensor δ_{jl} and the tensor $\tau_j\tau_l$, with coefficients depending on a unique invariant which can be made up of the components of τ, i.e., the length $\tau = |\tau|$. Changing over again to the coordinate system $0'x_1', ..., x_n'$ (in which $\tau_1 = \tau$, $\tau_2 = ... = \tau_n = 0$), we find from (4.13') that $B_{jl}'(\tau) = 0$ for $j \neq l$, $B_{11}'(\tau) = B_{LL}(\tau) = A_1(\tau)\tau^2 + A_2(\tau)$, and $B_{22}'(\tau) = ... = B_{nn}'(\tau) = B_{NN}(\tau) = A_2(\tau)$. Hence, it is clear that (4.13') is identical with (4.155).

[59]Equations (4.163) and (4.166) are due to Yaglom (1957). For the particular case where $n = 3$ and the functions $\Phi_1(k)$ and $\Phi_2(k)$ are absolutely continuous (i.e. the spectral densities $f_{jl}(k)$ exist), the general equations (4.173) for the correlation functions $B_{LL}(\tau)$ and $B_{NN}(\tau)$ have been previously obtained by Yaglom (1948) and Moyal (1952), in their study of isotropic turbulence in a compressible fluid.

Note also that, according to (4.161) and (4.162), $\Phi_1(0) = \Phi_2(0)$ (but, of course, addition of an arbitrary constant to $\Phi_1(k)$ and/or $\Phi_2(k)$ does not affect the functions $B_{LL}(\tau)$ and $B_{NN}(\tau)$). Moreover, the jumps $\Phi_1(+0) - \Phi_1(0) = \Phi_1(+0)$ and $\Phi_2(+0) - \Phi_2(0) = \Phi_2(+0)$ of the functions $\Phi_1(k)$ and $\Phi_2(k)$ at the point $k = 0$ add the same constant,

$$n^{-1}\{\Phi_1(+0) + (n-1)\Phi_2(+0)\}$$

to the values of $B_{LL}(\tau)$ and $B_{NN}(\tau)$. Thus, only the sum $\Phi_1(+0) + (n-1)\Phi_2(+0)$ is essential, and we can always assume that $\Phi_1(+0) =$

$\Phi_2(+\,0)$ (or that $\Phi_1(+\,0) = 0$, or that $\Phi_2(+\,0) = 0$). Besides, (4.163) and (4.166) imply that $B_{LL}(0) = B_{NN}(0) = n^{-1}[\Phi_1(\infty) + (n-1)\Phi_2(\infty)]$; hence $\langle|X(t)|^2\rangle = \Phi_1(\infty) + (n-1)\Phi_2(\infty)$.

[60]Equations (4.175), where $n = 3$ or $n = 2$, play an important part in the statistical theory of isotropic turbulence, and are most essential for this theory in the particular case where $f_1(k) = 0$ (the case of a solenoidal vector field $X(t)$; see Vol. I, pp. 380-383). In this particular case the first equation (4.175) for $n = 3$ is due to Heisenberg (1948), and for $n = 2$, to Y. Ogura (1952), while the general case of arbitrary n, under the same assumption that $f_1(k) = 0$, was considered by Tsuji (1955). Equations (4.175) for the case where $n = 3$ and both functions $f_1(k)$ and $f_2(k)$ do not vanish are given in Monin and Yaglom's book (1975).

[61]See, e.g., some related results of H. Ogura (1968), Ekhaguere (1979), and Malyarenko (1985a).

[62]For the particular case where $n = 3$, Eq. (4.176) is due to von Kármán (1937), and (4.177), to Obukhov (1954b); see also Monin and Yaglom (1975), Sec. 12.3. The representability of an arbitrary isotropic vector field $X(t)$ as a sum of mutually uncorrelated solenoidal and potential fields was proved by Obukhov (1954b); generalization of this result to the case of isotropic polyvector fields is due to Itô (1956).

[63]The identity of the class of all longitudinal correlation functions of isotropic solenoidal vector fields in $\mathbb{R}^n$ and the class of all lateral correlation functions of isotropic potential fields in the same space was proved by Yaglom (1957). Equations (4.184) - (4.186) for the special case where $n = 3$ are given in the book by Monin and Yaglom (1975). The model (2.139), where $\nu = 1/3$, of the longitudinal correlation function $B_{LL}(\tau)$ was proposed, for the case where $n = 3$, by von Kármán (1948), who also gave the formulae for the corresponding lateral correlation function $B_{NN}(\tau)$ and the spectral density $f_2(k)$ (see (4.185)). Von Kármán used this model for approximating experimental data on correlation functions and spectral densities of velocity fluctuations behind the grid in a wind tunnel. Therefore, in the mechanics of turbulence, these equations for $B_{LL}(\tau)$, $B_{NN}(\tau)$, and $f_2(k)$ are often called the *von Kármán equations*.

[64]See, e.g., Bochner (1941), Gel'fand (1950), and Krein (1949, 1950); cf. also Berg, Christensen, and Ressel (1984).

[65]The general representation (4.188) of invariant (i.e., satisfying (4.187)) positive definite kernels was given by Gel'fand (1950) for a very important class of so-called *symmetric homogeneous spaces* introduced by E. Cartan; see also Naimark (1959), Sec. 31.10. More general equations for invariant positive definite kernels on wide

classes of (nonsymmetric) homogeneous spaces can be found, e.g., in Yaglom's papers (1961, 1963).

[66]The spectral representation (4.189) of a homogeneous random field on an arbitrary symmetric homogeneous space R was obtained by Yaglom (1961), who derived also some more general forms of spectral representation for homogeneous fields on wide classes of nonsymmetric spaces R .   These spectral representations are implied, in fact, by equations for invariant positive definite kernels $B(t_1,t_2)$ by virtue of general addition theorems for spherical functions (which follow from the theory of group representation; see N. Vilenkin, 1968) and Karhunen's theorem formulated on pp. 39-41.   Also see in this connection Yaglom (1963), and Hannan (1965, 1969, 1970).

[67]The spectral representation (4.194) - (4.195) of a homogeneous random field on the sphere S^2 was given by Obukhov (1947); see also Yaglom (1961, 1963), Jones (1963), and Hannan (1969, 1970).   The estimation problem for the spectrum (i.e., the constants f_m) was considered by Jones (1963) under the condition that a single realization $x(\theta,\varphi)$ of the field $X(\theta,\varphi)$ is observed. The application of the spectral representation (4.194) to the investigation of the earth's magnetic field is, in fact, discussed by Strube (1985) (who does not properly use the concept of a random field in his paper).

[68]The spectral representation of a multidimensional field $X(\theta,\varphi) = \{X_1(\theta,\varphi), ..., X_s(\theta,\varphi)\}$, which has the constant mean value vector m and the correlation matrix $B(\theta_{12})$, was considered by Hannan (1970), Sec. II.10.   The spectral representation of a homogeneous vector field $X(\theta,\varphi)) = \{X_1(\theta,\varphi), X_2(\theta,\varphi), X_3(\theta,\varphi)\}$ on the sphere S^2 and of its correlation matrix $B(\theta_1,\varphi_1;\theta_2,\varphi_2)$ was studied by Yaglom (1955c, 1962b).

[69]The general form (4.196) of a positive definite kernel on the sphere S^{n-1}, depending only on the spherical distance between its two arguments, is due to Schoenberg (1942).   Note that the positive definiteness of the kernel (4.196) can easily be proved with the aid of the addition theorem for n-dimensional surface harmonics $S_m^l(t)$ referred to in Vol. I on p. 369 and p. 388.   This theorem enables one to write the kernel (4.196) in the form (4.197).   Now (4.197) implies at once that

$$\sum_{j,k=1}^{m} B(\theta_{jk})c_j c_k \geqslant 0$$

for any points t_1, ..., t_m of the sphere S^{n-1} and any real numbers c_1, ..., c_m.

On the other hand, if $B(\cos\theta)$ is an arbitrary continuous function of $\cos\theta$, then according to the well-known Funk-Hecke theorem (see, e.g., Erdélyi et al., 1953, Vol. 2, Sec. 11.4)

$$(4.14') \quad \int_{|t_2|=1} B(\cos\theta_{12})S^l_m(t_2)\,d\sigma(t_2) = b_m S^l_m(t_1)$$

where θ_{12} is the angle between unit vectors t_1 and t_2, $d\sigma(t_2)$ is an element of the $(n-1)$-dimensional area of the sphere $|t_2| = 1$, and b_m is a constant equal to

$$\int_{|t_1|=1} \int_{|t_2|=1} B(\cos\theta_{12})\,S^l_m(t_1)S^l_m(t_2)\,d\sigma(t_1)d\sigma(t_2).$$

If $B(\cos\theta)$ is a positive definite kernel on $\mathbf{S}^{n-1}$, then all the constants b_m will obviously be nonnegative. Moreover, according to (4.14'), the complete system of functions $S^l_m(t)$ coincides with the set of all the eigenfunctions of an integral equation with the kernel $B(\cos\theta_{12})$. Therefore, according to Mercer's theorem of the theory of integral equations, the positive definite kernel $B(\cos\theta_{12})$ can be represented in the form (4.197), which is equivalent to (4.196). Consequently, any positive definite kernel on the sphere $\mathbf{S}^{n-1}$, which depends on the spherical distance between its arguments, is representable in the form (4.196), where $f_m \geqslant 0$.

[70] The spectral representation (4.198) - (4.199) of the homogeneous random field on the $(n-1)$-dimensional sphere $\mathbf{S}^{n-1}$ was given by Yaglom (1961); see also Yadrenko (1983), Sec. 1.5.

[71] This remark is due to Yadrenko (1983), who noted that, if the correlation function $B(\theta)$ is given by (4.200), the equation

$$(4.15') \quad f_m = 2^{n-2}\,[\Gamma(n/2)]^2\,\left[1 + \frac{2m}{n-2}\right]\int_0^\infty \frac{J^2_{(n-2)/2+m}(kr)}{(kr)^{n-2}}\,d\Phi(k)$$

follows easily from the addition theorem for Bessel functions. Some additional examples of possible correlation functions of homogeneous fields on the sphere $\mathbf{S}^{n-1}$ can also be found in Yadrenko's book (1983), Sec. 1.5.

[72] The spectral representations (4.202) - (4.204) were given by Hannan (1969, 1970), who also showed that the homogeneous field $X(s,t)$ can be represented as a linear superposition of uncorrelated plane waves of various wavelengths and frequencies propagating isotropically in all spatial directions. A time-dependent (and statistically stationary) homogeneous random field on a sphere $X(s,t)$, where $s = (s_1,s_2)$, $s = (s_1^2 + s_2^2)^{1/2}$ = const, $-\infty < t < \infty$, was studied by Jones (1963). A more elementary example of a field $X(s,t)$, where $s = r\exp(i\varphi)$, $0 \leqslant \varphi < 2\pi$ (i.e. s is a point on a circle of fixed radius r), t takes integral values $0,\pm1,\pm2,\ \ldots$, and $\langle X(s,t)\rangle = 0, \langle X(r\exp(i\varphi_1),t_1) X(r\exp(i\varphi_2),t_2)\rangle = B(\varphi_1 - \varphi_2,\ t_1 - t_2)$, was considered by Roy (1972), who, in particular, inquired into the problem of spectral estimation for the field $X(s,t)$ when only observations $x(s,t) = x(r\exp(i\varphi),t)$, where $0 \leqslant \varphi \leqslant 2\pi$, $t = 0,1,\ \ldots,\ T-1$, are available. A more general

example of random fields in $\mathbb{R}^{n+m}$, homogeneous in respect to all variables and isotropic in respect to first n variables was studied (in a very general form) by Malyarenko (1985b).

[73]In particular, spectral representations for homogeneous random fields in an n-dimensional Lobachevsky space $\mathbf{L}^n$ (i.e., a hyperbolic space of constant negative curvature) were given by Yaglom (1961) (cf. also Wong, 1969), while the case of fields $X(s,t)$ depending on the spatial coordinates and the time, and homogeneous under all the Lorentz transformations of space-time continuum was studied by Hannan (1969) and Vovk (1981). As a rather simple but quite typical example of the above-mentioned spectral representations we present here the formulae referring to homogeneous fields in the Lobachevsky plane $\mathbf{L}^2$ (i.e., the two-dimensional hyperbolic space). In this case, the correlation function $B(r)$ depends on the non-Euclidean "Lobachevsky distance" r in $\mathbb{L}^2$ and has the form

$$(4.16')\qquad B(r) = \int_0^\infty P_{-1/2+(1/4-k)^{1/2}}(\cosh r)\,d\Phi(k)$$

where P_ν is Legendre's function of the first kind and $\Phi(k)$ is a bounded nondecreasing function. Similarly, the field $X(r,\varphi)$ itself, where (r,φ) are polar coordinates in $\mathbf{L}^2$, can be represented as

$$(4.17')\qquad X(r,\varphi) = \sum_{n=-\infty}^{\infty} e^{in\varphi}\int_0^\infty \gamma_n(k)P^n_{-1/2+(1/4-k)^{1/2}}(\cosh r)\,Z_n(dk),$$

where

$$\gamma_n(k) = \left\{(-1)^n\,\frac{\Gamma[(1/4-k)^{1/2}-|n|+1/2]}{\Gamma(1/4-k)^{1/2}+|n|+1/2]}\right\}^{1/2} > 0,$$

P^n_ν are associated Legendre's functions, and the complex-valued random measures $Z_n(\Delta)$ satisfy (4.143). (The representation (4.16') was given by Krein, 1949-50, while (4.17'), which follows easily from Krein's result, was obtained by Yaglom, 1961.)

[74]The structure functions were first used to describe statistical properties of random processes by Kolmogorov in his papers (1941b,c) on the theory of turbulence. The term "structure function" is due to Obukhov (1949a,b).

[75]The random process $X(t)$ is called a *process with strictly stationary increments* if the multidimensional probability distribution for the random variables $X(t + t_2) - X(t + t_1)$, $X(t + t_4) - X(t + t_3)$, ..., $X(t + t_{2n}) - X(t + t_{2n-1})$ for any n and $t, t_1, t_2, ..., t_{2n-1}, t_{2n}$ is independent of t. Some results about the moments of order p of the increments $\Delta_\tau X(t) = X(t + \tau) - X(t)$, where p is an arbitrary positive number and $X(t)$ is a process with strictly stationary increments, were given by Masani (1976).

The concept of a random process with stationary increments was introduced by Kolmogorov (1940a), who showed that in terms of the geometry of the Hilbert space H (see Vol. I, Sec. 5) a process with wide-sense stationary increments is equivalent to a *helix* (also called *spiral* or *screw curve*) in H (see also von Neumann and Schoenberg, 1941).   One can also consider the increment $\Delta_\tau X(t) = X(t + \tau) - X(t)$ as a random function of the interval $\Delta = [t + \tau, t) = \{t': t < t' \leqslant t + \tau\}$ and define processes with stationary increments as *stationary, additive interval functions* $X(\Delta)$. (This means that $X(\Delta)$ is a random function of time intervals Δ, having the first two moments or, for a strict sense concept, all the finite dimensional distributions invariant under time translation and satisfying the condition $X(\Delta_1 + \Delta_2) = X(\Delta_1) + X(\Delta_2)$, where Δ_1 and Δ_2 are adjoining time intervals and $\Delta_1 + \Delta_2$ is their union; see, e.g., Bochner, 1960, and Brillinger, 1972.) Note that for processes with stationary increments the values $X(t)$ at the given time t must not be considered at all and only differences $X(u) - X(t)$ make sense.

[76]This is due to the fact that, when we consider the differences $X(t + \tau) - X(t)$ instead of $X(t)$, we no longer have to deal with spectral components of the process $X(t)$ having the longest periods (much greater than τ) which introduce the largest errors for a relatively short averaging time. Some examples of the comparison of the correlation functions computed via the averaging of $X(t + \tau)X(t)$ and of $[X(t + \tau) - X(t)]^2$ are given by Schulz-DuBois and Rehberg (1981).

Note also that in many applied problems the quantity $D(\infty) = 2B(0)$ is only very approximately found from the measurements of the function $D(\tau)$ (since the approach of the function $D(\tau)$ to a constant value $D(\infty)$ is often very slow as $\tau \to \infty$). Therefore, the quantity $B(0) = \langle X^2(t)\rangle$ must usually be found directly by taking the time average of the square of the process.

[77]It is readily seen that (4.237) implies that

$$(4.18')\quad D(n\tau) \leqslant n^2 D(\tau)$$

for any integer n. (In fact, according to (4.237), $D(2\tau) \leqslant 4D(\tau)$ and also if (4.18') holds for $n = m - 1$, then $D(m\tau) \leqslant \{(m - 1)[D(\tau)]^{1/2} + [D(\tau)]^{1/2}\}^2 = m^2 D(\tau)$. Therefore, (4.18') must be valid for any n.) Replacing τ by τ/n in (4.18') we obtain

$$(4.19')\quad \frac{D(\tau)}{\tau^2} \leqslant \frac{D(\tau/n)}{(\tau/n)^2}.$$

Let now A be the maximum of the continuous function $D(\tau)/\tau^2$ in the interval $\tau_0 \leqslant \tau \leqslant 2\tau_0$. Recalling that any $\tau > \tau_0$ always lies between $n\tau_0$ and $2n\tau_0$, where n is an integer (so that $\tau_0 < \tau/n \leqslant 2\tau_0$), we find directly from (4.19') that

$$D(\tau) \leqslant A\tau^2 \quad \text{for } \tau > \tau_0.$$

The result (4.238) also follows at once from the spectral representation (4.250) of the structure function $D(\tau)$. In fact, if we use the first of the two elementary inequalities

$$|1 - \cos\omega\tau| < \frac{\omega^2\tau^2}{2}, \quad |1 - \cos\omega\tau| \leqslant 2,$$

for $0 < \omega \leqslant \omega_0$, and the second for $\omega > \omega_0$, we have from (4.250):

$$(4.20') \quad \frac{D(\tau)}{\tau^2} < 4 \int_0^{\omega_0} \omega^2 dF(\omega) + \frac{8}{\tau^2}\int_{\omega_0}^{\infty} dF(\omega) + A_1^2.$$

According to (4.251) the right-hand side of (4.20') is bounded, and consequently $D(\tau)$ cannot increase at infinity more rapidly than $A\tau^2$ where A is a constant. It is also easy to see that (4.20') implies the following result: If $A_1 = 0$, then the increase of $D(\tau)$ for $\tau \to \infty$ is always slower than τ^2, i.e., $\lim_{\tau \to \infty} \{D(\tau)/\tau^2\} = 0$.

Moreover, it follows also from (4.20') that for any $\omega_0 > 0$ the contribution of all the large frequencies in the range $\omega > \omega_0$ to $D(\tau)$ is bounded by a constant. Therefore, if $A_1 = 0$ and $D(\tau)$ increases without limit at infinity, then the rate of increase of $D(\tau)$ as $\tau \to \infty$ must be determined by the behavior of the spectral distribution function $F(\omega)$ for $\omega \to 0$. Consider, e.g., a process with stationary increments $X(t)$ which is nonstationary (and cannot be made stationary by adding some constant random variable to $X(t)$). Let $X(t)$ have the continuous spectral density $f(\omega) = F'(\omega)$ and $f(\omega) = \omega^\gamma g(\omega)$ in the neighborhood of $\omega = 0$, where $0 < g(0) < \infty$ and, evidently, $-3 < \gamma < -1$ (since condition (4.251a) must be valid and $X(t)$ is nonstationary). Then it is easy to show that $D(\tau)$ increases in proportion to $\tau^{-1-\gamma}$ as $\tau \to \infty$ (see Picinbono, 1974; cf., also Example 3 on p. 406 in Vol. I).

[78]In fact, Kolmogorov (1940a) gave ony a short sketch of the proof while von Neumann and Schoenberg (1941) considered only the real case. The full proof can be found, e.g., in the book by Doob (1953), Sec. XI.11, and in Masani's paper (1972); see also Itô (1954), Yaglom (1955b), and Pinsker (1955), where some more general results are proved. (The proof by Itô is outlined in Sec. 24 of this book.)
Note also that several different ways of writing the spectral representation of a process with stationary increments and of its structure functions are used by different authors. In particular, instead of (4.244), (4.245), and (4.246), formulae of the form (4.241), (4.242), and (4.243) are also sometimes used where, in general, $F_y(\omega)$ is a nondecreasing function on the line $(-\infty,\infty)$ such that

$$(4.21') \quad \int_{\omega_0}^{\infty} \frac{dF_y(\omega)}{\omega^2} < \infty, \quad \int_{-\infty}^{-\omega_0} \frac{dF_y(\omega)}{\omega^2} < \infty$$

(these conditions are clearly equivalent to (4.251)) and $Z_y(\omega)$ is a random function with uncorrelated increments satisfying the

relationship of the form (4.229) with $F(\omega)$ replaced by $F_y(\omega)$. (In
the real case $dZ_y(-\omega) = \overline{dZ_y(\omega)}$, $dF_y(-\omega) = dF_y(\omega)$ and hence only
the validity of the first condition (4.21') must be required.)
Moreover, the jump of the function $F_y(\omega)$ at $\omega = 0$ must be equal to
the numerical constant A_1, and the jump of $Z_y(\omega)$ at $\omega = 0$ is equal
to the constant random variable X_1.

It is also sometimes convenient to use a third way of writing the
spectral representations for $X(t)$, $D(\tau_1,\tau_2)$, and $D(\tau)$, which is
equivalent to the first two:

$$(4.22')\quad X(t) = \int_{-\infty}^{\infty}(e^{it\omega} - 1)\,\frac{i\omega + 1}{i\omega}\,dZ_2(\omega) + X_0,$$

$$(4.23')\quad D(\tau_1,\tau_2) = \int_{-\infty}^{\infty}(e^{i\tau_1\omega} - 1)(e^{-i\tau_2\omega} - 1)\,\frac{\omega^2 + 1}{\omega^2}\,dF_2(\omega),$$

$$(4.24')\quad D(\tau) = 2\int_{-\infty}^{\infty}(1 - \cos\omega\tau)\,\frac{\omega^2 + 1}{\omega^2}\,dF_2(\omega).$$

Here, $F_2(\omega)$ is a bounded nondecreasing function given on the line
$(-\infty, \infty)$, and $Z_2(\omega)$ is a random function with uncorrelated
increments, satisfying the relationship of the form (4.229) with $F(\omega)$
replaced by $F_2(\omega)$; see, e.g., Krein (1944) and Doob (1953).

[79]See, e.g., Karhunen (1947), p. 53, and Doob (1953), Sec. IX.4.

[80]See equation (10) in Uhlenbeck and Ornstein (1930), where it is
noted that this specific result was earlier found by L.S. Ornstein and
R. Fürth, independently of each other.

[81]See Kolmogorov (1940b) and also the subsequent papers by Hunt
(1951), Lamperti (1962), Mandelbrot and Van Ness (1968),
Mandelbrot (1982), and Pflug (1982), where many additional
references can be found.

[82]See Mandelbrot and Van Ness (1968). The simplest
representation of a fractional Brownian motion $X(t)$ as a
fractional derivative (or integral) of the ordinary Brownian
motion $W(t)$ was introduced by Lévy (1953); it has the form

$$(4.25')\quad X(t) = \frac{1}{\Gamma[(m + 1)/2]}\int_0^t (t - s)^{(m-1)/2}dW(s).$$

[83]See Kolmogorov (1940b). Note that since $a(h) = \{D(\tau)/D(h\tau)\}^{1/2}$
where $D(\tau)$ is nonnegative, even, and continuous, the function $a(h)$ is
also nonnegative, even, and continuous and hence $a(h) = a(|h|)$. Using
one similarity transformation and then the other, we obtain that
$a(h_1h_2) = a(h_1)a(h_2)$. It follows from this that $\log a(h)$ is a linear

function of $\log|h|$, i.e., $a(h) = \exp\{\alpha\log|h| + \beta\}$.   Since $a(1) = 1$, $\beta = 0$, and $a(h) = |h|^{\alpha}$ is a power function of $|h|$.   Hence $D(\tau) = D(1)/a^2(\tau) = C|\tau|^m$ where $C = D(1)$, $m = -2\alpha$.

It can be similarly shown that the self-similar stationary correlation function $B(\tau)$ must also be a power function of $|\tau|$. It is clear, however, that the non-zero stationary correlation function cannot be a power function (since $0 < B(0) < \infty$); therefore, self-similar correlation functions of stationary processes do not exist.

The considered concept of self-similarity is, of course, a wide-sense concept and it must be called, in fact, *wide-sense self-similarity.* (Note also that only the similarity of the increments of a process $X(t)$ with stationary increments is involved here. Thus, it is even more appropriate to call the studied random processes $X(t)$ *self-similar processes with stationary increments.* In general, the requirement of self-similarity can also be applied to the values of a random process, or, e.g., to its higher-order increments; cf. p. 429 in Vol. I.) The related strict-sense concept of self-similarity can be formulated as follows: A process with stationary increments $X(t)$ is said to be *strictly self-similar* if all the multidimensional probability distributions for the differences $X(t_2) - X(t_1)$, ..., $X(t_{2n}) - X(t_{2n-1})$, where n is an arbitrary integer and $t_1, t_2, ..., t_{2n}$ are arbitrary points of the time axis, are invariant under a similarity transformation $t \to th$, $X \to a(h)X$.   It is readily seen that, for Gaussian processes $X(t)$, the strict-sense self-similarity coincides with the wide-sense self-similarity; thus, fractional Brownian motions are strictly self-similar. However, there exist also non-Gaussian self-similar processes which often appear, e.g., in the study of the limiting distributions for the sum of random variables that exhibit a long-range dependence making the classical central limit theorem inapplicable (see, e.g., Rosenblatt, 1961, and Lamperti, 1962). The non-Gaussian self-similar random processes (and fields) also play an important part in the study of many geophysical (especially often, hydrological) processes, in economics, communication theory, and in some problems of modern theoretical physics; see, e.g., the reviews by Taqqu (1986a,b), where many additional references can be found.

Note, in conclusion, that self-similar processes are sometimes also called in the available literature *semi-stable, automodel,* or *scaling* processes.

[84]The spectral density estimates of the form (4.267) are considered by Brillinger (1972) for a more general case of multidimensional processes with stationary increments.

[85]Multidimensional random processes with stationary increments are studied in detail by Brillinger (1972). Special two-dimensional processes with stationary increments having applications in the theory of queues are considered in another paper by the same author (see Brillinger, 1974a).

[86]See Itô (1954) and Gel'fand (1955). A detailed treatment of this topic can be found in Gel'fand and Vilenkin's book (1964), Chap. III.

[87]A one-dimensional distribution function of the random variable $\alpha_1 X(h_1) + \ldots + \alpha_n X(h_n)$ clearly determines the corresponding characteristic function

$$(4.26')\quad \psi_{\alpha_1 \ldots \alpha_n}(t) = \langle \exp it\{\alpha_1 X(h_1) + \ldots + \alpha_n X(h_n)\}\rangle.$$

The knowledge of the function (4.26') for any $\alpha_1, \ldots, \alpha_n$, also allows us to determine the n-dimensional characteristic function

$$(4.27')\quad \psi(t_1, \ldots, t_n) = \langle \exp i\{t_1 X(h_1) + \ldots + t_n X(h_n)\}\rangle = \psi_{t_1 \ldots t_n}(1)$$

of the random vector $X(h_1, \ldots, h_n) = \{X(h_1), \ldots, X(h_n)\}$. Moreover the n-dimensional characteristic function (4.27') uniquely determines the n-dimensional probability distribution of the vector $X(h_1, \ldots, h_n)$.

The above simple proof of the fact that one-dimensional probability distributions of all linear combinations $\alpha_1 X_1 + \ldots + \alpha_n X_n$ uniquely determine the n-dimensional probability distribution of a random vector $X = (X_1, \ldots, X_n)$ is due to H. Cramér and H. Wold; see, e.g., Cramér (1962), Chap. IX. Explicit formulae for reconstructing the n-dimensional probability distribution functions of vectors X from probability distribution functions of linear combinations $\alpha_1 X_1 + \ldots + \alpha_n X_n$ are given by Khachaturov (1954).

[88]The functional (4.286a) is, of course, linear in h_1 and antilinear in h_2, i.e.,

$$B(h_1, \alpha_1 h_{21} + \alpha_2 h_{22}) = \bar{\alpha}_1 B(h_1, h_{21}) + \bar{\alpha}_2 B(h_1, h_{22}).$$

Note also that in the case of a complex Gaussian generalized process the functionals $m(h)$ and $B(h_1, h_2)$ do not determine the process $X(h)$ uniquely; cf. Note 15 to Chap. 1 on p. 24.

[89]See Schwartz (1950-51).

[90]The book by Gel'fand and Shilov (1968) is devoted to the study of various spaces K of functions $h(t)$ and the corresponding spaces of generalized functions. Note that, except for the trivial function which vanishes identically, the functions of K_∞ are all nonanalytic, since any analytic function vanishing outside an interval must vanish identically.

[91]See, e.g., Gel'fand and Shilov (1968), Sec. II.4, and Gel'fand and Vilenkin (1964), Sec. III.3. In general, an arbitrary linear functional $m(h)$ in the space $K = K_\infty$ is of the form

$$(4.28') \quad m(h) = \sum_{k=0}^{\infty} \int_{-\infty}^{\infty} f_k(t) h^{(k)}(t) dt,$$

where $f_k(t)$ are continuous functions, and only a finite number of functions $f_k(t)$ are different from zero on any finite interval. The condition (4.289) implies that all functions $f_k(t)$ are constant. Therefore all the terms on the right-hand side of (4.28'), involving derivatives $h^{(k)}(t)$, where $k \geqslant 1$, must vanish, so that (4.28') turns into (4.291), where $m = f_0(t)$.

[92]See, e.g., Itô (1954), Sec. 2, or Gel'fand and Vilenkin (1964), Sec. II.3. This result, in a slightly different form, is also given by Schwartz (1950-51).

[93]If $X(h)$ is of the form (4.281) (i.e., if it has point values), where $X(t)$ is a stationary random process, then

$$\begin{aligned} B(h_1, h_2) &= \Big\langle \int_{-\infty}^{\infty} X(t) h_1(t) dt \int_{-\infty}^{\infty} \overline{X(s) h_2(s)} ds \Big\rangle \\ &= \int_{-\infty}^{\infty} \int_{-\infty}^{\infty} B(t - s) h_1(t) \overline{h_2(s)} dt \, ds \\ &= \int_{-\infty}^{\infty} B(\tau) \Big\{ \int_{-\infty}^{\infty} h_1(\tau - \sigma) \overline{h_2(-\sigma)} d\sigma \Big\} d\tau \, , \end{aligned}$$

where $\tau = t - s$, $\sigma = -s$. This result clearly coincides with (4.293).

[94]See Schwartz (1950-51), Vol. I, Sec. VII.8, or Gel'fand and Vilenkin (1964), Sec. II.3.

[95]See, in particular, Gel'fand and Vilenkin (1964), Chap. II, Subsec. 3.4. The theory of generalized random processes $X(h)$ given on the space of entire analytic functions $h(t)$ which fall off exponentially at infinity and lead to spectral distribution functions of exponential growth is considered by Onoyama (1959).

[96]A rigorous proof of the existence of the spectral representation (4.299) can be found, e.g., in Itô's paper (1954) and in Gel'fand and Vilenkin's book (1964), Chap. III, Subsec. 3.4.

[97]A number of examples of generalized positive definite functions are given in Gel'fand and Vilenkin's book (1964), Sec. II.3.

[98]It is easy to see that (4.310) implies the following result: If $h(t)$ is an arbitrary function in K, then the mean value functional $m(h)$ has the form

$$m(h) = m_0 \, \alpha_0(h) + m_1 \alpha_1(h) = m_0 \int_{-\infty}^{\infty} h(t) dt + m_1 \int_{-\infty}^{\infty} t h(t) dt,$$

where m_0 and m_1 are some constants. The general form of the correlation functionals $B(h_1, h_2)$ of generalized processes with stationary increments for arbitrary h_1 and h_2 in K can be found in Gel'fand and Vilenkin's book (1964), Sec. III.3.

[99]It is easy to show that in the case of a process $X(h)$ with stationary increments of order n the mean value functional for an arbitrary h in K necessarily has the form

$$m(h) = \sum_{k=0}^{n} m_k \alpha_k(h) = \sum_{k=0}^{n} m_k \int_{-\infty}^{\infty} t^k h(t) dt$$

where m_k, $k = 0, 1, ..., n$, are some constants; see Gel'fand and Vilenkin (1964), Sec. III.3. The general form of the correlation functional $B(h_1, h_2)$ of such a process $X(h)$ for arbitrary h_1 and h_2 in K is considerably more complicated; see again the same section of Gel'fand and Vilenkin's book. The derivation of the spectral representation (generalizing (4.305)) of the process $X(h)$ itself for any h in K is outlined in Gel'fand's note (1955), containing the first sketch of the theory of generalized random processes with stationary increments of order n.

[100]The general theory of ordinary (nongeneralized) processes with stationary increments of order n was developed independently by Yaglom and by Pinsker; see Yaglom and Pinsker (1953), Pinsker (1955), Yaglom (1955b). (Note that the mentioned works also contain, along with the theory of one-dimensional processes with stationary increments of order n, the more general theory of multidimensional processes $\mathbf{X}(t) = \{X_1(t), ..., X_s(t)\}$ which have the same property. Moreover, the spectral representation of nth differences $\Delta_\tau^{(n)} X(t)$ and of structure functions $D^{(n)}(t, \tau_1, \tau_2)$ is written here not in the form (4.322), (4.324), but in a slightly different form generalizing (4.22') and (4.23') on p 138) Independently of the indicated authors, Itô (1954) also examined ordinary processes $X(t)$ with stationary increments of order n and showed that they can be defined as random processes whose nth derivative (which is usually a generalized process) is stationary. Starting from this, he obtained a spectral representation of the form (4.325) for $X(t)$. Itô also proved that the set of all generalized stationary processes $X(h)$ in the space $K = K_\infty$ coincides with the set of nth derivatives of all ordinary (nongeneralized) processes with stationary increments of order n, where n runs through all non-negative integers.

[101]See, e.g., Pinsker (1955). Self-similar processes with stationary increments of order n were also studied by Yaglom (1955b).

[102]Rigorous proofs of formulae (4.331) − (4.333) can be found in Yaglom's paper (1957) and in the book by Gel'fand and Vilenkin (1964). These proofs are, in fact, completely similar to those for the particular case where $n = 1$. Note, in this respect, that the general Bochner-Schwartz theorem, which was used above for deriving the general form of the correlation functional $B(h_1, h_2)$ of generalized

random processes, was proved by Schwartz (1950-51) for the general n-dimensional case and is given in the same general form in Gel'fand and Vilenkin's book (1964).

[103]The definition given on pp. 433-434 of Vol. I is naturally the definition of a *wide-sense locally homogeneous field*. Similarly, the random field $X(t)$ is said to be a *strict-sense* (or *strictly*) *locally homogeneous* if all the probability distributions for the field differences in a set of pairs of points are unaffected by any translation of all the points involved. Strictly locally homogeneous random fields appeared for the first time in Kolmogorov's paper (1941b) devoted to the local structure of turbulent fluid flows.

[104]See Yaglom (1957), where the proof of formulae (4.349) and (4.350) is also given. A brief outline of the spectral theory of locally homogeneous fields meant for physicists and fluid mechanicians can be found in the book by Monin and Yaglom (1975), Sec. 13.2; the theory of generalized locally homogeneous fields $X(h)$ is considered by Gel'fand and Vilenkin (1964), Sec. III.5. Note that all the above-mentioned sources also include a study of multidimensional locally isotropic fields $X(t) = \{X_1(t), ..., X_s(t)\}$ and $X(h) = \{X_1(h), ..., X_s(h)\}$.

[105]See Yaglom (1957), and Gel'fand and Vilenkin (1964).

[106]Some results concerning generalized random fields with homogeneous increments of higher orders can be found in Gel'fand and Vilenkin's book (1964), Sec. III.5.

[107]See Kolmogorov (1941b), where the notion of the locally isotropic field (in the strict sense) was first introduced, as well as Monin and Yaglom (1975), Chaps. 6 and 8.

[108]Yaglom (1957) showed that any locally isotropic random field is *wide-sense real*, i.e., such that its first and second moments are necessarily real.

[109]The spectral theory of locally isotropic random fields is due to Yaglom (1957); see also Gel'fand and Vilenkin (1964), Sec. III.5. Formula (4.358) was also derived by Neumann and Schoenberg (1941), who studied a special problem of Hilbert space geometry, which is, in fact, closely connected with the theory of locally isotropic random fields.

[110]Self-similar locally isotropic fields in $\mathbb{R}^n$ with power structure functions of the form (4.262), where $0 < m \leqslant 2$, have been considered by Yaglom (1957). The Gaussian random field in $\mathbb{R}^n$ with a correlation function of the form (4.363), where $m = 1$, has been studied in detail by Lévy (in a book published in 1965 and in a

number of other publications) under the name of *Brownian motion with* n *parameters.* The positive definiteness of the kernel (4.363) in $\mathbb{R}^n$ for $0 < m \leqslant 2$ was proved by Gangolli (1967a, b); this result naturally also follows from Yaglom's results (1957). Gangolli (1967a, b) also studied a special class of positive definite kernels related to the kernel (4.363) which are possible correlation functions of locally isotropic fields on a wide class of homogeneous spaces.

[111]The theory of multidimensional locally isotropic fields $X(t) = \{X_1(t), ..., X_s(t)\}$ having s scalar components $X_j(t)$, $j = 1, ..., s$, was given by Yaglom (1957); see also Monin and Yaglom (1975), Sec. 13.3.

[112]The theory of vector locally isotropic fields is discussed by Yaglom (1957), and Monin and Yaglom (1975), Sec. 13.3. In these sources one can also find a strict definition of such fields.

[113]Generalized locally isotropic (scalar and vector) fields are treated in Yaglom's paper (1957); see also Gel'fand and Vilenkin (1964), Sec. III.5.

[114]Some results relating to the form of the correlation functional $B(h_1, h_2)$ of generalized random fields with isotropic increments of higher orders can be found in Gel'fand and Vilenkin's book (1964), Subsec. III.5.4.

[115]See, e.g., Courant and Hilbert (1953) or Riesz and Sz.-Nagy (1955). Note that instead of a random process $X(t)$, where $a \leqslant t \leqslant b$, we may also consider an arbitrary random function $X(t)$ in any connected compact topological space T with a finite measure $m(dt)$ in T such that all open subsets of T are m-measurable. If the correlation function $\langle X(t)\overline{X(s)} \rangle = B(t,s)$ is continuous in both arguments, then the integral equation

$$(4.29')\quad \int_T B(t,s)\psi(s)m(ds) = \lambda\psi(t)\qquad \text{for all } t \text{ in } T$$

will have the same properties as (4.384) (but, of course, the eigenfunctions $\psi_j(t)$ now will be orthonormal with respect to the measure m). Therefore, in this more general case $X(t)$ can also be represented as the sum (4.387) where Z_j satisfy (4.388). In particular, $X(t)$ may be a random field given on a connected bounded region in $\mathbb{R}^n$, while the measure $m(dt)$ may, for instance, be given as $m(dt) = m(t)dt = m(t_1, ..., t_n)dt_1 ... dt_n$, where the density $m(t)$ is an arbitrary continuous positive function.

[116]Recall that we can use, instead of (4.384), an equation of the form (4.29'), where the measure $m(dt)$ is restricted only by some very general topological conditions. (In particular, dt in (4.384) may

be replaced by $m(t)dt$, where $m(t)$ is an arbitrary continuous positive function.)

For the process $X(t)$, $0 \leqslant t \leqslant 1$, having the correlation function of the form $B(t,s) = \min(t,s)$, two different expansions in the series with random uncorrelated coefficients are given by (4.393) and (4.30').

[117]See, e.g., Kosambi (1943), Loève (1945-46), Karhunen (1946), Pugachev (1953), and Obukhov (1954b). In the particular case where $X(t)$ is a function on a finite set T (i.e. a finite collection of random variables $\{X_1, ..., X_m\}$) the Karhunen-Loève expansion coincides with the well-known *principal component analysis* introduced by Hotelling (1933) and presently considered in all textbooks on multidimensional statistical analysis (see, e.g., Anderson, 1984, or C. R. Rao, 1973). Note that the derivation of (4.387) in the main text of the book implies only the mean square convergence of the series on the right-hand side of this equation. However, this series converges, under wide conditions, also almost surely (i.e., with probability one). In particular, there will be almost sure convergence in all cases where $X(t)$ is a real Gaussian process (or Gaussian function) with mean value zero (so that Z_n, $n = 1, 2, ...$ is a sequence of independent Gaussian random variables); see, e.g., Gihman and Skorohod (1969), Sec. V.2, and (1974), Sec. IV.3.

[118]The approximation of the random function $X(t)$ by the truncated Karhunen-Loève expansion, i.e., by the sum of a few special nonrandom functions multiplied by mutually uncorrelated random coefficients Z_n is used in nearly all the practical applications of Karhunen-Loève series. The number of applied works using this approximation runs into many hundreds; see, e.g., Lorenz (1956), Obukhov (1960), Kutzbach (1967), Lumley (1967, 1970), Meshcherskaya et al. (1970), Craddock and Flintoff (1970), and Fortus (1980). It should be noted that in practice one almost always has to use, instead of the unknown true correlation function $B(t,s)$, the "empirical correlation function", i.e. some estimate of $B(t,s)$ from observational data forming a finite sample. (The term "method of empirical orthogonal functions", which is widely used in applied works to denote Karhunen-Loeve expansion, is related to this fact.) Therefore the study of sampling errors in the estimated eigenfunctions ψ_n and eigenvalues λ_n and of the statistical reliability of the obtained conclusions becomes very important; see, e.g., North et al. (1982), Glukhovskii and Fortus (1982, 1984), and Storch and Hannoschöck (1985).

[119]The representation of the Wiener process $X(t)$ as the sum (4.393) was obtained by Wiener as far back as the early 1930s; see, e.g., Paley and Wiener (1934), Chap. 9. Recall that if $X(t)$ is a Wiener (i.e. Brownian motion) process (i.e. a real Gaussian process such that $\langle X(t) \rangle = 0$, $\langle X(t)X(s) \rangle = \min(t,s)$), then the series on the

right-hand side of (4.393) converges with probability one for all t. Gihman and Skorohod (1974) showed also that in this case the maximum deviation of $X(t)$, $0 \leqslant t \leqslant 1$, from the sum of the first n terms of the series on the right-hand side of (4.393) tends to zero in probability as $n \to \infty$.

Gihman and Skorohod (1969, 1974) gave still another expansion of the Wiener process $X(t)$, $0 \leqslant t \leqslant 1$, in a series of the form (4.387). They considered the Karhunen-Loève expansion of the process $X_1(t)$ $= X(t) - tX(1)$ which vanishes both at $t = 0$ and $t = 1$. It is easy to show that $\langle X_1(t)X_1(s) \rangle = B_1(t,s) = \min(t,s) - ts$ and that (4.384) with the kernel $B(t,s)$ replaced by $B_1(t,s)$ reduces again to the differential equation (4.392), but now with the boundary conditions $\psi(0) = \psi(1) = 0$. Hence it follows that here

$$\lambda_n = \frac{1}{n^2\pi^2}, \quad \psi_n(t) = \sqrt{2}\sin n\pi t, \quad n = 1, 2, \dots .$$

It is also easy to show that $\langle X^2(1) \rangle = 1$ and $\langle X(1)Z_n \rangle = 0$ for $Z_n = \int_0^1 X_1(t)\psi_n(t)dt$, $n = 1, 2, \dots$. Therefore, denoting $X(1) = Z_0$, we obtain the expansion

$$(4.30') \quad X(t) = tZ_0 + \sqrt{2} \sum_{n=1}^{\infty} \frac{\sin n\pi t}{n\pi} Z_n, \quad 0 \leqslant t \leqslant 1,$$

where $\langle Z_n Z_m \rangle = \delta_{nm}$, $n, m = 0, 1, 2, \dots$.

[120]In the applied literature, much prominence is given to the case where $X(t)$ is a stationary random process with rapidly decreasing correlation function $B(t,s) = B(t - s)$, which corresponds to the continuous spectral density $f(\omega)$; see, e.g., the books by Davenport and Root (1958), Middleton (1960), Pugachev (1965), Helstrom (1968), and Van Trees (1968). Particularly simple results are obtained if the spectral density $f(\omega)$ is a rational function of ω of the form (2.129) (or, what is the same, (2.130)). The point is that in the case of the stationary correlation function $B(\tau) = B(t - s)$ corresponding to the rational spectral density $f(\omega)$ the integral equation (4.384) can always be reduced to the so-called Sturm-Liouville problem (i.e. a problem with boundary conditions at both ends of a segment) for an ordinary differential equation with constant coefficients (see the above-cited books or the paper by Gel'fand and Yaglom, 1957). The explicit solution of the corresponding differential equation is expressed in terms of exponential and trigonometric functions, and the boundary conditions are reduced here to some transcendental equation for the eigenvalues λ_n (appearing in the expressions for $\psi_n(t)$), which must be solved numerically.

Let us consider in more detail the simplest specific case where $B(t - s) = C\exp(-\alpha|t-s|)$ while $[a,b]$ is the interval $[- T/2, T/2]$ symmetric about the point $t = 0$. (Since $X(t)$ is a stationary process, we can always assume that the midpoint of the interval $[a,b]$ coincides with the zero point.)

In this case (4.384) has the form

$$(4.31') \quad C\left\{\int_{-T/2}^{t} e^{\alpha(s-t)}\psi(s)ds + \int_{t}^{T/2} e^{\alpha(t-s)}\psi(s)ds\right\} = \lambda\psi(t),\; -T/2 \leqslant t \leqslant T/2.$$

Differentiating both sides of (4.31') twice with respect to t, we first obtain

$$(4.32') \quad \psi'(t) = \frac{C\alpha}{\lambda}\left\{-\int_{-T/2}^{t} e^{\alpha(s-t)}\psi(s)ds + \int_{t}^{T/2} e^{\alpha(t-s)}\psi(s)ds\right\}$$

and then

$$(4.33') \quad \psi''(t) + \frac{2C\alpha - \alpha^2\lambda}{\lambda}\,\psi(t) = 0.$$

Equation (4.32') for $t = T/2$ and $T = -T/2$ yields boundary conditions

$$(4.34') \quad \psi'(T/2) + \alpha\psi(T/2) = 0, \quad \psi'(-T/2) - \alpha\psi(-T/2) = 0.$$

The solutions of (4.33') satisfying (4.34') have the form

$$(4.35') \quad \psi_{n1}(t) = c_{n1}\cos b_{n1}t \quad \text{and} \quad \psi_{n2}(t) = c_{n2}\sin b_{n2}t,\; n = 1,2,\,\dots\,,$$

where b_{n1} and b_{n2} are the roots of the transcendental equations

$$(4.36') \quad b_{n1}\tan(b_{n1}T/2) = \alpha \quad \text{and} \quad b_{n2}\text{ctn}(b_{n2}T/2) = \alpha,\; n = 1,2,\,\dots\,.$$

Equations (4.36') may be solved numerically or graphically; see, e.g., Van Trees (1968), Sec. 3.4.1, or Fortus (1973). The eigenvalues λ_{n1} and λ_{n2} corresponding to the eigenfunctions (4.35') are expressed through b_{n1} and b_{n2} by formulae: $\lambda_{ni} = 2C\alpha/(\alpha^2 + b_{ni}^2)$, $i = 1, 2$. When b_{n1} and b_{n2} are determined, the coefficients c_{n1} and c_{n2} are found from the normalizing conditions

$$\int_{-T/2}^{T/2} \psi_{ni}^2(t)dt = 1, \quad i = 1, 2, \quad n = 1, 2, \,\dots\,.$$

In the case of more complicated correlation functions $B(t-s)$ corresponding to the rational spectral densities $f(\omega)$ similar reasoning may be applied, but this is often rather cumbersome. A simplified method of solving the integral equation (4.384) for these cases, which does not involve transition to the Sturm-Liouville problem, was developed independently by Slepian and Kadota (1969) and by Fortus (1973). The method is based on the application of the Fourier transform to (4.384) after replacing the integration limits by $-\infty$ and

$+\infty$ and the function $\psi(t)$ by $\hat{\bar{\psi}}(t) = \psi(t)$ for $|t| \leqslant T/2$ and $= 0$ for $|t| > T/2$. We assume for simplicity that all the roots $\alpha_1,\,\dots,\,\alpha_m,\bar{\alpha}_1,\,\dots,\,\bar{\alpha}_m$ of the polynomial in the denominator of (2.129) are distinct. (By

$\alpha_1, ..., \alpha_m$ we denote the roots with positive imaginary parts.) Then, according to the results of the above-mentioned papers, in order to find the eigenvalues and eigenfunctions of (4.384), where $a = -T/2$, $b = T/2$, one must first of all find m roots $\theta_1^2(\lambda), ..., \theta_m^2(\lambda)$ of the algebraic equation

$$(4.37') \qquad 2\pi f(x) = \lambda$$

of degree in m in x^2. We assume that all these roots are distinct. (The changes required in the rare cases where equation (4.37') has multiple roots have been analyzed by Slepian and Kadota, 1969.) Then the eigenvalues λ_{n1} and λ_{n2}, $n = 1, 2, ...,$ of (4.384) are the roots of the following two transcendental equations:

$$(4.38') \qquad \mathrm{Det}\|D_{jk}^{(1)}\| = 0 \quad \text{and} \quad \mathrm{Det}\|D_{jk}^{(2)}\| = 0$$

where

$$(4.39') \; D_{jk}^{(1)} = \frac{e^{-i\theta_k T/2}}{i(\alpha_j - \theta_k)} + \frac{e^{i\theta_k T/2}}{i(\alpha_j + \theta_k)} \, , \quad D_{jk}^{(2)} = \frac{e^{-i\theta_k T/2}}{i(\alpha_j - \theta_k)} - \frac{e^{i\theta_k T/2}}{i(\alpha_j + \theta_k)}.$$

The eigenfunctions corresponding to the roots of the first and, respectively, the second of equations (4.38') have the form

$$(4.40') \qquad \psi_{n1}(t) = \sum_{k=1}^{m} c_{n1}^{(k)}\cos\theta_k(\lambda_{n1})t \quad \text{and} \quad \psi_{n2}(t) = \sum_{k=1}^{m} c_{n2}^{(k)}\sin\theta_k(\lambda_{n2})t,$$

where the coefficients $c_{n1}^{(1)}, ..., c_{n1}^{(m)}$ and $c_{n2}^{(1)}, ..., c_{n2}^{(m)}$ are given as solutions to two systems of m homogeneous linear equations with a coefficient matrix $\|D_{jk}^{(1)}\|$ and, respectively, $\|D_{jk}^{(2)}\|$.

Some specific examples of evaluation of Karhunen-Loève expansions for stationary processes with rational spectral densities are given by Fortus (1973), who also studied the asymptotic behavior, as $T \to \infty$, of the eigenvalues and eigenfunctions of the corresponding equation (4.384). Fortus (1975, 1977) also developed a similar method for finding the Karhunen-Loève expansion of the homogeneous random field $X(t_1, t_2)$ given in a disc $t_1^2 + t_2^2 \leqslant R^2$ and having a spectral density $f(k_1, k_2)$ rational in k_1 and k_2. The principal attention in these works is given to the particular case where the field $X(t_1, t_2)$ is isotropic and

$$f(k_1, k_2) = \frac{A}{(k^2 + \alpha^2)(k^2 + \beta^2)}, \qquad k^2 = k_1^2 + k_2^2,$$

$$B(t_1, t_2) = \frac{2\pi A}{\alpha^2 - \beta^2} \left\{ K_0(\beta t) - K_0(\alpha t) \right\}, \qquad t_1^2 + t_2^2.$$

The Karhunen–Loève expansions for random fields on a two-dimensional lattice were considered by Jain (1981).

There is one more case of Karhunen-Loève expansion of a stationary process $X(t)$ in the interval $-T/2 \leqslant t \leqslant T/2$, which has been studied in the available literature, namely the case of band limited white noise having the spectral density (2.121) and the correlation function (2.122).   In this case the eigenfunctions of (4.384) are known as *prolate spheroidal wave functions*; these functions and the corresponding eigenvalues λ_n have been studied in detail and tabulated (see Slepian and Pollak, 1961, and Landau and Pollak, 1961, 1962; cf. also Van Trees, 1968, and Papoulis, 1965, p. 460, or 1984, p. 305).

[121]The well-known Plancherel theory of Fourier transforms for functions in the space L_2 (i.e., square integrable over $(-\infty,\infty)$) is given, e.g., in the books by Wiener (1933), Titchmarsh (1948), and Bochner (1959). Recall that in this case the improper integrals in (4.397) are to be understood as the mean square limits of integrals from $-A$ to A as $A \to \infty$.

[122]This result is due to Karhunen (1947), who showed also that all the increments $W(t + s) - W(t)$ of the Wiener process $W(t)$ can be determined uniquely from the values of the stationary process $X(t)$ if, and only if, $f(\omega) > 0$ almost everywhere (i.e. if $f(\omega)$ does not vanish on any set of positive measure). Similar results for stationary sequences $X(t)$ (i.e. for discrete time) were obtained by Kolmogorov (1941a); cf. Note 2 to Chap. 2 on pp. 27-28.

[123]See theorem XII in the book by Paley and Wiener (1934). Note that this theorem deals with functions $g(\omega)$, whose Fourier transform vanishes for $t \leqslant t_0$, where t_0 is an arbitrary given number. The conditions imposed on $f(\omega)$ clearly cannot depend on t_0, since if the Fourier transform of the function $g(\omega)$ vanishes for $t \leqslant t_0$, then the Fourier transform of the function $\exp(-i\omega t_0)g(\omega)$ (which has the same absolute value) will vanish for $t \leqslant 0$.

[124]This result is also due to Karhunen (1947), but it actually follows from the earlier results of Krein (1945). A similar result for stationary sequences (i.e. for discrete t) was obtained by Kolmogorov (1941a); cf. Note 2 on pp. 27-28.

[125]Representation (4.403) of a stationary process with the exponential correlation function was given, in particular, by McNeil (1972).

[126]Let $g_0(\omega)$ be a function analytic in the lower half-plane and decreasing rapidly enough at infinity. Then, according to the Cauchy theorem,

$$(4.41') \qquad \int_{-A}^{A} e^{i\omega t} g_0(\omega)\,d\omega = \int_{L_A} e^{i\omega t} g_0(\omega)\,d\omega$$

where L_A is a semicircle of radius A lying in the lower half-plane (with $[-A,A]$ as its base). Since $\exp(i\omega t)$ tends to zero at infinity in the lower half-plane if $t < 0$, the right-hand side of (4.41') tends to zero as $A \to \infty$ and $t < 0$; therefore, the Fourier transform of the function $g_0(\omega)$ vanishes for $t < 0$.   See also p. 177 in Vol. I, where similar assertions are made.

[127]The general theory of linear extrapolation of stationary sequences and processes was developed by Kolmogorov (1941a) and Krein (1945); see the brief review of it in Yaglom's book (1962a), Sec. 36, and its complete presentation in the books by Doob (1953), Rozanov (1967), and Gihman and Skorohod (1969, 1974).   The principal role in this general theory when one deals with stationary processes $X(t)$ is played by the representation of $X(t)$ in the form (4.396), where the increments $W(t_2) - W(t_1)$ for $t \geqslant t_2 > t_1$ may be determined uniquely from the data on $X(s)$, $s \leqslant t$.   It is rather difficult to find the function $g_0(\omega)$ corresponding to this representation of $X(t)$ in the general case; but if the spectral density $f(\omega)$ is rational, then finding $g_0(\omega)$ is trivial.   See, in this connection, Notes 31 (pp. 46-47) and 70 (pp. 61-62).

[128]See, e.g., Granger and Hatanaka (1964), Chap. 9, and Priestley (1965b) and (1981), Sec. 11.2.   The term "evolutionary spectral representations" (more precisely, "evolutionary spectra") and the term "oscillatory processes" (describing random processes which have at least one evolutionary spectral representation) are due to Priestley.

[129]See, e.g., Parzen (1959), Sec. 4, where it is shown that if there exists a representation of $X(t)$ as

$$X(t) = \int_{-\infty}^{\infty} \varphi(t,\omega) dZ(\omega)$$

where $Z(\omega)$ satisfies (2.78), then there is a multitude of such different representations of the process $X(t)$, each representation based on a different family of functions.   Hence it follows at once that if there is a representation of the process $X(t)$ in the form (4.406), then a multitude of such different representations necessarily exists.   The problem of choice of "the most appropriate" evolutionary spectral representations is briefly discussed in Note 131.

[130]Mandrekar (1972) presented a more general definition of oscillatory processes, which includes conditions (4.406) and (4.407), but does not impose any other conditions restricting the class of functions $A(t,\omega)$.   He showed then that the oscillatory processes defined in his paper may be described as "deformed stationary curves" in Hilbert space, which naturally generalize the concept of a "stationary curve" (i.e. of a "stationary random process").   Later on, Battaglia (1979) noted that the class of oscillatory random processes is inconvenient in that it is not closed with respect to the summation of independent elements (i.e., the sum of two independent oscillatory processes need no longer be an oscillatory process).   To remedy this situation, Battaglia introduced the concept of a sigma-oscillatory process, which admits the following spectral representation:

$$X(t) = \sum_{j=1}^{n} \int_{-\infty}^{\infty} A_j(t,\omega)e^{it\omega}dZ_j(\omega),$$

where n is an arbitrary integer,

$$\langle dZ_j(\omega)\overline{dZ_k(\omega')}\rangle = \delta_{jk}\delta(\omega - \omega')dF_j(\omega)d\omega',$$

and the functions $A_j(t,\omega)$ satisfy the same conditions as does the function $A(t,\omega)$ in (4.406) (but $F(\omega)$ in (4.407) must now be replaced by $F_j(\omega)$). The class of sigma-oscillatory processes naturally includes all the oscillatory processes, and it is already closed relative to the summation of the independent elements (and also has some other advantages over the class of oscillatory processes). At the same time many practically important results of the theory of oscillatory processes may be generalized rather easily to this new class.

[131]The following crude estimate of the length of the maximum time interval over which the process can be considered to be "approximately stationary" was proposed by Priestley (1965b, 1981) for one important subclass of oscillatory processes. It is clear that we may, without loss of generality, suppose that $A(0,\omega) = 1$ (since $A(0,\omega)$ may be included in the random measure $dZ(\omega)$, making $F(\omega)$ coincide with the value of the evolutionary spectral distribution function at $t = 0$). Then

$$\int_{-\infty}^{\infty} dK(\theta;\omega) = 1$$

for all ω. This allows us to introduce the following measure of the "width" of the Fourier transform of $A(t,\omega)$:

$$B_A(\omega) = \int_{-\infty}^{\infty} |\theta|\ |dK(\theta;\omega)|.$$

Priestley suggested that the oscillatory process $X(t)$ be called *semi-stationary* if the function $A(t,\omega)$ in the representation (4.406) of this process can be so chosen that $B_A(\omega)$ be bounded for all ω. Then the constant

$$T_A = [\sup_{\omega} B_A(\omega)]^{-1}$$

of the time dimension may be called the *characteristic width* of the family $\{A(t,\omega), -\infty < \omega < \infty\}$ of the functions of t. The constant T_A (or, more precisely, $2\pi T_A$) may, indeed, be regarded as a crude estimate of the time interval over which the frequency distribution (4.411) of the power of $X(t)$ is approximately independent of t.

Of course, the time constant T_A depends on the choice of the function $A(t,\omega)$ on the right-hand side of (4.406). Let $A_1(t,\omega)$ and $A_2(t,\omega)$ be two such choices and the corresponding constants T_{A_1} and

T_{A_2} satisfy the inequality $T_{A_2} < T_{A_1}$. Since the evolutionary spectral representations are intended for the description of slow variations in the spectral distribution of the power of the process, it would be natural in this case to consider the spectral representation based on the use of the function $A_1(t,\omega)$ as more appropriate than the representation based on the use of $A_2(t,\omega)$. It is therefore reasonable to introduce the quantity

$$T_x = \sup_A T_A$$

where supremum is taken over all the possible choices of the function $A(t,\omega)$. The constant T_x depends only on the process $X(t)$ and may be called the *characteristic width* of this process. Now the constant T_x (or, more precisely, $2\pi T_x$) may be interpreted as the length of the *maximum interval* over which the semi-stationary process $X(t)$ may be treated as "approximately stationary".

Denote by A_x the class of admissible functions $A(t,\omega)$ whose characteristic width is equal to T_x. If class A_x is not empty, then each evolutionary spectral representation of the process $X(t)$ based on the use of the function $A(t,\omega)$ from A_x may be regarded as a proper spectral representation of $X(t)$. In particular, for stationary processes the ordinary spectral representation (2.61) is a proper representation, and the characteristic width for such processes is infinite.

[132]See, e.g., Burford (1958), Shapiro and Silverman (1959), and Pugachev (1965); cf. also pp. 464 and 492 in Vol. I.

[133]The term *uniformly modulated process* is due to Priestley (1965b, 1981).

[134]See, e.g., Granger and Hatanaka (1964), Chap. 9 (examples of oscillatory processes, spectral analysis of such processes), Priestley (1981), Chap. 11 (the study of all the mentioned topics containing many additional references to earlier works of Priestley and his collaborators), Mandrekar (1972) (extrapolation of generalized oscillatory processes), and Brillinger and Hatanaka (1969) (spectral analysis of one-dimensional and multidimensional oscillatory processes).

[135]Regarding other approaches (which have, as a rule, little to do with the theorem on the generalized spectral representation) to the concept of the spectrum for a nonstationary random process, see, e.g., Priestley (1981), Sec. 11.3, Loynes (1968), Rytov (1976), Secs. 57 and 58, Sasaki and Sato (1977), and Bendat and Piersol (1986), Chap. 12, where many additional references may be found.

[136]Condition (4.418) may be formulated more precisely as follows:

$$(4.42') \qquad \sup_{i,j=1}^{n} |F(\Delta_i, \Delta_j')| = V_F < \infty$$

where the supremum is taken over all possible finite collections of disjoint frequency sets $\Delta_1, ..., \Delta_n$ and $\Delta_1', ..., \Delta_n'$ and $F(\Delta, \Delta')$ is defined by (4.419). The constant V_F is often called the *Vitali variation* of a function $F(\omega, \omega')$ (and a measure $F(\Delta, \Delta')$).

[137]See, e.g., Loève (1945-46, 1963) and Loève's appendix to Lévy's book (1965), which was first published in 1948.

[138]Rozanov (1959) showed that the Fourier-Stieltjes integral (4.415) has sense and determined a random process $X(t)$ also in the case of the spectral measure $F(\Delta, \Delta')$ which does not satisfy condition (4.42'), but satisfies the less restrictive condition

$$(4.43') \qquad \sup_{i,j=1}^{n} \alpha_i \bar{\alpha}_j F(\Delta_i, \Delta_j) = W_F < \infty$$

where supremum is taken over all finite collections of disjoint frequency sets $\Delta_1, ..., \Delta_n$ and complex numbers $\alpha_1, ..., \alpha_n$ not exceeding unity in absolute values (i.e. $|\alpha_i| \leqslant 1$ for $i = 1, ..., n$). The constant W_F is sometimes called the *Frechet variation* (or the *semi-variation*) of the measure in the plane $F(\Delta, \Delta')$. (Moreover, W_F is also called by some authors the *semi-variation of the corresponding random measure $Z(d\omega)$* with values in the Hilbert space H; see, e.g., Dunford and Schwartz 1958, Sec. IV.10, and Miamee and Salehi, 1978.) It can be shown that $W_F \leqslant V_F$, so that if Vitali's variation V_F is finite (i.e. (4.42') is valid), then the Frechet variation is also finite (i.e., (4.43') is also fulfilled). The finiteness of the Frechet variation, however, does not imply that Vitali's variation is finite as well; if $W_F < \infty$, one can only assert that the plane $-\infty < \omega, \omega' < \infty$ can be divided into a denumerable number of disjoint two-dimensional sets S_k, $k = 1, 2, ...,$ in such a way that Vitali's variation of F in any S_k is finite (i.e., the Vitali variation is σ-finite).

If the measure $F(\Delta, \Delta')$ satisfies condition (4.43'), but not (4.42'), then it is possible that the Fourier-Stieltjes integral on the right-hand side of (4.415) does not exist as the limit in the mean (2.62). However, this integral does exist in some more sophisticated sense, and the correlation function $B(t,s)$ of the process $X(t)$ is then again determined by (4.416); see Rozanov (1959), and also the papers by Miamee and Salehi (1978), Rao (1981, 1982, 1985), and Chang and Rao (1983, 1986) containing many additional details. Note, in particular, the following result of Miamee and Salehi: *The process $X(t)$, $-\infty < t < \infty$, with the values in the Hilbert space H is harmonizable* (in the Rozanov sense) *if and only if there exists a Hilbert space H_1 containing H and a stationary curve $X_1(t)$ in H_1 such that $X(t)$ is the projection of $X_1(t)$ onto H.* This result makes clear the relation between the classes of stationary and harmonizable processes.

We see that there are several distinct definitions of harmonizability (see also the final part of this Note). Therefore Gladyshev (1962, 1963) suggested that the narrower concept of harmonizability introduced by Loève be called "absolute harmonizability", and the term "harmonizability" (without an adjective) be applied only to the more inclusive concept of Rozanov's harmonizability. Later on Rao (1981, 1982, 1985) and Chang and Rao (1983, 1986) used the term "strong harmonizability" for the ordinary Loève harmonizability, and "weak harmonizability" for the Rozanov harmonizability.

Some further generalizations of the harmonizability concept were introduced by Hurd (1969) and Dragan (1980); see also Dragan and Yavorsky (1982), Chap. 2. These works will be referred to below.

[139]It should be noted that all Loève's publications and all the other numerous publications in which the theory of harmonizable random processes is developed in accordance with Loève's works, give the following, weaker, formulation: *A random process* X(t) *is harmonizable if, and only if, its correlation function* B(t,s) *has the form (4.416), where* $F(\omega,\omega')$ *is a correlation function* (i.e. a positive definite kernel) *of bounded variation in the plane.* However, the requirement that $F(\omega,\omega')$ be a correlation function is in fact unnecessary: it is easy to show that if a function of the form (4.416), where $F(\omega,\omega')$ is of bounded variation, is a correlation function, then the function $F(\omega,\omega')$ is necessarily also a correlation function; see Hurd (1973). (The proof of this fact is based on the inversion formula for two-dimensional Fourier-Stieltjes integrals which allows us to express $F(\omega,\omega')$ in terms of $B(t,s)$.) As for the proof of the harmonizability of any process with a correlation function $B(t,s)$ of the form (4.416), both the proofs of the spectral representation theorem for stationary processes $X(t)$ that are

based on Khinchin's formula (2.52) for $B(\tau) = \langle X(t + \tau)\overline{X(t)}\rangle$ and outlined in Note 17 to Chap. 2 can be generalized to this more general case. In particular, we can formally introduce the random function $Z(\omega)$ by formula (2.9') on p. 33 and then use (4.416) firstly to prove that the mean square limit on the right-hand side of (2.9') does exist, and secondly to verify the validity of (4.417). After this it is easy to show that for any t the integral on the right-hand side of (4.415) exists (as the mean square limit (2.62)) and satisfies (2.21'), i.e. it coincides with $X(t)$. This completes the first proof of harmonizability of the random process $X(t)$; cf. Loève (1963), Sec. 34.4.

Another proof of the harmonizability theorem is based on consideration, along with the Hilbert space H_X, of one more Hilbert space $L_2\{F(\omega,\omega')\} = L_2(F)$. This new space consists of all complex-valued function $\varphi(\omega)$, $-\infty < \omega < \infty$, satisfying the condition

$$(4.44') \quad \int_{-\infty}^{\infty}\int_{-\infty}^{\infty}\varphi(\omega)\overline{\varphi(\omega')}\, d^2F(\omega,\omega') < \infty$$

and is specified by the scalar product

$$(4.45') \quad (\varphi(\omega),\psi(\omega))_L = \int_{-\infty}^{\infty}\int_{-\infty}^{\infty}\varphi(\omega)\overline{\psi(\omega)}\, d^2F(\omega,\omega')$$

(cf. a similar discussion on p. 36).* Then, according to (4.416), the correspondence

$$(4.46') \quad X(t) \longleftrightarrow e^{it\omega}, \quad \sum_{k=1}^{m} c_k X(t_k) \longleftrightarrow \sum_{k=1}^{m} c_k e^{it\omega}$$

generates the linear isometric mapping A of the space H_X onto the subspace $L_2^{(1)}(F)$ of the space $L_2(F)$, which is the closure of the set of all trigonometric polynomials. It is not hard to verify that the set of all trigonometric polynomials is closed in $L_2(F)$, so that $L_2^{(1)}(F)$ coincides with the whole $L_2(F)$. Therefore A is the mapping of the space H_X onto the whole space $L_2(F)$. (This last circumstance implies the possibility of expressing $Z(\omega)$ in terms of $X(t)$, but it is not crucial for the proof of harmonizability of $X(t)$; see the footnote on p. 40 and the derivation of formula (4.51') from (4.49').) Let us now consider the indicator function $\chi_{\omega'}(\omega)$ of the half-line $(-\infty, \omega')$ defined by the relation $\chi_{\omega'}(\omega) = 1$ for $\omega \leqslant \omega'$ and $\chi_{\omega'}(\omega) = 0$ for $\omega > \omega'$. Clearly, the function $\chi_{\omega'}$ belongs to the space $L_2(F)$ for all ω'. We denote by $Z(\omega')$ the element of H_X (a random variable) which the mapping A transforms into $\chi_{\omega'}(\omega)$. Then the isometric property of A implies that

$$(4.47') \quad \langle Z(\omega)\overline{Z(\omega')}\rangle = \int_{-\infty}^{\omega}\int_{-\infty}^{\omega'} d^2F(\lambda,\lambda') = F(\omega,\omega').$$

(It is assumed here that the function $F(\omega,\omega')$ has the property that

$$(4.48') \quad \lim_{\omega \to -\infty} F(\omega,\omega') = \lim_{\omega' \to -\infty} F(\omega,\omega') = 0.$$

This assumption is legitimate for the function $F(\omega,\omega')$ of bounded variation specified only by (4.416).) Moreover, it is easy to show that for any function $\varphi(\omega)$ belonging to $L_2(F)$,

$$\int_{-\infty}^{\infty}\varphi(\omega)d\,Z(\omega)$$

is a well-defined random variable belonging to H_X, and A maps this element of H_X into $\varphi(\omega)$ (cf. the derivation of (2.29') on p. 38). By virtue of (4.46') this leads at once to (4.415). This is the second proof of the harmonizability theorem.

---

*The right-hand side of (4.45') can be taken for a scalar product in the Hilbert space of the functions $\varphi(\omega)$ (and the left-hand side of (4.44') for the square of the norm) since $F(\omega,\omega')$ is a correlation function (a positive definite kernel), and therefore the left-hand side of (4.44') is always nonnegative.

   As in the case of the spectral representation theorem for stationary processes, the second proof of the harmonizability theorem, which was outlined above, can also be substantially generalized. This generalization will be formulated here without pretense to the greatest possible generality. *Assume that the random process* X(t) *has the correlation function* B(t,s) = $\langle X(t)\overline{X(s)}\rangle$ *representable in the form*

$$(4.49') \quad B(t,s) = \int_{-\infty}^{\infty}\int_{-\infty}^{\infty}\varphi(t,\omega)\overline{\varphi(s,\omega')}d^2F(\omega,\omega').$$

Here F(ω,ω') *is a correlation function* (i.e., a positive definite kernel) *of finite variation in the plane, while* φ(t,ω) *is a family of functions of* t *depending on the parameter* ω (not necessarily a frequency) *and satisfying the condition that*

$$(4.50') \quad \int_{-\infty}^{\infty}\int_{-\infty}^{\infty}\varphi(t,\omega)\overline{\varphi(t,\omega')}d^2F(\omega,\omega') < \infty$$

for all t. *Then the random process* X(t) *is representable as an integral* (i.e., a mean square limit of the corresponding integral sums) *of the form*

$$(4.51') \quad X(t) = \int_{-\infty}^{\infty}\varphi(t,\omega)dZ(\omega)$$

where Z(ω) *is a random function of the parameter* ω, *which satisfies* (4.417). *Conversely, if a random process is representable in the form* (4.51'), *where* Z(ω) *satisfies* (4.417), *while* F(ω,ω') *is a function of bounded variation, then* B(t,s) = $\langle X(t)\overline{X(s)}\rangle$ *has the form* (4.49'). The above general statement may be called the *generalized harmonizability theorem;* it is due to Cramér (1951), who presented its proof in every detail (see also Rao, 1978, 1985, and Chang and Rao, 1986). This proof is quite similar to the above-outlined second proof of the ordinary harmonizability theorem (perhaps supplemented by arguments similar to those outlined in the footnote on p. 40); therefore, we will not pursue it here.

[140]Note that the boundedness of (Vitali's) variation of $F(\omega,\omega')$ implies that the limits $F(\omega + 0, \omega' + 0) = \lim_{\epsilon \to \infty} F(\omega + \epsilon, \omega' + \epsilon)$, and

$F(\omega + 0, \omega' - 0)$, $F(\omega - 0, \omega' + 0)$, and $F(\omega - 0, \omega' - 0)$ exist; see Loève (1963). The function $F(\omega,\omega')$ is commonly assumed to be continuous from the right (or left) in both variables; however, as was shown by Loève, it is more convenient in some respects to assume that the function $F(\omega,\omega')$ is normalized, i.e., is replaced by the function

$$(4.52') \quad \hat{F}(\omega,\omega') = \frac{1}{4}\{F(\omega + 0, \omega' + 0) + F(\omega + 0, \omega' - 0) + F(\omega - 0, \omega' + 0) + F(\omega - 0, \omega' - 0)\}.$$

It is clear that the right-hand side of (4.416) remains the same for any of the indicated three choices of the values of $F(\omega,\omega')$ at the discontinuity points, and the corresponding three functions of ω and ω' are restored uniquely from any one of them.

[141]See, e.g., Loève (1963), Sec. 34.4, and Rozanov (1959). Both these sources demonstate that the ordinary inversion formula

$$(4.53')\quad \Delta\Delta'F(\omega,\omega')$$
$$= \frac{1}{4\pi^2}\lim_{T\to\infty,S\to\infty}\int_{-T}^{T}\int_{-S}^{S}\frac{e^{-it\omega}(e^{-it\Delta}-1)}{-it}\frac{e^{-is\omega}(e^{-is\Delta'}-1)}{-is}B(t,s)\,dt\,ds$$

is valid for any four points $(\omega+\Delta,\omega'+\Delta'),(\omega+\Delta,\omega'),(\omega,\omega'+\Delta')$, and (ω,ω') provideed the function $F(\omega,\omega')$ is considered to be normalized (i.e., $F(\omega,\omega')$ is replaced by $\hat{F}(\omega,\omega')$; see (4.52')). It is also shown by Loève and Rozanov that

$$F(\omega+0,\omega'+0) - F(\omega+0,\omega'-0) - F(\omega-0,\omega'+0) + F(\omega-0,\omega'-0)$$

$$= \Delta_0\Delta_0'F(\omega,\omega')$$

(the contribution from the point (ω,ω') to the spectral measure) can be determined from the equation

$$(4.54')\quad \Delta_0\Delta_0'F(\omega,\omega') = \lim_{T\to\infty,S\to\infty}\frac{1}{4TS}\int_{-T}^{T}\int_{-S}^{S}e^{-i(t\omega-s\omega')}B(t,s)\,dt\,ds.$$

[142]The proofs of (4.421) and (4.422) can also be found in Loève (1963) and Rozanov (1959). Note also that (4.415) implies the relation

$$(4.55')\quad m(t) = <X(t)> = \int_{-\infty}^{\infty}e^{it\omega}\,dM(\omega)$$

where $M(\omega) = <Z(\omega)>$. Thus the mean value of a harmonizable process $X(t)$ is always representable as the Fourier-Stieltjes integral with respect to a functin $M(\omega)$, which is clearly of bounded variation. Moreover, it is easy to see that

$$(4.56')\quad \lim_{T\to\infty}\frac{1}{T}\int_{-T/2}^{T/2}m(t)\,dt = M(+0) - M(-0)$$

(cf. (4.422) and Note 170 below). Let now $b(t,s) = <\{X(t) - m(t)\}\{\overline{X(s) - m(s)}\}>$ be the correlation function of the centered process $\overset{\circ}{X}(t) = X(t) - m(t)$, and $\overset{\circ}{F}(\omega,\omega')$, the spectal distribution function of $\overset{\circ}{X}(t)$. Then evidently

$$b(t,s) = B(t,s) - m(t)\overline{m(s)},\overset{\circ}{F}(\omega,\omega') = F(\omega,\omega') - M(\omega)\overline{M(\omega')}.$$

It is also clear that (4.422) for $\omega = 0$ may be written as

$$(4.57') \quad \lim_{T \to \infty} \frac{1}{T} \int_{-T/2}^{T/2} X(t)dt = \{M(+0) - M(-0)\} + \{\mathring{Z}(+0) - \mathring{Z}(-0)\}$$

where $\mathring{Z}(\omega) = Z(\omega) - M(\omega)$. Formulae (4.56') and (4.57') imply that the mean square limit on the left-hand side of (4.423) exists for any harmonizable process $X(t)$ and that it is equal to the right-hand side of (4.423) if, and only if, $<|\mathring{Z}(+0) - \mathring{Z}(-0)|^2> = \Delta_0\Delta_0' \mathring{F}(0,0) = 0$ (cf. Nagabhushanam, 1969, where the corresponding proof for the case where t is discrete is considered at length.) We note in addition that according to (4.54') the necessary and sufficient condition for (4.423) to be valid may also be written as

$$(4.58') \quad \lim_{T \to \infty, S \to \infty} \frac{1}{4TS} \int_{-T}^{T} \int_{-S}^{S} b(t,s)dtds = 0$$

which is similar to Slutsky's condition (3.10a).

A number of papers are devoted to the study of the ordinary (*weak*) *law of large numbers* (given by (4.423), where the limit on the left-hand side is interpreted as the limit in mean), and of the *strong law of large numbers* (where the limit in mean is replaced by the limit with probability one) for special classes of nonstationary (not necessarily harmonizable) processes; see, e.g., Kawata (1965, 1973), Nagabhusanam and Bhagavan (1969) and Arimoto (1973), where many additional references can be found, and also pp. 485-486 and 492 in Vol. I and Notes 171 and 181.

[143]It is clear that

$$<|X(t+h) - X(t)|^2> = B(t+h,t+h) - B(t+h,t) - B(t,t+h) + B(t,t).$$

Hence is follows that the random process $X(t)$ is mean square continuous if its correlation function $B(t,s)$ is continuous at all the points of the diagonal $t = s$. It is also easy to show that the continuity of $B(t,s)$ at all the points of the diagonal $t = s$ implies the continuity of $B(t,s)$ in the whole place $-\infty < t,s < \infty$ (cf. pp. 64-65 in Vol I).

Thus, any random process $X(t)$ which is not mean square continuous (e.g., which has fixed discontinuity points) is necessarily nonharmonizable.

[144]Various forms of necessary and sufficient conditions for the representability of a function as a Fourier-Stieltjes integral have been given by Bochner (1939) (see also Krein, 1943, Eberlein, 1955, and Rudin, 1962), Cramér (1939) and Dominguez (1940). The most useful of them is the Bochner-Eberlein condition, which, when applied to functions of two variables, any be formulated as follows:

A function of two variables B(t,s) (not necessarily a correlation function) *can be represented as the Fourier-Stieltjes integral* (4.416) *with respect to a function* F(ω,ω') *of bounded variation in the plane if, and only if, for any integer* n, *any collection of points in the plane* $(t_j,s_j), j = 1, ..., n$, *and any collection of complex constants* $c_j, j = 1, ..., n$, *the following inequality if valid.*

$$(4.59') \qquad \left| \sum_{j=1}^{n} c_j B(t_j, s_j) \right| \leqslant M \sup_{\omega, \omega'} \left| \sum_{j=1}^{n} c_j e^{i(t_j - \omega - s_j \omega')} \right|$$

where the supremum is taken over all real numbers ω and ω', while $M < \infty$ coincides with the total variation of the function $F(\omega, \omega')$.

A number of applications of this condition have been given by Hurd (1973). Note that the necessity of condition (4.59') is obvious; only its sufficiency requires special proof.

[145]In particular, Gladyshev (1962, 1963) showed that the function $B(t,s) = G(t)G(s)$ (which is evidently a positive definite kernel, i.e. a correlation function, for any $G(t)$) is nonharmonizable (and not only in the Loève sense, but also in the weaker Rozanov sense; cf. Note 138), provided tht $G(t)$ is a continuous periodic function with nonabsolutely summable Fourier series. Moreover, Gladyshev also gave an example of a bounded and continuous correlation function $B(t,s)$ which is Rozanov harmonizable but is not Loève harmonizable (and also has the property that $B(t + T_0, s + T_0) = B(s,t)$ for some T_0).

[146]A useful survey of the spectral theory of linear transformations of harmonizable processes can be found in the book by Blanc-Lapierre and Fortet (1953), Chap. 8.

[147]Equations (4.415), (4.433) and (4.434) clearly imply that

$$(4.60') \qquad \begin{aligned} Y(t) &= \mathcal{L}X(t) = \int_{-\infty}^{\infty} e^{it\omega} H(t,\omega) dZ(\omega), \\ B_y(t,s) &= \langle Y(t)\overline{Y(s)} \rangle = \int_{-\infty}^{\infty}\int_{-\infty}^{\infty} e^{i(t\omega - s\omega')} H(t,\omega) H(s,\omega) d^2 F(\omega, \omega'). \end{aligned}$$

By virtue of (4.435) and of the second equation (4.60') the correlation function $B_y(t,s)$ satisfies the Bochner-Eberlein condition (4.59'); consequently, $Y(t)$ is a harmonizable process.

[148]See Cambanis and Liu (1970), Theorem 5.

[149]This result is due to Hurd (1973). It can be proved very easily by ascertaining that (4.437) implies the validity of condition (4.59') for the correlation function (4.436).

[150]This result can also be proved easily by checking the validity of condition (4.59') for the correlation function $B(t,s)$ of the random process (4.438); see Hurd (1973). A slightly weaker result had previously been proved by Cambanis and Liu (1970), Theorem 4, using a different method.

[151]See, e.g., Hurd (1973), which includes some additional applications of the Bochner-Eberlein condition (4.59') to the construction of examples of harmonizable processes.

[152]In particular, the following terms are also used to denote random processes $X(t)$ satisfying (4.443); *cyclostationary*, *periodically stationary*, *periodic nonstationary*, and *wide sense periodic*.

[153]See, e.g., Franks (1969) and Gardner and Franks (1975) (radio engineering applications), Rytov (1976), Sec. 59 (radiophysics), Monin (1963) and Jones and Brelsford (1967) (meteorology), Dragan and Yavorskii (1982), Dragan, Rozhkov and Yavorskii (1984), and Ortiz and Ruiz de Elvira (1985) (oceanography), Dubrovsky and Tumarkina (1967) (physiology), and Box and Jenkins (1970) (economics).

[154]See, e.g., Gudzenko (1959), Gladyshev (1961, 1963), Ogrua (1971), Papoulis (1973b), Gardner and Franks (1975), Dragan (1975) and (1980), Sec. 4.9, Rytov (1976), Sec. 59, Dragan and Yavorskii (1982), chap. 2, Honda (1983) and also the dissertations by Gladyshev (1962), Brelsford (1967), Hurd (1969) and Gardner (1972).

[155]A random process with equidistant discontinuities and adjoined random variables taking only values 1 and $-$ 1 with equal probabilities was introduced by Wiener (1930), Sec. 12, and (1933), p. 151; see also Papoulis (1965), p. 294, and (1984), p. 228.
A substantial generalization of the process in Example 2 was indicated by Kayatskas (1968). This author considered a periodically correlated process of the form

$$X(t) = S(t, \mathbf{X}^{(0)}(t))$$

where $S(t, \mathbf{X}^{(0)})$ is a determined (i.e., nonrandom) function of t and an n-dimensional vector $\mathbf{X}^{(0)} = (X_1^{(0)}, ..., X_n^{(0)})$ periodic in t with period T_0, and $\mathbf{X}^{(0)}(t) = \mathbf{X}_k = $ const for $kT_0 \leqslant t < (k+1)T_0$ where $\mathbf{X}_k, k = 0, \pm 1, ...$ is a sequence of independent identically distributed random vectors specified by a given n-dimensional probability distribution.

[156]See, e.g., Franks (1969), Papoulis (1973b, 1984), and Gardner and Franks (1975). Some authors (in particular, Fortet, 1954, and Rytov, 1976) considered also the pulse-position-modulated signals of the form

$$(4.61') \quad X(t) = \sum_{k=-\infty}^{\infty} \Gamma(t - kT_0 - \epsilon_k, X_k)$$

where $\{\epsilon_k\}$ and $\{X_k\}$ are two independent sequences of independent identically distributed random variables the second of which describes random variations in the pulse shape. Rytov (1976), Sec. 59, studied, in fact, more gteneral class of periodically correlated processes allowing the variables ϵ_k and X_k with the same index k to be dependent.

[157]See, e.g., the references cited in Note 154. In particular, Papoulis (1973b) and Rytov (1976) considered an example of a periodically correlated process $X(t)$ obtained by periodic repetition of the same segment (of length T_0) of a given stationary process $X(t)$.

Blanc-Lapierre and Fortet (1953), Sec. V.12, Papoulis (1965), p. 451, Hurd (1969), and Levin (1974), Sec. 4.2, studied a periodically correlated telegraph signal which differs from the random telegraphic signal described on p. 32 of Vol. I in that the points $\{t_n\}$ now form an inhomogeneous Poisson point system having a time-dependent density $\lambda(t) = \lambda_0 + \lambda_1 \cos\omega_0 t$. Kampé de Fériet and Frenkiel (1962) noted that the sum of a stationary and a strictly periodic random process is always a periodically correlated process. A broad class of periodically correlated processes may also be obtained by applying a stationary random process $X_0(t)$ to the input of a linear system $\mathfrak{L}$ with periodically varying parameters.

[158]This result is due to Gladyshev (1961, 1962). Recall that by virtue of (4.454) and the results of Sec. 2 (see, in particular, (1.18)) the function $C(t,\tau)$ is a correlation function of some stationary sequence if, and only if,

$$(4.62') \quad \sum_{p,q=1}^{M} C(t_q, t_p - t_q) a_p \bar{a}_q \geqslant 0$$

for any integer M and any collections of integral times $t_1, ..., t_M$ and complex numbers $a_1, ..., a_M$. Similarly, the given function matrix $\|B_{jk}(\tau)\|_{j,k=0,1,...,T_0-1}$ is a correlation matrix of some T_0-dimensional stationary sequence $X_j(t) = X(t,j)$, $j = 0,1, ..., T_0 - 1; t = 0,\pm 1, ...,$ if, and only if,

$$(4.63') \quad \sum_{p,q=1}^{M} B_{k_p k_q}(t_p - t_q) a_p \bar{a}_q \geqslant 0$$

for any integer M and any collection of indices $k_1, ..., k_M$ (which take on values from 0 to $T_0 - 1$), integral times $t_1, ..., t_M$, and complex numbers $a_1, ..., a_M$. Gladyshev showed by simple algebraic manipulations that by virtue of (4.457) condition (4.62') and (4,63') are equivalent to each other.

[159]The proof of harmonizability of any periodically correlated sequence and the determination of the set where it spectral measure is concentrated is also due to Gladyshev (1961, 1962). See also the papers by Jones and Brelsford (1967) and Troutman (1979) which contain, in particular, a number of interesting examples of periodically correlated sequences.

[160]To guarantee the convergence of the Fourier series of a continuous periodic function $C(s)$ to the function $C(s)$ at any s, one should assume that the function $C(s)$ satisfies some rather mild regularity condition. It will suffice, e.g., to require that $C(s)$ satisfy a Lipschitz condition of order α, where $0 < \alpha < 1$; see, e.g., Titchmarsh (1939).

[161]For the case of continuous correlation functions $C(t,\tau)$, the above result is due to Gladyshev (1962, 1963); its proof is similar to that of the related result for periodically correlated sequences (see Note 158). The fact that the function $C_0(\tau)$ belongs to the class of stationary correlation functions was noted by Gudzenko (1959). The validity of Gladyshev's result for discontinuous correlation functions $C(t,\tau)$ (normalized at discontinuities by the condition $C(s,\tau) = \{C(s + 0,\tau + 0) + C(s + 0,\tau - 0) + C(s - 0,\tau + 0) + C(s - 0,\tau - 0)\}/4)$ was proved by Hurd (1960, 1974b).

[162]See Hurd (1974b), where it is also proved that the continuity of the function $C_0(\tau)$ at $\tau = 0$ implies the validity of condition (4.470) (and hence the continuity of all the functions $C_k(\tau)$ $k = 0,\pm 1, ...,$ everywhere) and that in this case

$$\int_{-\infty}^{\infty}|dF_k(\omega)| \leqslant C_0(0) \text{ for } k = 0,\pm 1, ...$$

[163]This result is due to Gladyshev (1962, 1963).

[164]See Hurd's dissertation (1969), where three modes of convergence for Fourier partial sums are examined, namely pointwise convergence of $S_N(t,\tau)$ to $\{C(t - 0,\tau) + C(t + 0,\tau)\}/2$, mean square convergence, which means that

$$\lim_{N\to\infty} \int_0^{T_0} |S_N(t,\tau) - C(t,\tau)|^2 dt = 0$$

and convergence of arithmetic means (i.e., Cesaro means of order one), which means that

$$\lim_{N\to\infty} \frac{1}{N + 1} \sum_{k=0}^{n} S_k(t,\tau) = \{C(t + 0,\tau)\}/2$$

for any t. Any one of these three modes of convergence is valid under wide regularity conditions and any mode allows us to represent the correlation function $B(t,s)$ as

$$(4.64') \quad B(t,s) = \lim_{N\to\infty} \sum_{k=-N}^{N} e^{ik\omega_0 s} \int_{-\infty}^{\infty} e^{i\omega(t-s)} dF_k(\omega)$$

(with a suitable definition of the limit on the right-hand side depending on the employed mode of Fourier series convergence). Equation (4.64') clearly describes the *spectal representation* of $B(t,s)$, i.e., it shows that the correlation function $B(t,s)$ is *harmonizable* in some special sense.

Another generalization of the harmonizability concept, which also permits one to assume almost any periodically correlated process $X(t)$ of practical interest to be harmonizable, was introduced by Dragan (1975, 1980); see also Dragan and Yavorskii (1982), Chap. 2.

[165]This result was in fact known to Gladyshev (1962, 1963), but it

was not precisely formulated in his works. Later it was stated by H. Ogura (1971) without a strict proof; a formal proof was given by Hurd (1969) and Honda (1983).

[166]The spectral densities $f_k(\omega)$ were introduced by Gudzenko (1959). Some examples of evaluation of these densities for specific models of periodically correlated processes may be found in Hurd's dissertation (1969), Ogura's paper (1971), and Rytov's book (1976), Sec. 59.

[167]Some new representations of harmonizable periodically correlated processes $X(t)$ were derived by Ogura (1971) from (4.415), (4.417) and (4.473). (Only the case where spectral densities exist, i.e. where all the functions $F_k(\omega)$ are absolutely continuous, was considered in the mentioned paper.) In the first place Ogura suggested that the whole frequency range $-\infty < \omega < \infty$ on the right-hand side of (4.415) be divided into bands of width $\omega_0 = 2\pi/T_0$ with centres at $\omega = k\omega_0$, $k = 0, \pm 1, \ldots$. Such partition of the frequency domain leads to the representation

$$(4.65') \qquad X(t) = \sum_{k=-\infty}^{\infty} e^{i k\omega_0 t} \int_{-\omega_0/2}^{\omega_0/2} e^{i\omega t} dZ_k(\omega) = \sum_{k=-\infty}^{\infty} A_k(t) e^{ik\omega_0 t}$$

where $Z_k(\omega) = Z(k\omega_0 + \omega)$ and therefore by virtue of (4.473)

$$(4.66') \quad \langle dZ_k(\omega)dZ_m(\omega') \rangle = \delta(\omega - \omega')dF_{km}(\omega)d\omega', F_{km}(\omega) = F_{k-m}(k\omega_0 + \omega).$$

According to (4.66') the random processes $A_k(t)$ on the right-hand side of (4.65') are all stationary and stationarily correlated with each other.   It is also easy to show that a process $X(t)$ of the form (4.65') is periodically correlated with period $T_0 = 2\pi/\omega_0$ if, and only if, all processes $A_k(t)$, $k = 0, \pm 1, \ldots$, are stationary and stationarily correlated; see Gardner and Franks (1975), where some additional information on representastion (4.65') may also be found.   It is clear that representation (4.65') <u>implies</u> the representation of the corrrelation function $B(t,s) = \langle X(t)X(s) \rangle$ in the form

$$(4.67') \qquad B(t,s) = \sum_{k=-\infty}^{\infty} \sum_{m=-\infty}^{\infty} e^{i(kt-ms)\omega_0} \int_{-\omega_0/2}^{\omega_0/2} e^{i\omega(t-s)} dF_{km}(\omega),$$

which is equivalent to (4.472a).

Another useful class of representations of periodically correlated processes is formed by so-called *translation series representations*; see, e.g., K. Jordan (1961), Brelsford (1967), Gardner (1972) and Gardner and Franks (1975).   The simples translation series representation can be obtained by using the Karhunen-Loève expansion of a process $X(t)$ on the interval $0 \leqslant t \leqslant T_0$ and then translating the corresponding eigenfunctions $\psi_j(t)$, $j = 1, 2, \ldots$, along the time axis to the intervals $kT_0 \leqslant t \leqslant (k + 1)T_0$, $k = \pm 1, \pm 2, \ldots$. (Such a translation is justified by the second equation of (4.443).) Then we obtain the representation

(4.68') $\quad X(t) = \sum\limits_{k=-\infty}^{\infty} \sum\limits_{n=1}^{\infty} Z_n(k)\psi_n(t-kT_0), \quad Z_n(k) = \int_0^{T_0} X(t + kT_0)\overline{\psi_n(t)}dt,$

where $\psi_n(t), t = 1,2, ...,$ are the eigenfunctions of the integral operator on the interval $[0,T_0]$ with the kernel $B(t,s)$. (Any complete orthonormal system of functions on the interval $[0,T_0]$ can also be used instead of functions $\psi_n(t)$; see Brelsford, 1967, and Gardner and Franks, 1975.) It is easy to see that the random sequences $Z_n(k), n = 1,2, ...,$ are in this case stationary and stationarily correlated sequences. Several examples of application of translation series representations are given by Gardner and Franks (1975).

[168]Estimates (4.488) and (4.489) clearly coincide with the simplest unbiased estimate (3.5) of the mean value for stationary sequences $X_t(k)$ and $X_{ts}(k)$. Some additional results on the properties of the mean value and correlation function estimates for periodically correlated random process can be found in Hurd's dissertation (1969), Chap. V, and in Yavorskii's papers (1983, 1985a).

[169]See, e.g., Gudzenko (1959), Hurd (1969), Chap. VI, and Dragan, Mezentsev and Yavorskii (1984). The estimation of $f_0(\omega)$ and its Fourier transformation $B_0(\tau)$ is treated in Papoulis' paper (1973b). The estimation of the statistical characteristics of a periodically correlated process with the unknown period T_0 from a single observed realization is studied by Yavorskii (1985b).

[170]If $X(t)$ is a harmonizable process, then (4.55') is fulfilled, where $M(\omega) = <Z(\omega)>$ is a function of bounded variation. Hence

$$M_T = \frac{1}{T} \int_{-T/2}^{T/2} m(t)dt = \int_{-\infty}^{\infty} \varphi_{T/2}(\omega)dM(\omega),$$

(4.69')
$$M_T^{(1)} = \frac{1}{T} \int_0^T m(t)dt = \int_{-\infty}^{\infty} \varphi_{T/2}^{(1)}(\omega)dM(\omega),$$

where $\varphi_{T/2}(\omega) = (\omega T/2)^{-1}\sin\omega T/2, \varphi_{T/2}^{(1)}(\omega) = \exp(iT/2)\varphi_{T/2}(\omega)$ (cf. Note 22 to Chap. 2 on pp. 42-43). The asymptotic properties of $\varphi_{T/2}(\omega)$ and $\varphi_{T/2}^{(1)}(\omega)$ for $T \to \infty$ are the same and imply that both M_T and $M_T^{(1)}$ tend to $\Delta M(0) = M(+0) - M(-0)$ as $T \to \infty$ (cf. Note 22 again).

Blanc-Lapierre and Fortet (1953), Sec. IX.7, noted that for harmonizable processes

$$\lim_{T\to\infty} \frac{1}{T} \int_0^T [\quad] \, dt$$

can, in equations of the form (4.491) and (4.498), always be replaced by

$$\lim_{T\to\infty} \frac{1}{2T} \int_{-T}^{T} [\;]dt, \text{ or } \lim_{T\to\infty}\frac{1}{T} \int_a^{a+T} [\;]dt, \text{ or } \lim_{T\to\infty} \frac{1}{T} \int_{a-T}^{a} [\;]dt$$

where a is an arbitrary number. However, in general, the assumption that

$$\lim_{T \to \infty} \frac{1}{T} \int_a^{a+T} [\ ]\, dt = \lim_{T \to \infty} \frac{1}{T} \int_{a-T}^a [\ ]dt$$

substantially restricts the class of the random processes under consideration. Classes $\mathfrak{M}_1$ and $\mathfrak{M}_2$, which are introduced in this book are specified by the condition that there exist limits (4.491) and, respectively, (4.498), which depend exclusively on the values of $X(t)$ at $t \geqslant 0$. These conditions do not impose any restrictions on the behavior of $X(t)$ for $t < 0$ and they are fulfilled for many nonharmonizable processes.

[171]The general ergodic theorem for nonstationary random processes is given in Cramér and Leadbetter's book (1967), Sec. 5.5. Its reduction in the stationary case to Slutsky's theorem is obvious (cf. Note 5 to Chap. 3, p. 64). In the case of a harmonizable process $X(t)$, condition (4.497) is clearly equivalent to (4.58'); therefore, in this case the ergodic theorem mentioned turns into the statement formulated in Vol. I on p. 463.
A more general class of ergodic theorems, which is applicable to a wide class of random processes generalizing the class of periodically correlated processes, has been considered (for the case of a discrete time by Boyles and Gardner (1983).

[172]See, e.g., Cramér and Leadbetter (1967), Sec. 5.5, where one particular condition is given which is sufficient for the validity of the strong law of large numbers.

[173]Let $X_T(t) = X(t)$ for $0 \leqslant t \leqslant T$ and $X_T(t) = 0$ for $t > T$ and $t < 0$. Consider the function

$$(4.70')\quad B_T(\tau) = \frac{1}{T}\int_0^{T-\tau} \langle X(t+\tau)X(t)\rangle dt = \int_{-\infty}^{\infty} \langle X_T(t+\tau)X_T(t)\rangle dt$$

where $0 \leqslant \tau \leqslant T$, and extend it to all the values of τ by using relations $B_T(\tau) = 0$ for $\tau > 0$, $B_T(-\tau) = B_T(\tau)$. Then the positive definiteness of $B_T(\tau)$ may be proved in the same way as is used to prove the positive definiteness of $B_T^{**}(\tau)$ in Note 30 to Chap. 3. The positive definiteness of $B_T(\tau)$ clearly implies that $B^{(0)}(\tau) = \lim_{T \to \infty} B_T(\tau)$ is also positive definite (cf. (1.32) in Sec. 4).

[174]See Bunimovich (1951), Sec. 11, Blanc-Lapierre and Fortet (1953), Chap. IX, as well as, e.g., Fortet (1954), Rozanov (1959), Kharkevich (1957, 1960), Kampé de Fériet and Frenkiel (1959, 1962), Kampé de Fériet (1962), Johns (1960), Parzen (1962b), Yaglom (1963), Papoulis (1965), Nagabhushanam and Bhagavan (1968), Bhagavan (1974), and Rao (1978, 1981). Note that a single generally accepted term for the functions $B^{(0)}(\tau), F^{(0)}(\omega)$, and $f^{(0)}(\omega)$, and also for the class $\mathfrak{M}_2$ does not exist. Rozahov suggests that $\mathfrak{M}_2$ be called the "class of processes possessing a spectrum"; Kampé de Fériet and Frenkiel use the term "processes having a

correlation function" (and refer to $B^{(0)}(\tau)$ as the correlation function of a nonstationary process $X(t)$); Parzen says that $\mathfrak{M}_2$ is the "class of asymptotically stationary processes" (the same term is also used by Kampé de Fériet, 1962, and occasionally by Blanc-Lapierre and Fortet, 1953) or, alternatively, "the class of processes possessing a covariance function $B^{(0)}(\tau)$," while Rao (1978) and Chang and Rao (1986) calls $\mathfrak{M}_2$ "the class (KF)" in honor of Kampé de Fériet and Frenkiel. The terms "average correlation function" (more precisely, "average autocorrelation") and "average power spectrum" are applied by Papoulis (1965), p. 449, to $B^{(0)}(\tau)$ and $f^{(0)}(\omega)$, while Chang and Rao (1986) called the function $F^{(0)}(\omega)$ the "associated spectrum" of the process $X(t)$ and Dragan and Yavorskii (1982) and Dragan, Rozhkov and Yavorskii (1974) refer to the same function as "the Fortet-Karkevich-Rozanov spectrum".

[175]It should be noted, however, that some authros call all the processes of class $\mathfrak{M}_2$ asymptotically stationary; see the preceding Note.

[176]The derivation of (4.501), which is due to Rozanov (1959), is based on the obvious relation

$$\lim_{T\to\infty} \frac{1}{T} \int_0^T e^{i(\omega-\omega')t}dt = \begin{cases} 1 \text{ for } \omega = \omega' , \\ 0 \text{ for } \omega \neq \omega'. \end{cases}$$

A completely rigorous mathematical proof of the fact that all harmonizable processes $X(t)$ belong to class $\mathfrak{M}_2$ supplemented by the precise formulation of the relationship between $F(\Delta,\Delta')$ and $F^{(0)}(\omega)$ is given in Bhagavan's thesis (1974); see also Rao (1978). Rao proves also the following generalization of the statement that any harmonizable processes belongs to class $\mathfrak{M}_2$: *If the correlation function* $B(t,s)$ *of a process* $X(t)$ *can be represented in the form* (4.49'), *where* $\varphi(t,\omega)$ *is almost periodic in* t *for almost all* ω (the class where $\varphi(t,\omega) = \exp(it\omega)$ is, of course, automatically included), *then the process* $X(t)$ *necessarily belongs to class* $\mathfrak{M}_2$. Later Chang and Rao (1986) showed that class $\mathfrak{M}_2$ contains also some (but not all) processes harmonizable only in the wider sense due to Rozanov (i.e., weakly harmonizable, cf. Note 138).

[177]This topic is considered in more detail in the book by Blanc-Lapierre and Fortet (1953), Chap. IX.

[178]See Kampé de Fériet and Frenkiel (1959, 1962) and Parzen (1962b). Parzen also notes that if

$$\lim_{T\to\infty} \frac{1}{T^{1-\epsilon}} \int_0^T b^{(4)}(T,s,\tau)ds = 0$$

for some $\epsilon > 0$, then $B_T^{**}(\tau) \to B^{(0)}(\tau)$ for $T \to \infty$ not only in the mean, but also with probability one.

[179]See Parzen (1962b).

[180]This statement is also explained in Parzen's article (1962b). See also Papoulis' paper (1973b), where special attention is given to the case of a periodically correlated process $X(t)$.

[181]For the case where the time t is discrete (i.e., for random sequences having an average correlation function which can be defined similarly to (4.498)), this result has been proved by Nagahushanam and Bhagavan (1969). See also Bhagavan (1974) and Rao (1978), where the case of continuous t is studied.

[182]The case where $X(t) = A(t)X_0(t)$ is a moldulated stationary process was considered by Bunimovich (1951). See also Parzen (1962b), where equation (4.514) is given and it is also shown that if $X_0(t)$ is a stationary normal ergodic process, and $A(t)$ is a bounded function, then condition (4.506) is necessarily fulfilled for $X(t)$.

[183]Formula (4.519) was obtained by Wiener (1930), Sec. 12, and (1933), Sec. 21, for the case where X_n, $n = 0,\pm1,\pm2, ...$, take on values ±1 and -1 with equal probabilities. Later on, the same example of Wiener was studied at length by Kampé de Fériet and Frenkiel (1959, 1962). These authors found the bound for the rate of convergence of $B_T(\tau)$ (see (4.70')) to $B^{(0)}(\tau)$ and of $B_T^{**}(\tau)$ to $B^{(0)}(\tau)$; moreover, they also illustrated by numerical examples (constructed with the aid of the Monte-Carlo technique) the accuracy of approximation of the average correlation function $B^{(0)}(\tau)$ by the estimate $B_T^{**}(\tau)$ (or by the arithmetic mean of a number of such estimates calculated from independent realizations of the process $X(t)$).

[184]This subject was treated more comprehensively, e.g., by Middleton (1960), Kharkevich (1960), Franks (1969), and Levin (1974); see also Papoulis (1973b).

[185]See, e.g., Fortet (1954), Kharkevich (1957, 1960), Franks (1969) and Rytov (1976), Sec. 59, where some additional references may be found.

[186]The results (4.529) and (4.530) may be found in a number of sources; see, e.g., Parzen (1962) and also the short note by Wiener and Wintner (1958), who, however, considered only the stationarizable version of phase-modulated sine oscillation.

[187]See, e.g., Zadeh (1951), Kharkevich (1960), Middleton (1960), Chap. 14, and Levin (1974), Chap. 12. Many other aspects of the theory of randomly modulated (primarily phase-modulated) signals have been studied thoroughly in Papoulis' papers (1983a, b) where a number of additional references may be found.

[188]See, e.g., Middleton (1960), Kharkevich (1960), Franks (1969), Levin (1974) and Papoulis (1973, 1983a, b). The available radio engineering literature also includes a number of other books and papers containing calculations of the functions $B^{(0)}(\tau)$, $F^{(0)}(\tau)$ and $f^{(0)}(\tau)$ for various models of stochastic signals.

BIBLIOGRAPHY

Many of the Russian articles listed in the bibliography can be found in English in cover-to-cover translations of the corresponding Russian journals or collections of papers. Readers interested in reading a particular article in translation should enquire of their librarian.

The titles of Russian books and papers are given in English; the English translation of the titles of the cited Russian journals and collections of papers is often indicated in parentheses after the Russian titles. Moreover, in a few cases the second possible English spelling of Russian names is also given in parentheses after the first one. Note also that many Russian women's names terminate in the letter a (in constrast to the corresponding names of men); this letter is often omitted in the English translation but it is conserved in this bibliography.

Ables, J. G.
(1974). Maximum entropy spectral analysis. *Astron. Astrophys. Suppl. Series*, **15**, 383-393. Reprinted in Childers, 1978, pp. 23-31.
Abramowitz, M., and I. A. Stegun
(1964). *Handbook of Mathematical Functions*. Nat. Bureau Stand., Washington, D.C.
Adenstedt, R. K.
(1974). On large-sample estimation for the mean of a stationary random sequence. *Ann. Stat.*, **2**, No. 6, 1095-1107.
Agarwal, R. C., and J. W. Cooley
(1977). New algorithm for digital convolution. *IEEE Trans. Acoust., Speech, Signal Processing*, **25**, No. 5, 392-410.
Ahmed, N., and T. Natarajan
(1974). On logical and arithmetic autocorrelation functions. *IEEE Trans. Electromagnetic Compat.*, **16**, No. 3, 177-183.

Akaike, H.
(1969a). Fitting autoregressive models for prediction. *Ann. Inst. Stat. Math.*, **21**, 243-247.
(1969b). Power spectrum estimation through autoregressive model fitting. *Ann. Inst. Stat. Math.*, **21**, No. 3, 407-419.
(1970). Statistical predictor identification. *Ann. Inst. Stat. Math.*, **22**, No. 2, 203-217.
(1973). Information theory and an extension of the maximum likelihood principle. In *Proc. 2nd Int. Symp. Inform. Theory*, Suppl. to *Problems of Control and Inform. Theory* (B. N. Petrov and F. Csaki, eds.), pp. 267-281, Akademiai Kiado, Budapest.
(1974). A new look at the statistical model identification. *IEEE Trans. Autom. Control*, **19**, No. 6, 716-723. Reprinted in Childers, 1978, pp. 234-241.
(1977). On entropy maximization principle. In *Applications of Statistics* (P. Krishnaiah, ed.), pp. 27-41, North-Holland, Amsterdam.
(1978). A Bayesian analysis of the minimum AIC procedure. *Ann. Inst. Stat. Math.*, **30A**, No. 1, 9-14.
(1979). A Bayesian extension of the minimum AIC procedure of autoregressive model fitting. *Biometrika*, **66**, No. 2, 237-242.

Akhiezer, N. I.
(1956). *Theory of Approximations*. Ungar, New York.

Akhiezer, N. I., and I. M. Glazman
(1980). *Theory of Linear Operators in Hilbert Space. Vols. I and II.* Pitman Press, Boston.

Alekseev (Alekseyev), V. G.
(1965). Analysis of the performance of the instruments measuring time spectra of turbulent fluctuations. *Izv. Akad. Nauk SSSR, Ser. Fiz. Atmos. i Okeana* (Bull. Acad. Sci. USSR, Ser. Atmosph. and Oceanic Phys.), 1, No. 7, 688-695.
(1967). On fluctuations of spectrum analyzer readings. *Izv. Akad. Nauk SSSR, Ser. Fiz. Atmos. i Okeana*, **3**, No. 9, 928-935.
(1970). On the accuracy of the higher moment determinations for time series. *Izv. Akad. Nauk SSSR, Ser. Tekhn. Kibern.* (Bull. Acad. Sci. USSR, Ser. Techn. Cybernetics), No. 6, 171-176.
(1971a). On spectral estimates for some non-Gaussian random processes. *Teor. Veroyatn. i Mat. Stat.* (Probab. Theory and Math. Stat., Kiev), No. 5, 3-7.
(1971b). On choosing the spectral window when spectrum of a Gaussian stationary random process is estimated. *Problemy Peredachi Inform.* (Probl. Inform. Transmission), 7, No. 4, 46-54.
(1973a). Spectral analysis of some non-Gaussian random processes. *Teor. Veroyatn. i Mat. Stat.*, No. 9, 3-7.
(1973b). Some practical recommendations related to spectral analysis of Gaussian stationary random processes. *Problemy Peredachi Inform.*, **9**, No. 4, 42-48.
(1974a). Some questions of the spectral analysis for Gaussian random processes. *Teor. Veroyatn. i Mat. Stat.*, No. 10, 3-11.

(1974b). On the uniform convergence of the spectral density estimates for a Gaussian stationary random process. *Teor. Veroyatn. i Primen.* (Probab. Theory and Its Appl.), **19**, No. 1, 198-206.

(1980). The calculation of spectra of stationary random processes from large samples. *Problemy Peredachi Inform.*, **16**, No. 1, 42-49.

(1981). On large sample calculations of the spectral densities of random processes. *Vychisl. i Prikl. Mat.* (Comput. and Applied Math., Kiev), No. 44, 32-40.

(1985). Spectral density estimates for some models of stationary random processes. *Problemy Peredachi Inform.*, **21**, No. 2, 42-49.

Alekseev, V. G., and A. M. Yaglom
(1980). Nonparametric and parametric spectrum estimation methods for stationary time series. In *Time Series* (O. D. Anderson, ed.), pp. 401-422. North-Holland, Amsterdam.

Alisov, B. P., O. A. Drozdov, and E. S. Rubinstein
(1952). *A Course in Climatology*, Gidrometeoizdat, Leningrad.

An, H.-Z., and Z.-G. Chen
(1986). The identification of ARMA processes. *J. Appl. Probab.*, **A23**, 75-87.

An, H.-Z., Z.-G. Chen, and E. J. Hannan
(1982). Autocorrelation, autoregression and autoregressive approximation. *Ann. Stat.*, **10**, No. 3, 926-936.

Andersen, N.
(1974). On the calculation of filter coefficients for maximum entropy spectral analysis. *Geophysics*, **39**, No. 1, 69-72. Reprinted in Childers, 1978, pp. 252-255.

(1978). Comments on the performance of maximum entropy algorithms. *Proc. IEEE*, **66**, No. 11, 1581-1582.

Anderson, O. D.
(1979a). Formulae for the expected values of the sampled variances and covariances from series generated by general autoregressive integrated moving average processes of order (p, d, q). *Sankhyā*, Ser. B, **41**, Nos. 3-4, 177-195.

(1979b). On the bias of autocorrelations. *Math. Operationsforsch. Stat.*, Ser. Statistics, **10**, No. 2, 299-305.

Anderson, O. D., and J. G. De Gooijer
(1982). The covariances between sampled autocovariances and between serial correlations for finite realizations from ARUMA time series models. In *Time Series Analysis: Theory and Practice*, 1 (O. D. Anderson, ed.), pp. 7-22. North-Holland, Amsterdam.

Anderson, P. W.
(1954). A mathematical model for the narrowing of spectral lines by exchange of motions. *J. Phys. Soc. Japan*, **9**, No. 3, 316-339.

Anderson, T. W.
(1971). *The Statistical Analysis of Time Series*. Wiley, New York.

(1984). *An Introduction to Multivariate Statistical Analysis*. 2nd edition. Wiley, New York.

Anderson, T. W., and A. M. Walker
(1964). On the asymptotic distribution of the autocorrelations of a sample from a linear stochastic process. *Ann. Math. Stat.*, **35**, No. 3, 1296-1303.

Andrews, L. C.
(1980). Analysis of a cross correlator with a clipper in one channel. *IEEE Trans. Inform. Theory*, **26**, No. 6, 743-746.

Anishin, N. S., and A. M. Tivkov
(1978). A fast algorithm for the computation of correlation functions and investigation of its error. *Avtometriya* (Autometry, Novosibirsk), No. 3, 37-40.

Antonov, V. S.
(1963). Will the Caspian Sea level be decreasing? *Morskoy Flot* (The Marine), **23**, No. 11, 42-43.

Arimoto, A.
(1973). Some theorems for harmonizable stochastic processes. *Keio Math. Seminar Reports, No. 1* (Rep. Sem. "Stochastic Processes", Keio Univ., Yokohama), 24-30.

Askey, R.
(1975). Some characteristic functions of unimodal distributions. *J. Math. Anal. and Appl.*, **50**, No. 3, 465-469.

Åström, K. J.
(1970). *Introduction to Stochastic Control Theory.* Academic Press, New York.

Azencott, R., and D. Dacunha-Castelle
(1986). *Series of Irregular Observations. Forecasting and Model Building.* Springer, New York.

Babić, H., and G. C. Temes
(1976). Optimum low-order windows for discrete Fourier transform systems. *IEEE Trans. Acoust., Speech, Signal Process.*, **24**, No. 6, 512-517.

Bachelier, L.
(1900). Théorie de la spéculation. *Ann. Sci. École Norm. Sup.*, 17, 21-86.

Bain, W. C.
(1976). The power spectrum of temperatures in central England. *Quart. J. Roy. Meteor. Soc.*, **102**, No. 432, 464-466.

Balakrishnan, A. V.
(1957). A note on the sampling principle for continuous signals, *IRE Trans. Inform. Theory*, **3**, No. 2, 143-146.

Ball, G. A.
(1968). *Instrumental Correlation Analysis for Random Processes.* Energiya, Moscow.

Barber, N. F.
(1961). *Experimental Correlograms and Fourier Transforms.* Pergamon Press, Oxford.

Barnett, T. P., and K. E. Kenyon
(1975). Recent advances in the study of wind waves. *Rep. Progr. Phys.*, **38**, No. 6, 667-729.

Barrett, J. F., and D. G. Lampard
(1955). An expansion for some second-order probability distributions and its application to noise problems. *IRE Trans. Inform. Theory*, 1, No. 1, 10-15.

Bartels, C. P. A., and R. H. Ketellapper, eds.
(1979). *Exploratory and Explanatory Statistical Analysis of Spatial Data*. M. Nijhoff, Boston, Mass.

Bartlett, M. S.
(1946). On the theoretical specification and sampling properties of autocorrelated time series. *J. Roy. Stat. Soc., Suppl.* **8**, No. 1, 27-41.
(1948). Smoothing periodograms from time series with continuous spectra. *Nature*, **161**, No. 4091, 486-487.
(1950). Periodogram analysis and continuous spectra. *Biometrika*, **37**, No. 1, 1-16.
(1963). Statistical estimation of density functions. *Sankhyā, Ser. A*, **25**, No. 3, 245-254.
(1967). Some remarks on the analysis of time-series. *Biometrika*, **54**, Nos. 1-2, 25-38.
(1975). *The Statistical Analysis of Spatial Pattern*. Chapman and Hall, London.
(1978). *An Introduction to Stochastic Processes*. 3rd edition. Cambridge University Press, Cambridge.

Bass, J.
(1962). *Les fonctions pseudo-aléatoire*. Gauthier-Villars, Paris.

Batchelor, G. K.
(1982). *The Theory of Homogeneous Turbulence*. Cambridge University Press, Cambridge. (The book was first published in 1953.)

Båth, M.
(1974). *Spectral Analysis in Geophysics*. Elsevier, Amsterdam.

Battaglia, F.
(1979). Some extensions in the evolutionary spectral analysis of a stochastic process. *Boll. Unione Mat. Ital.*, **B16**, No. 5, 1154-1166.

Baum, R. F.
(1969). The correlation function of Gaussian noise passed through nonlinear devices. *IEEE Trans. Inform. Theory*, **15**, No. 4, 448-456.

Bayley, G. V., and J. M. Hammersley
(1946). The "effective" number of independent observations in an autocorrelated time series. *J. Roy. Stat. Soc., Suppl.* **8**, 184-197.

Beamish, N., and M. B. Priestley
(1981). A study of autoregressive and window spectral estimation. *Appl. Stat.*, **30**, No. 1, 41-58.

Belyaev, Yu. K.
(1959) Analytical random processes. *Teor. Veroyatn. i Primen.* (Probab. Theory and Its Appl.), **4**, No. 4, 437-444.

Bendat, J. S.
(1958). *Principles and Applications of Random Noise Theory*. Wiley, New York.

Bendat, J. S., and A. G. Piersol
(1986). *Random Data: Analysis and Measurement Procedures*. 2nd edition. Wiley-Interscience, New York.

Bennett, R. J.
(1979). *Spatial Time Series. Analysis, Forecasting, Control*. Pion-Academic Press, London.

Bentkus, R. J.
(1972). Asymptotic normality of the spectral distribution function estimate. *Litovsk. Mat. Sbornik* (Lithuan. Math. Collect.), 12, No. 3, 5-18.
(1977). Cumulants of polylinear forms of a stationary sequence. *Litovsk. Mat. Sbornik*, 17, No. 1, 27-46.
(1985a). On asymptotic behaviour of the minimax mean square risk for statistical estimates of a spectral density in L_2 space. *Litovsk. Mat. Sbornik*, 25, No. 1, 23-42.
(1985b). Uniform convergence rate of the spectral density estimators in the space of differentiable functions. *Litovsk. Mat. Sbornik*, 25, No. 3, 17-31.
Bentkus, R. J., and R. A. Rudzkis ·
(1982). On distributions of some statistical estimates of a spectral density. *Teor. Veroyatn. i Primen.* (Probab. Theory and Its Appl.), 27, No. 4, 739-756.
Bentkus, R. J., and B. Rutkauskas
(1973). On asymptotic behaviour of two first moments of second-order spectral estimates. *Litovsk. Mat. Sbornik*, 13, No. 1, 29-45.
Berg, Ch., J. P. R. Christensen, and P. Ressel
(1984). *Harmonic Analysis on Semigroups. Theory of Positive Definite and Related Functions*. Springer, New York.
Bergland, G. D.
(1969). A guided tour of the fast Fourier transform. *IEEE Spectrum*, 6, No. 7, 41-52.
Berk, K. N.
(1974). Consistent autoregressive spectral estimates. *Ann. Stat.*, 2, No. 3, 489-502.
Berndt, H.
(1968). Correlation function estimation by a polarity method using stochastic reference signals. *IEEE Trans. Inform. Theory*, 14, No. 6, 796-801.
Berry, B. L. J., and D. F. Marble, eds.
(1968). *Spatial Analysis: A Reader in Statistical Geography*. Prentice-Hall, Englewood Cliffs, N. J.
Besag, J. E.
(1972). On the correlation structure of some two-dimensional stationary processes. *Biometrika*, 59, No. 1, 43-48.
Beutler, F. J.
(1961a). On stationarity conditions for certain periodic random processes. *J. Math. Anal. Appl.*, 3, No. 1, 25-36.
(1961b). Sampling theorems and bases in a Hilbert space. *Inform. and Control*, 4, Nos. 2-3, 97-117.
Beveridge, W. H.
(1921). Weather and harvest cycles. *Econ. J.*, 31, 429-452.
(1922). Wheat prices and rainfall in Western Europe. *J. Roy. Stat. Soc.*, 85, 412-459.

Bhagavan, C. S. K.
(1974). *Nonstationary Processes, Spectra and Some Ergodic Theorems.* Thesis, Andhra Univ., Waltair, India.

Bhansali, R. J.
(1986a). The criterion autoregressive transfer function of Parzen. *J. Time Series Anal.*, 7, No. 2, 79-104.
(1986b). Asymptotically efficient selection of the order by the criterion autoregressive transfer function. *Ann. Stat.*, 14, No. 1, 315-325.
(1986c). A derivation of the information criteria for selecting autoregressive models. *Adv. Appl. Probab.*, 18, No. 2, 360-387.

Bingham, C., M. D. Godfrey, and J. W. Tukey
(1967). Modern techniques of power spectrum estimation. *IEEE Trans. Audio Electroacoust.*, 15, No. 2, 56-66, Reprinted in Childers, 1978, pp. 6-16.

Birkhoff, G. D.
(1931). Proof of the ergodic theorem. *Proc. Nat. Acad. Sci. USA*, 17, 656-660.

Blackman, R. B., and J. W. Tukey
(1959). *The Measurement of Power Spectra.* Dover, New York.

Blanc-Lapierre, A., and R. Brard
(1946). Les fonctions aléatoires stationnaires et la loi des grands nombres. *Bull. Soc. Math. France*, 74, 102-115.

Blanc-Lapierre, A., and R. Fortet
(1946a). Sur la décomposition spectrale des fonctions aléatoires stationnaires d'ordre deux. *C.R. Acad. Sci.*, 222, No. 9, 467-468.
(1946b). Résultats sur la décomposition spectrale des fonctions aléatoires stationnaires d'ordre 2. *C.R. Acad. Sci.*, 222, No. 13, 713-714.
(1947a). Sur une propriété fundamentale des fonctions de corrélation. *C.R. Acad. Sci.*, 224, 786-788.
(1947b). Les fonctions aléatoires de plusieurs variables. *Revue Sci.*, 85, No. 7, 419-422.
(1953). *Théorie des fonctions aléatoires.* Masson, Paris. English translation: *Theory of Random Functions, Vols. I and II*, Gordon and Breach, New York, 1965 and 1968.

Blanc-Lapierre, A., and A. Tortrat
(1968). Sur la loi forte des grands nombres pour les fonctions aléatoires stationnaires du second ordre. *C.R. Acad. Sci.*, Ser. A, 267, No. 20, 740-743.

Blankinship, W. A.
(1974). Note on computing autocorrelations. *IEEE Trans. Acoust., Speech, Signal Process.*, 22, No. 1, 76-77.

Blomqvist, Å.
(1979). Figures of merit of windows for power density spectrum estimation with the DFT. *Proc. IEEE*, 67, No. 3, 438-439.

Bloomfield, P.
(1973). An exponential model for the spectrum of a scalar time series. *Biometrika*, 60, No. 2, 217-226.

(1976). *Fourier Analysis of Time Series: An Introduction*, Wiley, New York.

Bochkov, G. N., and A. A. Dubkov
(1974). Correlation analysis of nonlinear stochastic functionals. *Izv. VUZov, Ser. Radiofiz.* (Bull. Colleges and Univ. USSR, Ser. Radiophys.), 17, No. 3, 376-382.

Bochner, S.
(1934). A theorem on Fourier-Stieltjes integrals. *Bull. Amer. Math. Soc.*, **40**, 271-276.
(1941). Hilbert distances and positive definite functions. *Ann. Math.*, **42**, No. 3, 647-656.
(1959). *Lectures on Fourier Integrals*. Princeton University Press, Princeton, N. J.
(1960). *Harmonic Analysis and the Theory of Probability*. University of California Press, Berkeley and Los Angeles.

Bogner, R. E.
(1965). New relay correlation method. *Electronic Letters IEE*, 1, No. 3, 53-54.

Borgioli, R. C.
(1968). Fast Fourier transform correlation versus direct discrete time correlation. *Proc. IEEE*, **56**, No. 9, 1602-1604.

Box, G. E. P., and G. M. Jenkins
(1970). *Time Series Analysis. Forecasting and Control*. Holden-Day, San Francisco.

Boyles, R. A., and W. A. Gardner
(1983). Cycloergodic properties of discrete-parameter nonstationary stochastic processes. *IEEE Trans. Inform. Theory*, **29**, No. 1, 105-114.

Bradshaw, P.
(1971). *An Introduction to Turbulence and Its Measurement*. Pergamon Press, Oxford.

Brand, L.
(1966). *Differential and Difference Equations*. Wiley, New York.

Bras, R. L., and I. Rodriguez-Iturbe
(1985). *Random Functions and Hydrology*. Addison-Wesley, Reading, Mass.

Brelsford, W. M.
(1967). *Probability Predictions and Time Series with Periodic Structure*. Ph.D. Dissertation, Johns Hopkins University, Baltimore, Md.

Brigham, E. O.
(1974). *The Fast Fourier Transform*. Prentice-Hall, Englewood Cliffs, N. J.

Brillinger, D. R.
(1968). Estimation of the cross-spectrum of a stationary bivariate Gaussian process from its zeros. *J. Roy. Stat. Soc., Ser. B*, **30**, No. 1, 145-159.
(1969). Asymptotic properties of spectral estimates of second order. *Biometrika*, **56**, No. 2, 375-390.
(1970). The frequency analysis of relation between stationary spatial series. In *Proc. 12th Biennial Seminar of the Canad. Math. Congress* (R. Pyke, ed.), pp. 39-81, Canad. Math. Congress, Montreal.

(1972). The spectral analysis of stationary interval functions. *Proc. 6th Berkeley Symp. Math. Stat. Probab.*, 1, 483-513.

(1974a). Cross-spectral analysis of processes with stationary increments including the stationary $G/G/\infty$ queue. *Ann. Probab.*, 2, No. 5, 815-827.

(1974b). Fourier analysis of stationary processes. *Proc. IEEE*, 62, No. 12, 1628-1643.

(1975). *Time Series. Data Analysis and Theory.* Holt, Rinehart and Winston, New York.

(1979). Confidence intervals for the crosscovariance function. *Selecta Statistica Canadiana*, V, 3-16.

(1981). The key role of tapering in spectrum estimation. *IEEE Trans. Acoust., Speech, Signal Processing*, 29, No. 5, 1075-1076.

Brillinger, D. R., and M. Hatanaka

(1969). Harmonic analysis of nonstationary multivariate economic processes. *Econometrica*, 37, No. 1, 131-141.

Brillinger, D. R., and M. Rosenblatt

(1967). Computation and interpretation of k-th order spectra. In *Spectral Analysis of Time Series* (B. Harris, ed.), pp. 189-232. Wiley, New York.

Brissaud, A., and U. Frisch

(1971). Theory of Stark broading-2. Exact line profile with model microfield. *J. Quantum Spectrosc. Radiat. Transfer*, 11, No. 12, 1767-1784.

(1974). Solving linear stochastic differential equations. *J. Math. Phys.*, 15, No. 5, 524-534.

Brown, J. L.

(1957). On a cross-correlation property for stationary random processes. *IRE Trans. Inform. Theory*, 3, No. 1, 28-31.

Brown, R.

(1828). A brief account of microscopical observations made in the months of June, July, and August, 1827, on the particles contained in the pollen of plants; and on the general existence of active molecules in organic and inorganic bodies. *Phil. Mag., N.S.*, 4, 161-173.

(1829). Additional remarks on active molecules. *Phil. Mag., N.S.*, 6, 161-166.

Brown, R. G.

(1963). *Smoothing, Forecasting, and Prediction of Discrete Time Series.* Prentice-Hall, Englewood Cliffs, N. J.

Brown, T. J., and J. C. Hardin

(1973). A note on Kendall's autoregressive series. *J. Appl. Probab.*, 10, No. 2, 475-478.

Brown, W. M.

(1963). *Analysis of Linear Time-Invariant Systems.* McGraw-Hill, New York.

Bruckner, L. A.

(1971). Moving averages of homogeneous random fields. *Ann. Math. Stat.*, 42, No. 6, 2147-2149.

Buckingham, M. J.
(1983). *Noise in Electronic Devices and Systems.* E. Horwood, Chichester, England.
Bulinskii, A. V., and I. G. Zhurbenko
(1976a). The central limit theorem for random fields. *Dokl. Akad. Nauk SSSR* (Repts. Acad. Sci. USSR), 226, No. 1, 23-25.
(1976b). The central limit theorem for additive random functions. *Teor. Veroyatn. i Primen.* (Probab. Theory and Its Appl.), 21, No. 4, 707-714.
Bunimovich, V. I.
(1951). *Fluctuation Processes in Radio-Receiving Devices.* Soviet Radio, Moscow.
Burford, T. M.
(1958). Nonstationary velocity estimation. *Bell System Tech. J.*, 37, 1009-1021.
Burg, J. P.
(1967). Maximum entropy spectral analysis. Paper presented at *37th Ann. Intern. Meeting Soc. Explor. Geophys.*, *Oklahoma City, Oklahoma*, 8 pp. Reprinted in Childers, 1978, pp. 34-41.
(1968). A new analysis technique for time series data. Paper presented at *NATO Adv. Study Inst. on Signal Process. with Emph. on Underwater Acoust.*, *Enschede, Netherlands*, 7 pp. Reprinted in Childers, 1978, pp. 42-48.
(1975). *Maximum Entropy Spectral Analysis.* Ph.D. Dissertation, Dept. Geophysics, Stanford University, Stanford, Calif.
Buslenko, N. P., D. I. Golenko, Yu. A. Shreider, I. M. Sobol', and V. G. Sragovich
(1966). *The Monte Carlo Method: The Method of Statistical Trials.* Pergamon Press, Oxford.
Bussgang, J. J.
(1952). *Cross Correlation Function of Amplitude Distorted Gaussian Signals.* Techn. Rep. No. 216, Res. Lab. Elec., Mass. Inst. Technology, Cambridge, Mass.
Byrne, C. L., and R. M. Fitzgerald
(1984). Spectral estimators that extend the maximum entropy and maximum likelihood methods. *SIAM J. Appl. Math.*, 44, No. 2, 425-442.
Cacopardi, S.
(1983). Applicability of the relay correlator to radar signal processing. *Electron. Lett.*, 19, No. 18, 722-723.
Cadzow, J. A.
(1982). Spectral estimation: An overdetermined rational model equation approach. *Proc. IEEE*, 70, No. 9, 907-939.
Cambanis, S., and B. Liu
(1970). On harmonizable stochastic processes. *Information and Control*, 17, No. 2, 183-202.
Cambanis, S., and E. Masry
(1978). On the reconstruction of the covariance of stationary Gaussian processes observed through zero-memory nonlinearities. *IEEE Trans. Inform. Theory*, 24, No. 4, 485-494.

Campbell, N.
(1909a). The study of discontinuous phenomena. *Proc. Cambr. Phil. Soc.*, **15**, 117-136.
(1909b). Discontinuities in light emission. *Proc. Cambr. Phil. Soc.*, **15**, 310-328.

Capon, J.
(1969). High-resolution frequency-wavenumber spectrum analysis. *Proc. IEEE*, **57**, No. 8, 1408-1418. Reprinted in Childers, 1978, pp. 119-129.

Carter, G. C., and A. H. Nuttall
(1980a). On the weighted overlapped segment-averaging method for power spectral estimation. *Proc. IEEE*, **68**, No. 10, 1352-1354.
(1980b). A brief summary of a generalized framework for power spectral estimation. *Signal Processing*, **2**, No. 4, 387-390.
(1983). Analysis of a generalized framework for spectral estimation. Parts 1 and 2. *IEE Proc., Part F*, **130**, No. 3, 239-245.

Cauer, W.
(1958). *Synthesis of Linear Communication Networks. Vols. 1 and 2.* McGraw-Hill, New York.

Chan, Y. T., and R. P. Langford
(1982). Spectral estimation via high-order Yule-Walker equations. *IEEE Trans. Acoust., Speech, Signal Process.*, **30**, No. 5, 689-698.

Chan, Y. T., and J. C. Wood
(1984). A new order determination technique for ARMA processes. *IEEE Trans. Acoust., Speech, Signal Process.*, **32**, No. 3, 517-521.

Chandrasekhar, S.
(1943). Stochastic problems in physics and astronomy. *Rev. Mod. Phys.*, **15**, No. 1, 1-89. Reprinted in Wax, 1954, pp. 3-91.

Chang, D. K., and M. M. Rao
(1983). Bimeasures and sampling theorems for weakly harmonizable processes. *Stochast. Anal. and Appl.*, **1**, No. 1, 21-55.
(1986). Bimeasures and nonstationary processes. In *Real and Stochastic Analysis* (M. M. Rao, ed.), pp. 7-118, Wiley, New York.

Chang, K.-Y., and D. Moore
(1970). Modified correlator and its estimation errors. *IEEE Trans. Inform. Theory*, **16**, No. 6, 699-706.

Chapman, S., and J. Bartels
(1940). *Geomagnetism. Vol. 1.* Clarendon Press, Oxford.

Childers, D. G., ed.
(1978). *Modern Spectrum Analysis.* IEEE Press, New York.

Chistyakov, S. P.
(1986). On a regression problem for random processes. *Teor. Veroyatn. i Primen.* (Probab. Theory and Its Appl.), **31**, No. 1, 142-145.

Chiu, W.-C.
(1967). On the interpretation of energy spectrum. *Amer. J. Phys.*, **35**, No. 7, 642-648.

Choi, B.-S.
(1986). An algorithm for solving the extended Yule-Walker equations of an autoregressive moving-average time series. *IEEE Trans. Inform. Theory*, **32**, No. 3, 417-419.
Choi, B.-S., and T. M. Cover
(1984). An information-theoretic proof of Burg's maximum entropy spectrum. *Proc. IEEE*, **72**, No. 8, 1094-1095.
Chover, J.
(1961). On normalized entropy and the extension of a positive-definite function. *J. Math. Mech.*, **10**, No. 6, 927-945.
Christian, E., and E. Eisenmann
(1967). *Filter Design, Tables, and Graphs*. Wiley, New York.
Cook, C. E., and M. Bernfield
(1967). *Radar Signals*. Academic Press, New York.
Cooley, J. W., P. A. W. Lewis, and P. D. Welch
(1967a). Historical notes on the fast Fourier transform. *IEEE Trans. Audio Electroacoust.*, **15**, No. 2, 76-79.
(1967b). Application of the fast Fourier transform to computation of Fourier integrals, Fourier series, and convolution integrals. *IEEE Trans. Audio Electroacoust.*, **15**, No. 2, 79-84.
(1967c). The fast Fourier transform algorithm and its applications. *IBM Memorandum RC1743*.
(1970). The application of the fast Fourier transform algorithm to the estimation of spectra and cross-spectra. *J. Sound Vibr.*, **12**, No. 3, 339-352.
Cooley, J. W., and J. W. Tukey
(1965). An algorithm for the machine calculation of complex Fourier series. *Math. Comp.*, **19**, No. 90, 297-301. Reprinted in Oppenheim, 1969, pp. 146-150.
Courant, R., and D. Hilbert
(1953). *Methods of Mathematical Physics. Vol. 1*. Interscience, New York.
Cox, D. R.
(1962). *Renewal Theory*. Methuen, London.
(1981). Statistical analysis of time series: Some recent developments. *Scand. J. Statist.*, **8**, No. 2, 93-115.
Cox, D. R., and P. A. W. Lewis
(1966). *The Statistical Analysis of Series of Events*. Methuen, London.
Craddock, J. M., and S. Flintoff
(1970). Eigenvector representations of Northern Hemispheric fields. *Quart. J. Roy. Meteor. Soc.*, **96**, No. 407, 124-129.
Cramér, H.
(1939). On the representation of a function by certain Fourier integrals. *Trans. Amer. Math. Soc.*, **46**, 191-201.
(1940). On the theory of stationary random processes. *Ann. Math.*, **41**, 215-230.
(1942). On harmonic analysis in certain functional spaces. *Arkiv för Mat. Astr. Fys.*, **28B**, No. 12, 1-7.
(1946). *Mathematical Methods of Statistics*. Princeton University Press, Princeton, N. J.

(1951). A contribution to the theory of stochastic processes. *Proc. Second Berkeley Symp. Math. Stat. Prob.*, 329-339.

(1962). *Random Variables and Probability Distributions.* 2nd edition. Cambridge University Press, Cambridge.

Cramér, H., and M. R. Leadbetter

(1967). *Stationary and Related Stochastic Processes.* Wiley, New York.

Currie, R. G.

(1973). Fine structure in the sunspot spectrum − 2 to 70 years. *Astrophys. Space Sci.*, **20**, No. 2, 509-518.

Dahlhaus, R.

(1985a). On a spectral density estimate obtained by averaging periodograms. *J. Appl. Probab.*, **22**, No. 3, 598-610.

(1985b). Asymptotic normality of spectral estimates. *J. Multivar. Anal.*, **16**, No. 3, 412-431.

Dang, Dyk Khaw

(1982). On asymptotic behaviour of the variance of the estimate for the unknown mean value of a homogeneous random field. *Dokl. Akad. Nauk Ukrain. SSR, Ser. A, Phys-Math. and Eng. Sci.*, No. 9, 3-5.

(1984). Asymptotic behaviour of the variance of the arithmetic mean estimate for the unknown mean value of a homogeneous random field. *Dokl. Akad. Nauk Ukrain. SSR, Ser. A, Phys.-Math. and Eng. Sci.*, No. 2, 8-10.

Daniell, P. J.

(1946). Discussion of paper by M. S. Bartlett. *J. Roy. Stat. Soc., Suppl.*, **8**, No. 1, 88-90.

Daniels, H. E.

(1962). The estimation of spectral densities. *J. Roy. Stat. Soc., Ser. B*, **24**, No. 1, 185-198.

Davenport, W. B., Jr., R. A. Johnson, and D. Middleton

(1952). Statistical errors in measurement of random functions. *J. Appl. Phys.*, **23**, No. 4, 377-388.

Davenport, W. B., Jr., and W. L. Root

(1958). *An Introduction to the Theory of Random Signals and Noise.* McGraw-Hill, New York.

Davidovich, Yu. S.

(1970). Asymptotic behaviour of the variances of spatially averaged values of a homogeneous random field. *Teor. Veroyatn. i Mat. Stat.* (Probab. Theory and Math. Stat., Kiev), No. 3, 35-49.

Davidovich, Yu. S., and Yu. V. Kartashov

(1968). On the estimation of the mean of a homogeneous random field. *Vestnik Kiev Univ., Ser. Mat. Mekh.* (Kiev Univ. Herald, Ser. Math. Mech.), No. 10, 119-123.

Davis, B. R., and W. G. Cowley

(1982). Bias and variance of spectral estimates from an all-pole digital filter. *IEEE Trans. Acoust., Speech, Signal Process.*, **30**, No. 2, 322-329.

Davis, J. C.

(1973). *Statistics and Data Analysis in Geology.* Wiley, New York.

De Bruijn, N. G.

(1967). Uncertainty principle in Fourier analysis. In *Inequalities* (O. Shisha, ed.), pp. 57-71. Academic Press, New York.

De Gooijer, J. G.
(1981). An investigation of the moments of the sample autocovariances and autocorrelations for general ARMA processes. *J. Stat. Comput. Simul.*, **12**, Nos. 3-4, 175-192.
De Gooijer, J. G., B. Abraham, A. Gould, and L. Robinson
(1985). Methods for determining the order of an autoregressive-moving average process: a survey. *Intern. Statistical Rev.*, **55**, No. 3, 301-329.
Deo, C. M.
(1975). A functional central limit theorem for stationary random fields. *Ann. Probab.*, **3**, No. 4, 708-715.
Deutsch, R.
(1962). *Nonlinear Transformations of Random Processes.* Prentice-Hall, Englewood Cliffs, N. J.
Dickinson, B. W.
(1978). Two recursive estimates of autoregressive models based on maximum likelihood. *J. Stat. Comput. Simul.*, **7**, No. 2, 85-92.
(1980). Two-dimensional Markov spectrum estimates need not exist. *IEEE Trans. Inform. Theory*, **26**, No. 1, 120-121.
Dodds, C. J., and J. D. Robson
(1973). The description of road surface roughness. *J. Sound Vibr.*, **31**, 175-183.
Dominguez, A. G.
(1940). The representation of functions by Fourier integrals. *Duke Math. J.*, **6**, 246-255.
Donsker, M. D.
(1964). On function space integrals. In *Analysis in Function Space* (W. T. Martin and I. Segal, eds.), pp. 17-30. MIT Press, Cambridge, Mass.
Doob, J. L.
(1942). The Brownian movement and stochastic equations. *Ann. Math.*, **43**, No. 2, 351-369. Reprinted in Wax, 1954, pp. 319-337.
(1944). Elementary Gaussian processes. *Ann. Math. Stat.*, **15**, 229-282.
(1949). Time series and harmonic analysis. In *Proc. (First) Berkeley Symp. Math. Stat. Probab.* (J. Neyman, ed.), pp. 303-344. Univ. California Press, Berkeley.
(1953). *Stochastic Processes.* Wiley, New York.
(1966). Wiener's work in probability theory. *Bull. Amer. Math. Soc.*, **72**, No. 1, Part II, 69-72.
Dragan, Ya. P.
(1975). On the representation of periodically correlated random process in terms of stationary components. *Otbor i Peredacha Inform.* (Selection and Transmiss. of Inform., Kiev), No. 45, 7-20.
(1980). *Structure and Representations of Models of Stochastic Signals.* Naukova Dumka, Kiev.
Dragan, Ya. P., V. P. Mezentsev, and I. N. Yavorskii
(1984). Computation of the estimates for spectral characteristics of periodically correlated random processes. *Otbor i Peredacha Inform.*, No. 69, 29-35.

Dragan, Ya. P., V. A. Rozhkov, and I. N. Yavorskii
(1984). Applications of the theory of periodically correlated random processes to the probabilistic analysis of oceanological time series. In *Probabilistic Analysis and Modelling of Oceanological Processes* (V. A. Rozhkov, ed.), pp. 4-23. Gidrometeoizdat, Leningrad.

Dragan, Ya. P., and I. N. Yavorskii
(1982). *Rhythmics of Sea Waves and Underwater Acoustic Signals.* Naukova Dumka, Kiev.

Drozdov, O. A., and T. V. Pokrovskaya
(1961). On the contribution of random variations of water balance to the level oscillations for closed lakes. *Meteorologiya i Gidrologiya* (Meteor. and Hydrol.), No. 8, 43-48.

Dubrovskii, N. A., and L. N. Tumarkina
(1967). Investigation of the human perception of amplitude-modulated noise. *Akust. Zhurn.* (Acoust. J.), 13, No. 1, 51-59.

Dunford, N., and J. T. Schwartz
(1956). Convergence almost everywhere of operator averages. *J. Rat. Mech. Anal.*, 5, 129-178.
(1958). *Linear Operators. Part I: General Theory.* Interscience, New York.

Durbin, J.
(1960). The fitting of time series models. *Rev. Inst. Intern. de Stat.*, 28, 233-244.

Durrani, T. S., ed.
(1983). *Papers on Spectral Analysis.* Special issue of *IEE Proc., Part F*, 130, No. 3.

Dykhovichny, A. A.
(1983). On the estimation of the correlation function of a homogeneous and isotropic Gaussian random field. *Teor. Veroyatn. i Mat. Stat.* (Probab. Theory and Math. Stat., Kiev), No. 29, 37-40.
(1985). On the estimates of the correlation and spectral distribution functions of a homogeneous and isotropic Gaussian random field. *Teor. Veroyatn. i Mat. Stat.*, No. 32, 17-27.

Dym, H.
(1966). Stationary measures for the flow of a linear differential equation driven by white noise. *Trans. Amer. Math. Soc.*, 123, No. 1, 130-164.

Dzhaparidze, K. O.
(1986). *Parameter Estimation and Hypothesis Testing in Spectral Analysis of Stationary Time Series.* Springer, New York.

Dzhaparidze, K. O., and A. M. Yaglom
(1983). Spectrum parameter estimation in time series analysis. In *Development in Statistics, Vol. 4* (P. R. Krishnaiah, ed.), pp. 1-96. Academic Press, New York.

Eberhard, A.
(1973). An optimal discrete window for the calculation of power spectra. *IEEE Trans. Audio Electroacoust.*, 21, No. 1, 37-43.

Eberlein, W. F.
(1955). Characterization of Fourier-Stieltjes transforms. *Duke Math. J.*, **22**, 465-468.

Eberlein, E., and M. S. Taqqu, eds.
(1986). *Dependence in Probability and Statistics. A Survey of Recent Results*. Birkhäuser, Boston.

Edward, J. A., and M. M. Fitelson
(1973). Notes on maximum entropy processing. *IEEE Trans. Inform. Theory*, **19**, No. 2, 232-234. Reprinted in Childers, 1978, pp. 94-96.

Egau, P. C.
(1984). Correlation systems in radio astronomy and related fields. *IEE Proc., Part F*, **131**, No. 1, 32-39.

Egorov, I. S.
(1968). Commutator correlograph. *Radiotekhn. i Elektr.* (Radio Eng. and Electronics), **13**, No. 8, 1509-1512.

Einstein, A.
(1914). Méthode pour la détermination de valeurs statistiques d'observations concernant des grandeurs soumises à des fluctuations irrégulières. *Arch. Sci. Phys. et Natur.*, Ser. *4*, 37, 254-256.
(1956). *Investigations on the Theory of the Brownian Movement* (R. Fürth, ed.). Dover, New York.

Ekhaguere, G. O. S.
(1979). A class of Gaussian fields of Markov type. *Physica*, **A99**, No. 3, 545-568.

Elliott, D. F., and K. R. Rao
(1982). *Fast Transforms, Algorithms, Analysis, Applications*. Academic Press, New York.

Elliott, J. P., and P. G. Dawber
(1979). *Symmetry in Physics. Vols. 1 and 2*. MacMillan Press, London.

Erdélyi, A., W. Magnus, F. Oberhettinger, and F. G. Tricomi
(1953). *Higher Transcendental Functions. Vols. 1-3*. McGraw-Hill, New York.
(1954). *Tables of Integral Transforms. Vols. 1 and 2*. McGraw-Hill, New York.

Erickson, R. V.
(1971). Constant coefficient linear differential equations driven by white noise. *Ann. Math. Stat.*, **42**, No. 2, 820-823.

Fama, E. F., and R. Roll
(1968). Some properties of symmetric stable distributions. *J. Amer. Statist. Assoc.*, **63**, No. 323, 817-836.

Farrell, R. H.
(1979). Asymptotic lower bounds for the risk of estimators of the value of a spectral density function. *Zeitschr. Wahrscheinlichkeitstheorie verw. Gebiete*, **49**, No. 2, 221-234.

Favre, A., J. Gaviglio, and R. Dumas
(1954). Quelques fonctions d'autocorrélation et de répartitions spectrales d'energie, pour la turbulence en aval des diverses grilles. *C.R. Acad. Sci.*, **238**, No. 15, 1561-1563.

Feller, W.
(1966). *Introduction to Probability Theory and Its Applications. Vol. 2.*
Wiley, New York.

Fine, T. L., and W. G. Hwang
(1979). Consistent estimation of system order. *IEEE Trans. Autom.
Control*, **24**, 389-402.

Fine, T., and N. Johnson
(1965). On the estimation of the mean of a random process. *Proc.
IEEE*, **65**, No. 2, 187-188.

Fishman, G. S.
(1969). *Spectral Methods in Econometrics.* Harvard University Press,
Cambridge, Mass.

Fomin, S. V.
(1949). On the theory of dynamic systems with a continuous
spectrum. *Doklady Akad. Nauk SSSR* (Repts. Acad. Sci. USSR), **67**,
No. 3, 435-437.
(1950). On dynamic systems in function space. *Ukr. Matem. Zhurn.*
(Ukrain. Math. J.), **2**, No. 2, 25-47.

Fortet, R.
(1954). Spectre moyen d'une suite d'impulsions en principe
périodiques et identiques mais déplacées aléatoirement. *L'onde
electrique*, **34**, Nos. 329-330, 683-687.

Fortus, M. I.
(1973). Statistically orthogonal functions for finite segments of a
random process. *Izv. Akad. Nauk SSSR, Ser. Fiz. Atm. i Okeana*
(Bull. Acad. Sci. USSR, Ser. Atmosph. and Oceanic Phys.), **9**, No. 4,
34-46.
(1975). Statistically orthogonal functions for a random field given
in a bounded planar domain. *Izv. Akad. Nauk SSSR, Ser. Fiz.
Atmosf. i Okeana*, **11**, No. 11, 1107-1112.
(1977). The eigenfunctions of the Wiener-Hopf integral operator in
a circular domain. *Teor. Veroyatn. i Mat. Stat.* (Probab. Theory and
Math. Stat., Kiev), No. 16, 135-141.
(1980). Method of empirical orthogonal functions and its
meteorological applications. *Meteorologiya i Gidrologiya*
(Meteorology and Hydrology, Moscow), No. 4, 113-119.

Foster, G. A. R.
(1946). Some instruments for the analysis of time-series and their
application to textile research. *J. Roy. Stat. Soc. Suppl.*, **8**, 42-61.

Franks, L. E.
(1969). *Signal Theory.* Prentice-Hall, Englewood Cliffs, N. J.

Freiberger, W., and U. Grenander
(1959). Approximate distributions of noise power measurements.
Quart. Appl. Math., **17**, No. 3, 271-283.

Frenkiel, F. N.
(1953). Turbulent diffusion: Mean concentration distribution in a
flow field of homogeneous turbulence. *Adv. Appl. Mech.*, **3**, 61-107.

Friedlander, B.
(1982). A recursive maximum likelihood algorithm for ARMA spectral estimation. *IEEE Trans. Inform. Theory*, **28**, No. 4, 639-646.
(1983). Efficient algorithm for ARMA spectral estimation. *IEE Proc., Part F*, **130**, No. 3, 195-201.
Friedlander, S. K., and L. Topper, eds.
(1961). *Turbulence. Classical Papers on Statistical Theory.* Interscience, New York.
Fuchs, A.
(1968). Note sur un théoreme de Pólya. *Publ. Inst. Stat. Univ. Paris*, **17**, No. 3, 7-10.
Furutsu, K.
(1963). On the statistical theory of electromagnetic waves in a fluctuating medium. *J. Res. Nat. Bureau Stand., Sec. D*, **67**, No. 3, 303-323.
Gabriel, K. J.
(1983). Comparison of three correlation coefficient estimators for Gaussian stationary processes. *IEEE Trans. Acoustics, Speech, Signal Process.*, **31**, No. 4, 1023-1025.
Gaile, G. L., and C. J. Willmott, eds.
(1984). *Spatial Statistics and Models.* D. Reidel, Dordrecht.
Gan, R. G., K. E. Eman, and S. M. Wu
(1984). An extended FFT algorithm for ARMA spectral estimation. *IEEE Trans. Acoust., Speech, Signal Process.*, **32**, No. 1, 168-170.
Gangolli, R.
(1967a). Abstract harmonic analysis and Lévy's Brownian motion of several parameters. *Proc. Fifth Berkeley Symp. Math. Stat. and Probab.*, **2**, Part I, 13-30.
(1967b). Positive definite kernels on homogeneous spaces and certain stochastic processes, related to Lévy's Brownian motion of several parameters. *Ann. Inst. H. Poincaré, Sec. B*, No. 2, 121-225.
Gaposhkin, V. F.
(1973). On the strong law of large numbers for wide sense stationary processes and sequences. *Teor. Veroyatn. i Primen.* (Probab. Theory and Its Appl.), **18**, No. 2, 388-392.
(1977a). A theorem on the almost-sure convergence of a sequence of measurable functions and its applications to sequences of stochastic integrals. *Mat. Sbornik* (Math. Collection), **104**, No. 1, 3-21.
(1977b). Conditions for applicability of strong law of large numbers to some classes of wide sense stationary random processes and homogeneous random fields. *Teor. Veroyatn. i Primen.*, **22**, No. 2, 295-319.
(1980). Almost sure convergence of the spectral density estimates for stationary processes. *Teor. Veroyatn. i Primen.*, **25**, No. 1, 172-178.
(1981a). The dependence of the convergence rate in the strong law of large numbers for stationary processes on the rate of decrease of the correlation function. *Teor. Veroyatn. i Primen.*, **26**, No. 4, 720-733.

(1981b). Multiparameter law of strong numbers for homogeneous random fields. *Uspekhi Mat. Nauk* (Adv. Math. Sci.), **36**, No. 6, 197-198.

(1984). On strong law of large numbers for isotropic random fields. *Teor. Veroyatn. i Mat. Statistika* (Probab. Theory and Math. Stat., Kiev), No. 30, 34-38.

Gardner, W. A.

(1972). *Representation and Estimation of Cyclostationary Processes.* Ph.D. Dissertation, Univ. of Massachusetts, Amherst; also Univ. Mass. Res. Inst. Tech. Rep. TR-2.

(1978). Stationarizable random processes. *IEEE Trans. Inform. Theory*, **24**, No. 1, 8-22.

Gardner, W. A., and L. E. Franks

(1975). Characterization of cyclostationary random signal processes. *IEEE Trans. Inform. Theory*, **21**, No. 1, 4-14.

GARP Publication Series

(1975). *The Physical Basis of Climate and Climate Modelling.* Report of the Intern. Study Confer. in Stockholm, 29 July - 10 August 1974. Global Atmosph. Res. Program Publ. Ser. No. 16, World Meteor. Organization, Geneva, Switzerland.

Geckinli, N. C., and D. Yavuz

(1978). Some novel windows and a concise tutorial comparison of window families. *IEEE Trans. Acoust., Speech, Signal Process.*, **26**, 501-507.

Gel'fand, I. M.

(1950). Spherical functions on symmetric Riemann spaces. *Doklady Akad. Nauk SSSR*, **70**, No. 1, 5-8.

(1955). Generalized random processes. *Doklady Akad. Nauk SSSR*, **100**, No. 5, 853-856.

Gel'fand, I. M., and G. E. Shilov

(1964). *Properties and Operations. (Generalized Functions, Vol. 1).* Academic Press, New York.

(1968). *Spaces of Fundamental and Generalized Functions. (Generalized Functions, Vol. 2).* Academic Press, New York.

Gel'fand, I. M., and N. Ya. Vilenkin

(1964). *Applications of Harmonic Analysis. (Generalized Functions, Vol. 4).* Academic Press, New York.

Gel'fand, I. M., and A. M. Yaglom

(1957). Calculation of the amount of information about a random function contained in another such function. *Uspekhi Mat. Nauk* (Adv. Math. Sci.), **12**, No. 1(73), 3-52. English transl. in *Amer. Math. Soc. Translations*, **12**, 192-246, 1959.

Gel'fond, A. O.

(1967). *Finite Difference Calculus.* 3rd edition. Nauka, Moscow.

Gentleman, W. M., and G. Sande

(1966). Fast Fourier transforms for fun and profit. *1966 Fall Joint Computer Conference, AFIPS Proc.*, **28**, 563-578, Spartan, Washington, D.C.

Geranin, V. A.
(1972). Effect of the non-Gaussianity of a random process on the accuracy of correlation function measurements. In *Proc. 5th USSR Symp. on Represent. Methods and Instr. Analysis of Random Processes and Fields*, Sec. *1*, pp. 80-86. USSR Res. Inst. Electr. Measur. Instr., Leningrad.

German, D. Ya., N. F. Morozov, and Yu. L. Svalov
(1974). Errors in the estimates of the mean value, variance, and correlation function of stationary random processes. In *Proc. 7th USSR Symp. on Represent. Methods and Instr. Analysis of Random Processes and Fields*, Sec. *2*, pp. 128-133, USSR Res. Inst. Electr. Measur. Instr., Leningrad.

Gersch, W.
(1970). Spectral analysis of EEG's by autoregressive decomposition of time series. *Math. Biosci.*, 7, 205-222.

Gershman, S. G., and E. L. Feinberg
(1955). On the measurement of the correlation coefficient. *Akust. Zhurn.* (Acoust. J.), 1, No. 4, 326-338. English transl. in *Sov. Phys. —Acoust.*, 1, 340-352.

Getis, A., and B. Boots
(1978). *Models of Spatial Processes.* Cambridge University Press, Cambridge.

Gihman (Gikhman), I. I., and A. V. Skorohod (Skorokhod)
(1969). *Introduction to the Theory of Random Processes.* Saunders, Philadelphia.
(1974, 1975, 1979). *The Theory of Stochastic Processes.* Vols. 1-3. Springer, Berlin.

Gladyshev, E. G.
(1961). On periodically correlated random sequences. *Doklady Akad. Nauk SSSR* (Repts. Acad. Sci. USSR), 137, No. 5, 1026-1030.
(1962). *Theory of Periodically Correlated Sequences and Processes.* Candidate Dissertation, Inst. Atmosph. Phys. Acad. Sci. USSR, Moscow.
(1963). Periodically and almost periodically correlated random processes with continuous time parameter. *Teor. Veroyatn. i Prim.* (Probab. Theory and Its Appl.), 8, No. 2, 184-189.

Glukhovskii, A. B., and M. I. Fortus
(1982). Estimation of statistical reliability for empirical orthogonal functions. *Izv. Akad. Nauk SSSR, Ser. Fiz. Atmosf. i Okeana* (Bull. Acad. Sci. USSR, Ser. Atmosph. and Oceanic Phys.), 18, No. 5, 451-459.
(1984). Statistical reliability of analysis of meteorological variable profiles. *Izv. Akad. Nauk SSSR, Ser. Fiz. Atmosf. i Okeana*, 20, No. 12, 1139-1149.

Gnedenko, B. V.
(1962). *The Theory of Probability.* Chelsea, New York.

Goering, H., ed.
(1958). *Sammelband zur Statistischen Theorie der Turbulenz.* Akademie-Verlag, Berlin.

Goldberg, R. R.
 (1961). *Fourier Transforms*. Cambridge University Press, London.
Goldberg, S.
 (1958). *Introduction to Difference Equations*. Wiley, New York.
Goldie, C. M., and G. J. Morrow
 (1986). Central limit questions for random fields. In *Dependence in Probability and Statistics* (E. Eberlein and M. S. Taqqu, eds.), pp. 275-289, Birkhäuser, Boston.
Goldstein, R. M.
 (1962). *Radar Exploration of Venus*. Ph.D. Thesis, Calif. Inst. of Technology, Pasadena, Calif.
Goodman, J. W.
 (1968). *Introduction to Fourier Optics*. McGraw-Hill, New York.
Gorbatsevich, E. D.
 (1971). *Approximating Correlometers*. Energiya, Moscow.
Gorodetskii, V. V.
 (1984). The central limit theorem and the invariance principle for weakly dependent random fields. *Doklady Akad. Nauk SSSR* (Repts. Acad. Sci. USSR), **276**, No. 3, 528-531.
Gradshteyn, I. S., and I. M. Ryzhik
 (1980). *Tables of Integrals, Series, and Products: Corrected and Enlarged Edition*. Academic Press, New York.
Grandell, J., M. Hamrud, and P. Toll
 (1980). A remark on the correspondence between the maximum entropy method and the autoregressive model. *IEEE Trans. Inform. Theory.*, **26**, No. 6, 750-751.
Granger, C. W. J., and M. Hatanaka
 (1964). *Spectral Analysis of Economic Time Series*. Princeton University Press, Princeton, N. J.
Granger, C. W. J., and A. O. Hughes
 (1971). A new look at some old data: the Beveridge wheat price series. *J. Roy. Stat. Soc., Ser. A*, **134**, No. 3, 413-428.
Grenander, U.
 (1950). Stochastic processes and statistical inference. *Ark. Mat.*, **1**, 195-277.
 (1951). On empirical spectral analysis of stochastic processes. *Ark. Mat.*, **1**, No. 35, 503-531.
 (1958). Bandwidth and variance in estimation of the spectrum. *J. Roy. Stat. Soc., Ser. B*, **20**, No. 1, 152-157.
 (1963). *Probabilities on Algebraic Structures*. Wiley, New York.
Grenander, U., H. O. Pollak, and D. Slepian
 (1959). The distribution of quadratic forms in normal variates: A small sample theory with applications to spectral analysis. *J. Soc. Ind. Appl. Math.*, **7**, 374-401.
Grenander, U., and M. Rosenblatt
 (1953). Statistical spectral analysis of time series arising from stationary stochastic processes. *Ann. Math. Stat.*, **24**, No. 4, 537-558.
 (1956a). *Statistical Analysis of Stationary Time Series*. Almqvist and Wiksell, Stockholm; also Wiley, New York, 1957, and (2nd edition) Chelsea, New York, 1984.

(1956b). Some problems in estimating the spectrum of a time series. *Proc. Third Berkeley Symp. Math. Stat. Probab.*, 1, 77-93.

Grenier, Y., ed.
(1986). *Major Trends in Spectral Analysis.* Special issue of *Signal Processing*, 10, No. 1.

Gribanov, Yu. I., and V. L. Mal'kov
(1974). *Spectral Analysis of Random Processes.* Energiya, Moscow.
(1978). *Sample Estimates for Spectral Characteristics of Stationary Random Processes.* Energiya, Moscow.

Gribanov, Yu. I., G. P. Veselova, and V. N. Andreev
(1971). *Automatic Digital Correlators.* Energiya, Moscow.

Gudzenko, L. I.
(1959). On periodically nonstationary processes. *Radiotekhnika i Elektronika* (Radio Eng. and Electronics), 4, No. 6, 1062-1064.

Hall, F.
(1950). Communication theory applied to meteorological measurements. *J. Meteor.*, 7, No. 2, 121-129.

Halmos, P. R.
(1950). *Measure Theory.* Van Nostrand, Princeton, N. J.
(1951). *Introduction to Hilbert Space and the Theory of Spectral Multiplicity.* Chelsea, New York.

Hamming, R. W.
(1977). *Digital Filters.* Prentice-Hall, Englewood Cliffs, N. J.

Hannan, E. J.
(1957). The variance of the mean of a stationary time series. *J. Roy. Stat. Soc., Ser. B*, 19, No. 2, 282-285.
(1960). *Time Series Analysis.* Methuen, London, and Wiley, New York.
(1965). *Group Representations and Applied Probability.* Methuen, London (also *J. Appl. Probab.*, 2, No. 1, 1-68).
(1969). Fourier methods and random processes. *Bull. Intern. Stat. Inst.*, 42, Book 1, 475-496.
(1970). *Multiple Time Series.* Wiley, New York.
(1974). The uniform convergence of autocovariances. *Ann. Stat.*, 2, No. 4, 803-806.
(1976). The asymptotic distribution of serial covariances. *Ann. Stat.*, 4, No. 2, 396-399.
(1980). The estimation of the order of an ARMA process. *Ann. Stat.*, 8, No. 5, 1071-1081.

Hannan, E. J., and C. C. Heyde
(1972). On limit theorems for quadratic functions of discrete time series. *Ann. Math. Stat.*, 43, No. 6, 2058-2066.

Hannan, E. J., and L. Kavalieri
(1984). A method for autoregressive-moving average estimation. *Biometrika*, 71, No. 2, 273-283.

Hannan, E. J., and B. G. Quinn
(1979). The determination of the order of an autoregression. *J. Roy. Stat. Soc., Ser. B*, 41, No. 2, 190-195.

Hannan, E. J., and J. Rissanen
(1982). Recursive estimation of mixed autoregressive-moving average order. *Biometrika*, **69**, No. 1, 81-94; erratum, *ibid.*, **70**, No. 2, 303, 1983.

Harbaugh, J. W., and F. W. Preston
(1968). Fourier series analysis in geology. In Berry and Marble, 1968, pp. 218-238.

Harris, B., ed.
(1967). *Spectral Analysis of Time Series (Proc. Adv. Seminar at the Univ. of Wisconsin, Madison, October 3-5, 1966).* Wiley, New York.

Harris, F. J.
(1978). On the use of windows for harmonic analysis with the discrete Fourier transform. *Proc. IEEE*, **66**, No. 1, 51-83.

Hartman, P., and A. Wintner
(1940). On the spherical approach to the normal distribution law. *Amer. J. Math.*, **62**, No. 4, 759-779.

Haubrich, R. A.
(1965). Earth noise, 5 to 500 millicycles per second. *J. Geophys. Res.*, **70**, No. 6, 1415-1427.

Hausdorff, F.
(1923). Momentenprobleme für ein endliches Intervall. *Math. Zeitschr.*, **16**, 220-248.

Haykin, S., ed.
(1983). *Nonlinear Methods of Spectral Analysis.* 2nd edition. Springer, Berlin.

Haykin, S., and J. A. Cadzow, eds.
(1982). *Spectral Estimation.* Special issue of *Proc. IEEE*, **70**, No. 9.

Haykin, S., B. W. Currie, and S. B. Kesler
(1982). Maximum-entropy spectral analysis of radar clutter. *Proc. IEEE*, **70**, No. 9, 953-962.

Heideman, M. T., D. H. Johnson, and C. S. Burrus
(1984). Gauss and the history of the Fast Fourier Transform, *IEEE Acoust., Speech, Signal Process. Magazine*, **1**, No. 4, 14-21.

Heine, V.
(1955). Models for two-dimensional stationary stochastic processes. *Biometrika*, **42**, Nos. 1-2, 170-178.

Heisenberg, W.
(1948). Zur statistischen Theorie der Turbulenz. *Zeitschr. Physik*, **124**, Nos. 7-12, 628-657.

Helms, H. D., J. F. Kaiser, and L. R. Rabiner, eds.
(1975). *Literature in Digital Signal Processing: Author and Permuted Title Index.* IEEE Press, New York.

Helstrom, C. W.
(1968). *Statistical Theory of Signal Detection.* 2nd edition. Pergamon Press, New York.

Hepple, L. W.
(1974). *Spatial Series and Regression Theory.* Pion, London.

Herglotz, G.
(1911). Über Potenzreihen mit positiven reelen Teil im Einheitskreis. *Ber. Verh. Sächs. Akad. Wiss. Leipzig, Math.-Phys. Kl.*, **63**, 501-511.
Hertz, D.
(1982). A fast digital method of estimating the autocorrelation of a Gaussian stationary process. *IEEE Trans. Acoust., Speech, Signal Process.*, **30**, No. 2, 329.
Hext, G. R.
(1966). *A New Approach to Time Series with Mixed Spectra.* Tech. Rep. No. 14, Dept. Stat., Stanford University, Stanford, Calif.
Hinich, M.
(1967). Estimation of spectra after hard clipping of Gaussian processes. *Technometrics*, **9**, 391-400.
Hinze, J. O.
(1975). *Turbulence.* 2nd ed. McGraw-Hill, New York.
Hoffman, K.
(1962). *Banach Spaces of Analytic Functions.* Prentice-Hall, Englewood Cliffs, N. J.
Honda, I.
(1983). On the Fourier series and some sample properties of periodic stochastic processes. *Repts. Stat. Appl. Res. Japan Union Sci. Eng.*, **30**, No. 1, 1-17.
Hotelling, H.
(1933). Analysis of a complex of statistical variables into principal components. *J. Educ. Psychology*, **24**, 417-441; 498-520.
Huang, T. S., ed.
(1981). *Two-Dimensional Digital Signal Processing II.* Springer, Berlin.
Hunt, G. A.
(1951). Random Fourier transforms. *Trans. Amer. Math. Soc.*, **71**, No. 1, 38-69.
Hurd, H. L.
(1969). *An Investigation of Periodically Correlated Stochastic Processes.* Ph.D. Dissertation, Duke Univ., Durham, N.C.
(1973). Testing for harmonizability. *IEEE Trans. Inform. Theory*, **19**, No. 3, 316-320.
(1974a). Stationarizing properties of random shifts. *SIAM J. Appl. Math.*, **26**, No. 1, 203-212.
(1974b). Periodically correlated processes with discontinuous correlation functions. *Teor. Veroyatn. i Primen.* (Probab. Theory and Its Appl.), **19**, No. 4, 834-838.
Hurt, J.
(1973). On a simple estimate of correlations of stationary random sequences. *Aplicace Matem.* (Prague), **18**, No. 3, 176-187.
Huzii, M.
(1962). On a simplified method of the estimation of the correlogram for a stationary Gaussian process. *Ann. Inst. Stat. Math.*, **14**, 259-268.

(1964). On a simplified method of the estimation of the correlogram for a stationary Gaussian process, II. *Kodai Math. Sem. Reps.*, **16**, 199-212.

Ibragimov, I. A.

(1963). On the estimate of the spectral distribution function of a stationary Gaussian process. *Teor. Veroyatn. i Primen.* (Probab. Theory and Its Appl.), **8**, No. 4, 391-430.

Ibragimov, I. A., and Yu. V. Linnik

(1971). *Independent and Stationary Sequences of Random Variables.* Walters-Noordhoff, Groningen.

Isserlis, L.

(1918). On a formula for product-moment coefficient in any number of variables. *Biometrika*, **12**, 134-139.

Itô, K.

(1954). Stationary random distributions. *Mem. Coll. Sci. Univ. Kyoto, Ser. A*, **28**, No. 3, 209-223.

(1956). Isotropic random current. *Proc. Third Berkeley Symp. Math. Stat. and Probab.*, **2**, 125-132.

Itô, K., and H. P. McKean

(1965). *Diffusion Processes and Their Sample Paths.* Academic Press, New York.

Itskovich, E. L., and I. A. Tsaturova

(1969). Selection of the realization parameters for a random stationary process when the error of the correlation function estimate is calculated. *Avtomatika i Telemekh.* (Automatics and Remote Control), No. 2, 25-30.

Ivanov, A. V., and N. N. Leonenko

(1980). On an invariance principle for the estimator of the correlation function of a homogeneous random field. *Ukrain. Mat. Zhurn.* (Ukrain. Math. J.), **32**, No. 3, 323-331.

(1981a). On an invariance principle for the estimator of the correlation function of a homogeneous random field. *Dokl. Akad. Nauk Ukrain. SSR, Ser. A, Phys.-Math. and Eng. Sci.*, No. 1, 17-19.

(1981b). On an invariance principle for the estimator of the correlation function of a homogeneous and isotropic random field. *Ukrain. Mat. Zhurn.*, **33**, No. 3, 313-323.

Iwase, K.

(1973). On the formula of the variance of a simplified estimator of the covariogram for a stationary normal process. *Repts. Stat. Appl. Res. Japan Union Sci. Eng.*, **20**, No. 4, 113-117.

Jackson, L. C.

(1950). *Wave Filters.* Wiley, New York.

Jacovitti, G., and R. Cusani

(1985). Performances of hybrid-sign correlation coefficient estimator for Gaussian stationary processes. *IEEE Trans. Acoust., Speech, Signal Process.*, **33**, No. 3, 731-733.

Jacovitti, G., A. Neri, and R. Cusani
(1984). On a fast digital method of estimating the autocorrelation of a Gaussian stationary process. *IEEE Trans. Acoust., Speech, Signal Process.*, **32**, No. 5, 968-976.

Jahnke, E., F. Emde, and F. Lösch
(1960). *Tables of Higher Functions.* McGraw-Hill, New York.

Jain, A. K.
(1981). Advances in mathematical models for image processing. *Proc. IEEE*, **69**, No. 5, 502-528.

Jajte, R.
(1967). On stochastic processes over an Abelian locally compact group. *Colloquium Math.*, **17**, No. 2, 351-355.

James, H. M., N. B. Nichols, and R. S. Phillips, eds.
(1947). *Theory of Servomechanisms.* McGraw-Hill, New York, and Dover, New York.

Jaynes, E. T.
(1982). On the rationale of maximum-entropy methods. *Proc. IEEE*, **70**, No. 9, 939-952.

Jeffreys, H., Sir, and Lady Jeffreys (B. Swirles)
(1950). *Methods of Mathematical Physics.* Cambridge University Press, Cambridge.

Jenkins, G. M.
(1961). General considerations in the analysis of spectra. *Technometrics*, **3**, No. 2, 133-166.
(1965). A survey of spectral analysis. *Appl. Stat.*, **14**, No. 1, 2-32.

Jenkins, G. M., and D. G. Watts
(1968). *Spectral Analysis and Its Applications.* Holden-Day, San Francisco.

Jerri, A. J.
(1977). The Shannon sampling theorem − its various extensions and applications: A tutorial review. *Proc. IEEE*, **65**, No. 11, 1565-1596.

Jespers, P., P. T. Chu, and A. Fettweis
(1962). A new method to compute correlation functions. *IRE Trans. Inform. Theory*, **8**, No. 5, S106-S107.

Jolley, L.
(1925). *Summation of Series.* Chapman and Hall, London; also Dover, New York, 1961.

Jones, R. H.
(1963). Stochastic processes on a sphere. *Ann. Math. Stat.*, **34**, No. 1, 213-218.
(1965). A reappraisal of the periodogram in spectral analysis. *Technometrics*, **7**, No. 4, 531-542.
(1974). Identification and autoregressive spectrum estimation. *IEEE Trans. Automat. Contr.*, **19**, No. 6, 894-898. Reprinted in Childers, 1978, pp. 301-304.
(1978). Multivariate autoregressive estimation using residuals. In *Applied Time Series Analysis* (D. F. Findley, ed.), pp. 139-162. Academic Press, New York.

Jones, R. H., and W. M. Brelsford
(1967). Time series with periodic structure. *Biometrika*, **54**, Nos. 3-4, 403-408.
Jordan, J. R.
(1986). Correlation algorithms, circuits, and measurement applications. *IEEE Proc., Part G*, **133**, No. 1, 58-74.
Jordan, K. L.
(1961). *Discrete Representations of Random Signals.* Ph.D. Dissertation. MIT, Cambridge, Mass.; also MIT-RLE Tech. Rep. 378.
Kaganov, E. I., and A. M. Yaglom
(1976). Errors in wind-speed measurements by rotation anemometers. *Boundary-Layer Meteor.*, **10**, No. 1, 15-34.
Kailath, T.
(1980). *Linear Systems.* Prentice-Hall, Englewood Cliffs, N. J.
Kailath, T., ed.
(1977). *Linear Least-Squares Estimation.* (*Benchmark Papers in Electrical Engineering and Computer Sciences, Vol. 17.*) Dowden, Hutchington, and Ross, Stroudsburg, Penn.
Kamash, K. M. A., and J. D. Robson
(1977). Implications of isotropy in random surfaces. *J. Sound. Vibr.*, **54**, No. 1, 131-145.
(1978). The application of isotropy in road surface modelling. *J. Sound Vibr.*, **57**, No. 1, 89-100.
Kampé de Fériet, J.
(1948). Analyse harmonique des fonctions aléatoires stationnaires d'ordre 2 définies sur un groupe abélien localement compact. *C.R. Acad. Sci.*, **226**, No. 11, 868-870.
(1962). Correlations and spectra of asymptotically stationary random functions. *Math. Student*, **30**, Nos. 1-2, 55-67.
Kampé de Fériet, J., and F. N. Frenkiel
(1959). Estimation de la corrélation d'une fonction aléatoire non stationnaire. *C.R. Acad. Sci.*, **249**, 348-351.
(1962). Correlations and spectra for non-stationary random functions. *Math. Comput.*, **16**, No. 77, 1-21.
Karhunen, K.
(1946). Zur Spektraltheorie stochastischer Prozesse. *Ann. Acad. Sci. Fennicae, Ser. A*, I, No. 34, 3-7.
(1947). Ueber lineare Methoden in der Wahrscheinlichkeitsrechnung. *Ann. Acad. Sci. Fennicae, Ser. A*, I, No. 37, 3-79.
Kármán, T., von
(1937). The fundamentals of the statistical theory of turbulence. *J. Aeronaut. Sci.*, **4**, No. 4, 131-138.
(1948). Progress in the statistical theory of turbulence. *Proc. Nat. Acad. Sci. USA*, **34**, No. 11, 530-539. Reprinted in Friedlander and Topper, 1961, pp. 162-174.
Kashyap, R. L.
(1980a). Inconsistency of the AIC rule for estimating the order of autoregressive models. *IEEE Trans. Autom. Control*, **25**, No. 5, 996-998.

(1980b). Consistent estimation of the order of autoregressive models. In *Proc. 19th IEEE Conf. Decis. and Contr. incl. Symp. Adapt. Process., Albuquerque, N.M., 1980, Vol. 2*, pp. 986-987. IEEE Press, New York.

Kaveh, M.
(1979). High resolution spectral estimation for noisy signals. *IEEE Trans. Acoust., Speech, Signal Process.*, 27, No. 3, 286-287.
(1983). Statistical efficiency of correlation-based methods for ARMA spectral estimation. *IEE Proc., Part F*, 130, No. 3, 211-217.

Kawata, T.
(1965). Fourier analysis of nonstationary stochastic processes. *Trans. Amer. Math. Soc.*, 118, No. 6, 276-302.
(1973). Nonstationary stochastic processes. *Keio Math. Sem. Repts.*, No. 1 (Rep. Sem. "Stoch. Processes", Keio Univ., Yokohama), 1-23.

Kay, S. M.
(1980). A new ARMA spectral estimator. *IEEE Trans. Acoust., Speech, Signal Process.*, 28, No. 5, 585-588.

Kay, S. M., and S. L. Marple, Jr.
(1981). Spectral analysis — A modern perspective. *Proc. IEEE*, 69, No. 11, 1380-1419; erratum, *ibid.*, 1982, 70, No. 10, 120; comments, *ibid.*, 1983, 71, No. 6, 776-779 and No. 11, 1324-1325.

Kayatskas, A. A.
(1968). Periodically correlated random processes. *Radiotekhnika* (Radio Engineering, Moscow), 23, No. 9, 91-97.

Kedem, B.
(1980). *Binary Time Series*. M. Dekker, New York.
(1984). On the sinusoidal limit of stationary time series. *Ann. Stat.*, 12, No. 2, 665-674.

Kendall, M. G.
(1944). On autoregressive time series. *Biometrika*, 33, 105-122.
(1946). *Contributions to the Study of Oscillatory Time Series*. Cambridge University Press, Cambridge.

Kendall, M. G., and A. Stuart
(1966). *The Advanced Theory of Statistics. Vol. 2. Inference and Relationship*. Griffin, London.
(1968). *The Advanced Theory of Statistics. Vol. 3. Design and Analysis, and Time-Series*. Griffin, London.

Kendall, W. B.
(1974). A new algorithm for computing correlations. *IEEE Trans. Comp.*, 23, No. 1, 88-90.

Kesler, S. B., and S. Haykin
(1978). The maximum entropy method applied to the spectral analysis of radar clutter. *IEEE Trans. Inform. Theory*, 24, No. 2, 269-272.

Khachaturov, A. A.
(1954). Determination of the values of a measure for a region in n-dimensional Euclidean space from its values for all half-spaces. *Uspekhi Mat. Nauk* (Adv. Math. Sci.), 9, No. 3, 205-212.

Khachaturova, T. V.

(1972). Spectral estimates for stationary random processes. *Zapiski Nauchn. Sem. LOMI* (Trans. Res. Seminars, Leningrad Branch Math. Inst. Acad. Sci. USSR), **29**, 42-50.

Khalidov, I. A.

(1978). Some problems from the theory of correlation functions. *Vestnik LGU* (Leningrad State Univ. Herald), No. 13, 63-68.

Khalilov, A.

(1982). A central limit theorem for continuous time random fields. *Teor. Veroyatn. i Mat. Stat.* (Probab. Theory and Math. Stat., Kiev), No. 26, 151-155.

Kharkevich, A. A.

(1957). Spectrum evaluation for random processes. *Radiotekhnika* (Radio Engineering, Moscow), **12**, No. 5, 5-11.

(1960). *Spectra and Analysis.* Consultants Bureau, New York Enterprises Inc., New York.

Kharybin, A. E.

(1957). Analysis of errors in determination of the mean value and mean square of a random variable due to finiteness of the observation time. *Avtomat. i Telemekh.* (Automatics and Remote Control), No. 4, 304-314.

Khavkin, V. P., and A. I. Grinberg

(1970). On the errors in the experimental determination of the relay correlation functions. *Zavodsk. Laborat.* (Factory Laboratory), **36**, No. 10, 1236-1238.

Khinchin, A. Ya.

(1932). Zur Birkhoffs Lösung des Ergodenproblems. *Math. Ann.*, **107**, 485-488.

(1933). Ueber stationäre Reihen zufälliger Variabeln. *Matem. Sbornik* (Math. Collect., Moscow), **40**, 124-128.

(1934). Korrelationstheorie der stationären stochastischen Prozesse. *Math. Ann.*, **109**, 604-615.

(1949). *Mathematical Foundations of Statistical Mechanics.* Dover, New York.

Kholmyansky, M. Z.

(1972). Measurements of the microturbulent fluctuations of wind velocity derivative in the atmospheric surface layer. *Izv. Akad. Nauk SSSR, Ser. Fiz. Atmosf. i Okeana* (Bull. Acad. Sci. USSR, Ser. Atmosph. and Oceanic Phys.), **9**, No. 8, 812-828.

Khomyakov, V. A.

(1968). On a method for determination of the correlation function of a random process. *Avtomat. i Telemekh.* (Automatics and Remote Control), No. 6, 75-78.

Khudyakov, G. I.

(1974). The spectral-correlation method of random function representations. *Radiotekhn. i Electr.* (Radio Eng. and Electronics), **19**, No. 4, 715-720.

Khurgin, Ya. I., and V. P. Yakovlev
 (1971). *Finite Functions in Physics and Engineering*. Nauka, Moscow.
Khusu, A. P., Yu. P. Witenberg, and V. A. Pal'mov
 (1975). *Surface Roughness. A Probabilistic Approach*. Nauka, Moscow.
King, L. J.
 (1969). *Statistical Analysis in Geography*. Prentice-Hall, Englewood
 Cliffs, N. J.
Kinkel, J. F., J. Perl, L. Scharf, and A. Stubberub
 (1979). A note on covariance-invariant digital filter design and
 autoregressive moving average spectral estimation. *IEEE Trans.
 Acoust., Speech, Signal Process.*, 27, No. 2, 200-202.
Kiriyanov, K. G.
 (1971). Determination of auto- and cross-correlation functions from
 the dependence of the conditional mean value on time shift.
 Radiotekhn. i Elektr. (Radio Eng. and Electronics), 16, No. 4,
 617-620.
Kisiel, C. C.
 (1969). Time series analysis of hydrologic data. *Advances in
 Hydrosci.*, 5, 1-119.
Klesov, O. I.
 (1981). The strong law of large numbers for homogeneous random
 fields. *Teor. Veroyatn. i Mat. Stat.* (Probab. Theory and Math. Stat.
 Kiev), No. 25, 29-40.
Knowles, J. B., and H. T. Tsui
 (1967). Correlation devices and their estimation errors. *J. Appl.
 Phys.*, 38, No. 2, 607-612.
Koh, T., and E. J. Powers
 (1985). Efficient methods to estimate correlation functions of
 Gaussian processes and their performance analysis. *IEEE Trans.
 Acoust., Speech, Signal Process.*, 33, No. 4, 1032-1035.
Kolmogorov, A. N.
 (1931). Ueber die analytischen Methoden in der Wahrscheinlichkeitsrechnung. *Math. Ann.*, 104, 415-458.
 (1935). La transformation de Laplace dans les espaces lineaires. *C.R.
 Acad. Sci.*, 200, 1717-1718.
 (1940a). Curves in Hilbert space which are invariant with respect to
 a one-parameter group of motions. *Dokl. Akad. Nauk SSSR* (Repts.
 Acad. Sci. USSR), 26, No. 1, 6-9.
 (1940b). Wiener's spiral and some other interesting curves in Hilbert
 space. *Dokl. Akad. Nauk SSSR*, 26, No. 2, 115-118.
 (1941a). Stationary sequences in Hilbert space. *Bull. MGU, Ser. Mat.*
 (Bull. Moscow State Univ., Ser. Math.), 2, No. 6, 1-40. English
 translation in Kailath, 1977, pp. 66-89.
 (1941b). Local structure of turbulence in an incompressible fluid at
 very high Reynolds numbers. *Dokl. Akad. Nauk SSSR*, 30, No. 4,
 299-303. English translation in Friedlander and Topper, 1961, pp.
 151-155.

(1941c). Energy dissipation in locally isotropic turbulence. *Dokl. Akad. Nauk SSSR*, **32**, No. 1, 19-21. English translation in Friedlander and Topper, 1961, pp. 159-161.

(1947). Statistical theory of oscillations with a continuous spectrum. In *Jubilee Collection, Part 1*, pp. 242-254. Izd. Akad. Nauk SSSR (Acad. Sci. USSR Press), Moscow.

(1956). *Foundations of the Theory of Probability.* 2nd English edition. Chelsea, New York. (The book was first published in German in 1933.)

Konyaev, K. V.

(1973). *Spectral Analysis of Random Processes and Fields.* Nauka, Moscow.

(1981). *Spectral Analysis of Random Oceanological Fields.* Gidrometeoizdat, Leningrad.

Koopmans, L. H.

(1974). *The Spectral Analysis of Time Series.* Academic Press, New York.

Kopilovich, L. E.

(1966). On the theory of the polarity-coincidence correlator. *Izv. VUZov, Ser. Radiotekhn.* (Bull. Colleges and Univ. USSR, Ser. Radio Eng.), **9**, No. 6, 719-725.

Korn, G. A.

(1966). *Random-Process Simulation and Measurements.* McGraw-Hill, New York.

Kosambi, D. D.

(1943). Statistics in function space. *J. Indian Math. Soc.*, **7**, No. 3, 76-88.

Koslov, J. W., and R. H. Jones

(1985). A unified approach to confidence bounds for the autoregressive spectral estimator. *J. Time Series Anal.*, **6**, No. 3, 141-151.

Kosyakin, A. A., and G. F. Filaretov

(1972). Theory of relay correlator. In *Proc. 5th USSR Symp. on Repres. Methods and Instr. Analysis of Rand. Process. and Fields, Sec. 3*, pp. 144-150. USSR Res. Inst. Electr. Measur. Instr., Leningrad.

Kotel'nikov, V. A.

(1933). On the capacity of the "either" and wire in radio communication. In *Materials for the 1st USSR Congr. Techn. Reconstruction in Communication Eng. and Developm. Weak Current Industry*, 19pp. USSR Energy Committee, Moscow.

Kovasznay, L. S. G., M. S. Uberoi, and S. Corrsin

(1949). The transformation between one- and three-dimensional power spectra in isotropic scalar fluctuation field. *Phys. Rev.*, **76**, No. 8, 1263-1264.

Kozhevnikov, V. A., and R. M. Meshcherskii

(1963). *Modern Methods of Electroencephalogram Analysis.* Medgiz, Moscow.

Kozhevnikova, I. A.

(1979). A comparison of nonparametric and parametric spectral density estimates. In *Stochastic Control Systems* (A. A. Medvedev, ed.), pp. 92-98, Nauka, Novosibirsk.

Kozulyaev, P. A.
(1941). On extrapolation and interpolation problems for stationary sequences. *Doklady Akad. Nauk SSSR* (Repts. Acad. Sci. USSR), 30, No. 1, 13-17.

Krein, M. G.
(1943). On representations of functions by Fourier-Stieltjes integrals. *Uchen. Zap. Kuibyshev Pedagog. Inst.* (Sci. Reports of Kuibyshev Teachers College), 7, 123-148.
(1944). On the problem of continuing helical arcs in Hilbert space. *Doklady Akad. Nauk SSSR*, **45**, No. 4, 147-150.
(1945). On an extrapolation problem of A. N. Kolmogorov. *Doklady Akad. Nauk SSSR*, **46**, 339-342.
(1949-50). Hermitian-positive kernels on homogeneous spaces. I and II. *Ukrain. Mat. Zhurn.* (Ukrain. Math. J.), 1, No. 4, 64-98; 2, No. 1, 10-59.

Kritsky, S. N., and M. F. Menkel
(1964). On level oscillations for closed lakes. *Meteorologiya i Gidrologiya* (Meteorology and Hydrology), No. 7, 32-36.

Kromer, R. E.
(1969). *Asymptotic Properties of the Autoregressive Spectral Estimator.* Tech. Rep. No. 13, Stanford Univ., Stanford, California.

Krylov, Yu. M.
(1966). *Spectral Methods in Sea Wave Investigations and Computations.* Gidrometeoizdat, Leningrad.

Kubo, R.
(1954). Note on the stochastic theory of resonance absorption. *J. Phys. Soc. Japan*, 9, No. 6, 935-944.

Kurochkin, S. S.
(1972). *Multichannel Computing Systems and Correlators.* Energiya, Moscow.

Kutin, B. N.
(1957). Calculation of the correlation function of a stationary random process from experimental data. *Avtomatika i Telemekh.* (Automatics and Remote Control), **18**, No. 3, 201-222.

Kutzbach, J. E.
(1967). Empirical eigenvectors of sea-level pressure, surface temperature, and precipitation complexes over North America. *J. Appl. Meteor.*, **6**, No. 5, 791-802.

Lambert, P. J., and D. S. Poskitt
(1983). *Stationary Processes in Time Series Analysis: The Mathematical Foundations.* Vandenhoeck and Ruprecht, Göttingen.

Lamperti, J.
(1962). Semi-stable processes. *Trans. Amer. Math. Soc.*, **104**, 62-78.
(1977). *Stochastic Processes. A Survey of the Mathematical Theory.* Springer, New York.

Landau, H. J., and H. O. Pollak
(1961). Prolate spheroidal wave functions, Fourier analysis and uncertainty — II. *Bell System Tech. J.*, **40**, No. 1, 65-84.

(1962). Prolate spheroidal wave functions, Fourier analysis and uncertainty — III. *Bell System Tech. J.*, **41**, No. 4, 1295-1336.

Lang, S. W., and J. H. McClellan
(1982). Multidimensional MEM spectral estimation. *IEEE Trans. Acoust., Speech, Signal Process.*, **30**, No. 6, 880-887.

Lange, F. H.
(1959). *Korrelationselektronik*. Verlag Technik, Berlin. English translation was published under the title *Correlation Techniques* by Iliffe Books, London, 1967.

Langevin, P.
(1908). Sur la theorie du mouvement brownien. *C.R. Acad. Sci.*, **146**, 530-533.

Laning, J. H., Jr., and R. H. Battin
(1956). *Random Processes in Automatic Control*. McGraw-Hill, New York.

Larimore, W. E.
(1977). Statistical inference on stationary random fields. *Proc. IEEE*, **65**, No. 6, 961-970.

Larsen, H.
(1976). Comments on "A fast algorithm for the estimation of autocorrelation functions". *IEEE Trans. Acoust., Speech, Signal Process.*, **24**, No. 5, 432-434.

Lawson, J. L., and G. E. Uhlenbeck
(1950). *Threshold Signals*. McGraw-Hill, New York.

LeCam, L.
(1947). Un instrument d'étude des fonctions aléatoires: la fonctionnelle caractéristique. *C.R. Acad. Sci.*, **224**, 710-711.

Leith, C. E.
(1973). The standard error of time averaged estimates of climatic means. *J. Appl. Meteor.*, **12**, No. 6, 1066-1069.

Leonenko, N. N., and A. V. Ivanov
(1986). *Statistical Analysis of Random Fields*. Izd. pri Kievskom Univ. Obyedin. Vishcha Shkola (Kiev Univ. Division of Vishcha Shkola Press), Kiev.

Leonov, V. P.
(1960). Applications of the characteristic functional and cumulants to ergodic theory of stationary processes. *Doklady Akad. Nauk SSSR* (Repts. Acad. Sci. USSR), **133**, No. 3, 523-526.
(1961). On the variance of time average of a stationary stochastic process. *Teor. Veroyatn. i Primen.* (Probab. Theory and Its Appl.), **6**, No. 1, 93-101.
(1964). *Some Applications of Higher Cumulants to the Theory of Stationary Stochastic Processes*. Nauka, Moscow.

Levin, B. R.
(1974,1975,1976). *Theoretical Foundations of Statistical Radioengineering. Parts 1-3*. Soviet Radio, Moscow.

Levin, V. A.
(1986). On window choice in problems of digital signal processing. *Radiotekhn. i Elektr.* (Radio Eng. and Electronics), **31**, No. 3, 515-525.

Levinson, N.
  (1947). The Wiener RMS (root mean square) error criterion in filter design and prediction. *J. Math. Phys.*, 25, No. 4, 261-278. Reprinted in N. Wiener, 1949, pp. 129-148.
Lévy, P.
  (1925). *Calcul des probabilités.* Gauthier-Villars, Paris.
  (1953). Random functions: General theory with special reference to Laplacian random functions. *Univ. Calif. Publ. Stat.*, 1, 331-390.
  (1965). *Processus stochastiques et mouvement brownien.* Gauthier-Villars, Paris.
Lifermann, J.
  (1977). *Theorie et applications de la transformation de Fourier rapide.* Masson, Paris.
Lighthill, M. J.
  (1968). *Fourier Analysis and Generalized Functions.* Cambridge University Press, Cambridge.
Lim, J. S., and F. V. Dowla
  (1983). A new algorithm for high-resolution two-dimensional spectral estimation. *Proc. IEEE*, 71, No. 2, 284-285.
Lindgren, C.
  (1974). Spectral moment estimation by means of level crossings. *Biometrika*, 61, No. 3, 401-418.
Livshits, N. A., and V. N. Pugachev
  (1963). *Probabilistic Analysis of Automatic Control Systems. Vols. 1 and 2.* Sovjet Radio, Moscow.
Lloyd, S. P.
  (1959). A sampling theorem for stationary (wide sense) stochastic processes. *Trans. Amer. Math. Soc.*, 92, No. 1, 1-12.
Loève, M.
  (1945-46). Sur le fonctions aléatoire de second ordre. *Rev. Sci.*, 83, 297-303; 84, 195-206.
  (1963). *Probability Theory*, 3rd edition. Van Nostrand, Princeton, N. J. The 4th edition was published in two volumes by Springer-Verlag, New York, 1977, 1978.
Lomnicki, Z. A., and S. K. Zaremba
  (1955). Some applications of zero-one process. *J. Roy. Stat. Soc., Ser. B*, 17, No. 2, 243-255.
Longuet-Higgins, M. S.
  (1957a). The statistical analysis of a randomly moving surface. *Phil. Trans. Roy. Soc.*, A249, 321-387.
  (1957b). Statistical properties of an isotropic random surface. *Phil. Trans. Roy. Soc.*, A250, 157-174.
Lopresti, P. V., and H. L. Suri
  (1974). A fast algorithm for the estimation of autocorrelation functions. *IEEE Trans. Acoustics, Speech, Signal Process.*, 22, No. 6, 449-453.
Lord, R. D.
  (1954). The use of the Hankel transform in statistics. I. *Biometrika*, 41, Nos. 1-2, 44-55.

Lorenz, E. N.
(1956). *Empirical Orthogonal Functions and Statistical Weather Prediction.* Sci. Rep. No. 1, Stat. Forecasting Proj., Dept. Meteorology, Mass. Inst. Technol. (MIT), Cambridge, Mass., 49 pp.

Loynes, R. M.
(1968). On the concept of the spectrum for non-stationary processes. *J. Roy. Stat. Soc., Ser. B,* **30,** No. 1, 1-30.

Lukacs, E.
(1970). *Characteristic Functions.* Griffin, London.

Lumley, J. L.
(1967). The structure of inhomogeneous turbulent flows. In *Atmospheric Turbulence and Radio Wave Propagation* (A. M. Yaglom and V. I. Tatarskii, eds.), pp. 166-178. Nauka, Moscow.
(1970). *Stochastic Tools in Turbulence.* Academic Press, New York.
(1972). Application of central limit theorems to turbulence problems. In *Statistical Models and Turbulence,* ed. by M. Rosenblatt and C. Van Atta (*Lecture Notes in Phys.,* 12), pp. 1-26, Springer, Berlin.

Lumley, J. L., and H. A. Panofsky
(1964). *The Structure of Atmospheric Turbulence.* Interscience Publ., New York.

Lütkepohl, H.
(1985). Comparison of criteria for estimating the order of a vector autoregressive process. *J. Time Series Anal.,* **6,** No. 1, 35-52.

McClellan, J. H.
(1982). Multidimensional spectral estimation. *Proc. IEEE,* **70,** No. 9, 1029-1039.

McClellan, J. H., and S. W. Lang
(1983). Duality for multidimensional MEM spectral analysis. *IEEE Proc., Part F,* **130,** No. 3, 230-235.

McCombie, C. W.
(1953). Fluctuation theory in physical instruments. *Rep. Progress Phys.,* **16,** 266-320.

MacDonald, D. K. C.
(1962). *Noise and Fluctuations: An Introduction.* Wiley, New York.

McFadden, J. A.
(1965). An alternate proof of Nuttall's theorem on output cross-covariances. *IEEE Trans. Inform. Theory,* **11,** No. 2, 306-307.

McKean, H. P.
(1969). *Stochastic Integrals.* Academic Press, New York.

McNeil, D. R.
(1967). Estimating the covariance and spectral density functions from a clipped stationary time series. *J. Roy. Stat. Soc., Ser. B,* **29,** No. 1, 180-195.
(1972). Representations for homogeneous processes. *Z. Wahrscheinlichkeitstheorie verw. Geb.,* **22,** No. 4, 333-339.

Malakhov, A. N.
(1968). *Fluctuations in Self-Oscillating Systems.* Nauka, Moscow.

Malevich, T. L.
(1965). Some properties of spectral estimates of a stationary process. *Teor. Veroyatn. i Prim.* (Probab. Theory and Its Appl.), **10,** No. 3, 500-509.

Malik, N. A., and J. S. Lim
(1982). Properties of two-dimensional maximum entropy power spectrum estimates. *IEEE Trans. Acoust., Speech, Signal Process.*, 30, No. 5, 788-798.
Malinvaud, E.
(1969). *Méthodes statistiques de l'économétrie.* Dunod, Paris.
Malyarenko, A. A.
(1985a). Spectral representation of multidimensional homogeneous and isotropic random fields. *Doklady Akad. Nauk Ukrain. SSR* (Repts. Acad. Sci. Ukrain. SSR), *Ser A, Phys.-Math. and Eng. Sci.*, No. 7, 20-22.
(1985b). Spectral representation of general multidimensional homogeneous random fields which are isotropic with respect to some variables. *Teor. Veroyatn. i Mat. Stat.* (Probab. Theory and Math. Stat., Kiev), No. 32, 66-72.
Mandelbrot, B. B.
(1977). *Fractals: Forms, Chance, and Dimension.* Freeman, San Francisco.
(1982). *The Fractal Geometry of Nature.* Freeman, San Francisco.
Mandelbrot, B. B., and J. W. Van Ness
(1968). Fractional Brownian motions, fractional noises and applications. *SIAM Rev.*, 10, No. 4, 422-437.
Mandrekar, V.
(1972). A characterization of oscillatory processes and their prediction. *Proc. Amer. Math. Soc.*, 32, No. 1, 280-284.
Mann, H. B., and A. Wald
(1943). On the statistical treatment of linear stochastic difference equations. *Econometrica*, 11, 173-220.
Marple, S. L., Jr.
(1980). A new autoregressive spectrum analysis algorithm. *IEEE Trans. Acoust., Speech, Signal Process.*, 28, No. 4, 441-454.
Marple, S. L., Jr., and A. H. Nuttall
(1983). Experimental comparison of three multichannel linear prediction spectral estimators. *IEEE Proc., Part F*, 130, No. 3, 218-229.
Martinelli, G., G. Orlandi, and P. Burraccano
(1986). Pole identification of ARMA processes by extended Levinson recursion. *IEEE Trans. Circuits Syst.*, 33, No. 3, 348-350.
Maruyama, G.
(1949). The harmonic analysis of stationary stochastic processes. *Mem. Fac. Sci. Kyūsyū Univ., Ser. A*, 4, No. 1, 45-106.
Masani, P.
(1972). On helixes in Hilbert space. I. *Teor. Veroyatn. i Prim.* (Probab. Theory and Its Appl.), 17, No. 1, 3-20.
(1976). On helixes in Banach spaces. *Sankhyā*, A38, No. 1, 1-27.
Masry, E.
(1972). On covariance function of unit processes. *SIAM J. Appl. Math.*, 23, No. 1, 28-33.

Matérn, B.
(1960). Spatial variation. *Medd. Statens Skogsforskn. Inst.*, **49**, No. 5, 3-144.

Matheron, G.
(1970). Structure aléatoires et géologie mathématiques. *Intern. Stat. Rev.*, **38**, 1-11.

Max, J.
(1981). *Méthodes et techniques de traitement du signal et applications aux mesures physiques. Tomes 1 et 2.* Masson, Paris.

Meshcherskaya, A. V., L. V. Ruchovets, M. I. Yudin, and N. I. Yakovleva
(1970). *Natural Components of Meteorological Fields.* Gidrometeoizdat, Leningrad.

Miamee, A. G., and H. Salehi
(1978). Harmonizability, V-boundedness and stationary dilation of stochastic processes. *Indiana Univ. Math. J.*, **27**, No. 1, 37-50.

Middleton, D.
(1960). *Introduction to Statistical Communication Theory.* McGraw-Hill, New York.

Milatz, J. M. W., and J. J. van Zolingen
(1953). The Brownian motion of electrometers. *Physica*, **19**, No. 3, 181-194.

Mirsky, G. Ya.
(1972). *Instrumental Determination of Random Process Characteristics.* 2nd edition. Energiya, Moscow.
(1978). Measurement of the normalized correlation function by the method which uses a conditional mean value of the sign function. In *Proc. 8th USSR Symp. on Represent. Methods and Instr. Anal. Random Process. and Fields*, Sec. 2, pp. 10-15. USSR Res. Inst. Electr. Measur. Instr., Leningrad.

Monin, A. S.
(1963). Stationary and periodic time series in the general circulation of the atmosphere. In *Time Series Analysis* (M. Rosenblatt, ed.), pp. 144-151, Wiley, New York.

Monin, A. S., and R. V. Ozmidov.
(1985). *Turbulence in the Ocean.* Reidel, Dordrecht.

Monin, A. S., and A. M. Yaglom
(1971,1975). *Statistical Fluid Mechanics, Vols. 1 and 2.* The MIT Press, Cambridge, Mass.

Moran, P. A. P.
(1949). The spectral theory of discrete stochastic processes. *Biometrika*, **36**, Nos. 1-2, 63-70.
(1950). The oscillatory behaviour of moving averages. *Proc. Cambr. Phil. Soc.*, **46**, No. 2, 272-280.
(1975). The estimation of standard errors in Monte-Carlo simulation experiments. *Biometrika*, **62**, No. 1, 1-4.

Morf, M., A. Vieira, D. T. L. Lee, and T. Kailath
(1978). Recursive multichannel maximum entropy spectral estimation. *IEEE Trans. Geosci. Electron.*, **16**, No. 1, 85-94. Reprinted in Childers, 1978, pp. 172-181.
Morris, M. D., and S. F. Ebey
(1984). An interesting property of the sample mean under a first-order autoregressive model. *Amer. Stat.*, **38**, No. 2, 127-129.
Moyal, J. E.
(1952). The spectra of turbulence in a compressible fluid: Eddy turbulence and random noise. *Proc. Cambr. Phil. Soc.*, **48**, No. 2, 329-344.
Munroe, M. E.
(1953). *Introduction to Measure and Integration.* Addison-Wesley, Cambrige, Mass.
Nachbin, L.
(1965). *The Haar Integral.* Van Nostrand, Princeton, N. J.
Nagabhushanam, K.
(1969). Harmonizable processes and mean ergodic theorem. *Sankhyā*, *Ser. A*, **31**, No. 4, 413-420.
Nagabhushanam, K., and C. S. K. Bhagavan
(1968). Non-stationary processes and spectrum. *Canad. J. Math.*, **20**, 1203-1206.
(1969). A mean ergodic theorem for a class of non-stationary processes. *Sankhyā*, *Ser. A*, **31**, No. 4, 421-424.
Nahapetian, B. S.
(1980). The central limit theorem for random fields with mixing property. In *Multicomponent Random Systems* (R. L. Dobrushin and Ya. G. Sinai, eds.), pp. 531-542, M. Dekker, New York.
Naimark, M. A.
(1959). *Normed Rings.* P. Noordhoff, Groningen.
Nash, R. A., Jr., and S. K. Jordan
(1978). Statistical geodesy - An engineering perspective. *Proc. IEEE*, **66**, No. 5, 532-550.
Neaderhouser, C. C.
(1978). Limit theorems for multiply indexed random variables, with application to Gibbs random fields. *Ann. Probab.*, **6**, No. 2, 207-215.
Neave, H. R.
(1970a). Spectral analysis of a stationary time series using initially scarce data. *Biometrika*, **57**, 111-122.
(1970b). An improved formula for the asymptotic variance of spectrum estimates. *Ann. Math. Stat.*, **41**, No. 1, 70-77.
(1971). The exact error in spectrum estimates. *Ann. Math. Stat.*, **42**, No. 3, 961-975.
(1972). A comparison of lag window generators. *J. Amer. Stat. Assoc.*, **67**, No. 337, 152-158.
Nelson, E.
(1967). *Dynamical Theories of Brownian Motion.* Princeton University Press, Princeton, N. J.

Nemirovsky, A. S.
(1964). *Probabilistic Methods in Measurement Technique. Measurements of Stationary Random Processes.* Izdat. Standartov (Standardization Publ. House), Moscow.
Neumann, J. von
(1932a). Proof of the quasi-ergodic hypothesis. *Proc. Nat. Acad. Sci. USA*, **18**, No. 1, 70-82.
(1932b). Zur Operatorenmethode in der klassischen Mechanik. *Ann. Math.*, **33**, No. 4, 587-642.
Neumann, J. von, and I. J. Schoenberg
(1941). Fourier integrals and metric geometry. *Trans. Amer. Math. Soc.*, **50**, No. 2, 226-251.
Newton, H. J., and M. Pagano
(1984). Simultaneous confidence bound for autoregressive spectra. *Biometrika*, **71**, No. 1, 197-202.
Nidekker, I. G.
(1968). Problems on accuracy increase in computing the spectral density for a random process. *Soobshch. po Vychisl. Mat.* (Repts. on Comput. Math.), No. 3, 1-24, Comput. Center Acad. Sci. USSR, Moscow.
Nikias, Ch. L., and M. R. Raghuveer
(1985). Multi-dimensional parametric spectral estimation. *Signal Process.*, **9**, No. 3, 191-205.
North, G. R., Th. L. Bell, R. F. Cahalan, and F. J. Moeng
(1982). Sampling errors in the estimation of empirical orthogonal functions. *Monthly Weather Rev.*, **110**, No. 7, 699-706.
Novikov, E. A.
(1959). On the predictability of synoptic processes. *Izv. Akad. Nauk SSSR, Ser. Geofiz.* (Bull. Acad. Sci. USSR, ser. Geophys.), No. 11, 1721-1724.
(1964). Functionals and random force method in turbulence theory. *Zhurn. Exper. i Teor. Fiz.* (J. Exper. Theor. Phys.), **47**, No. 5(11), 1919-1926.
Nussbaumer, H. J.
(1982). *Fast Fourier Transform and Convolution Algorithms.* 2nd corrected and updated edition. Springer, Berlin.
Nuttall, A. H.
(1958). *Theory and Application of the Separable Class of Random Processes.* Techn. Rep. No. 343, Res. Lab. Elec., Mass. Inst. Technology, Cambridge, Mass.
(1981). Some windows with very good sidelobe behavior. *IEEE Trans. Acoust., Speech, Signal Process.*, **29**, No. 1, 84-91.
Nuttall, A. H., and G. C. Carter
(1980). A generalized framework for power spectral estimation. *IEEE Trans. Acoust., Speech, Signal Process.*, **28**, No. 3, 334-335.
(1982). Spectral estimation using combined time and lag weighting. *Proc. IEEE*, **70**, No. 9, 1115-1125.
Oberhettinger, F.
(1973). *Fourier Transforms of Distributions and Their Inverses. A Collection of Tables.* Academic Press, New York.

Obukhov, A. M.
(1947). Statistical homogeneous random fields on a sphere. *Uspekhi Mat. Nauk* (Adv. Math. Sci.), 2, No. 2, 196-198.
(1949a). Structure of the temperature field in a turbulent flow. *Izv. Akad. Nauk SSSR, Ser. Geogr. i Geofiz.* (Bull. Acad. Sci. USSR, Ser. Geogr. and Geophys.), 13, No. 1, 58-69.
(1949b). Local structure of atmospheric turbulence. *Doklady Akad. Nauk SSSR* (Repts. Acad. Sci. USSR), 67, No. 4, 643-646.
(1954a). Probabilistic description of continuous fields. *Ukrain. Mat. Zhurn.* (Ukrain. Math. J.), 6, No. 1, 37-42.
(1954b). Statistical description of continuous fields. *Trudy Geofiz. Inst. Akad. Nauk SSSR* (Proc. Geophys. Inst. Acad. Sci. USSR), No. 24(151), 3-42. German translation in Goering, 1958, pp. 1-42. English translation by Liaison Office, Technical Information Center, Wright-Patterson AFB F-TS-9295/v.
(1960). Statistically orthogonal expansion of empirical functions. *Izv. Akad. Nauk SSSR, Ser. Geofiz.*, No. 3, 432-439.
Ogura, H.
(1966). Polar spectral representation of a homogeneous and isotropic random field. *J. Phys. Soc. Japan*, 21, No. 1, 154-166.
(1968). Spectral representation of vector random field. *J. Phys. Soc. Japan*, 24, No. 6, 1370-1380.
(1971). Spectral representation of periodic non-stationary random processes. *IEEE Trans. Inform. Theory*, 17, No. 2, 143-149.
Ogura, Y.
(1952). The structure of two-dimensional isotropic turbulence. *J. Meteor. Soc. Japan*, 30, No. 2, 59-64.
Olberg, M.
(1972). Bemerkungen zur Verwendung der effektiven Anzahl unabhängiger Stichprobenwerte bei statistischen Testverfahren in der Meteorlogie und Geophysik. *Gerlands Beitr. Geophys.*, 81, No. 1/2, 57-64.
Olshen, R. A.
(1967). Asymptotic properties of the periodogram of a discrete stationary process. *J. Appl. Probab.*, 4, No. 3, 508-528.
Olson, H. F.
(1958). *Dynamical Analogies.* Van Nostrand, Princeton, N. J.
Onoyama, T.
(1959). Note on random distributions. *Mem. Faculty Sci. Kyushu Univ.*, Ser. A, 13, No. 2, 208-213.
Oppenheim, A. V., ed.
(1969). *Papers on Digital Signal Processing.* MIT Press, Cambridge, Mass.
Oppenheim, A. V., and R. W. Schafer
(1975). *Digital Signal Processing.* Prentice-Hall, Englewood Cliffs, N. J.
Ortiz, J., and A. Ruiz de Elvira
(1985). A cyclo-stationary model of sea surface temperatures in the Pacific Ocean. *Tellus*, A37, No. 1, 14-23.

Otnes, R. K., and L. Enochson
(1972). *Digital Time Series Analysis.* Wiley, New York.
(1978). *Applied Time Series Analysis. Vol. 1. Basic Techniques.* Wiley, New York.
Owens, A. J.
(1978). On detrending and smoothing random data. *J. Geophys. Res., Ser. A,* **83,** No. 1, 221-224.
Ozaki, T.
(1977). On the order determination of ARIMA models. *Appl. Stat.,* **26,** No. 3, 290-301.
Pagano, M.
(1971). Some asymptotic properties of a two-dimensional periodogram. *J. Appl. Probab.,* **8,** No. 4, 841-847.
Palenskis, V., J. Laučidus, and Z. Šoblickas
(1985). The choice of lag windows when the spectrum of a random process is calculated. *Litovsk. Fiz. Sbornik* (Lithuan. Phys. Coll.), **25,** No. 3, 98-104.
Paley, R. E. A. C., and N. Wiener
(1934). *Fourier Transforms in the Complex Domain.* Amer. Math. Soc., Providence, R. I.
Palmer, D. F.
(1969). Bias criteria for the selection of spectral windows. *IEEE Trans. Inform. Theory,* **15,** No. 5, 613-615.
Panofsky, H. A., and J. A. Dutton
(1984). *Atmospheric Turbulence.* Wiley, New York.
Papoulis, A.
(1965). *Probability, Random Variables, and Stochastic Processes.* McGraw-Hill, New York.
(1973a). Minimum-bias windows for high resolution spectral estimates. *IEEE Trans. Inform. Theory,* **19,** No. 1, 9-12.
(1973b). Spectral estimation of a certain class of non-stationary processes. *Trans. 6th Prague Conf. Inform. Theory, Stat. Decis. Funct., Random Process.,* 683-694.
(1981). Maximum entropy and spectral estimation: A review. *IEEE Trans. Acoust., Speech, Signal Process.,* **29,** No. 6, 1176-1186.
(1983a). Spectra of stochastic FM signals. *Trans. 9th Prague Conf. Inform. Theory, Stat. Decis. Funct., Rand. Process., Vol. B,* 105-111.
(1983b). Random modulation: A review. *IEEE Trans. Acoust., Speech, Signal Process.,* **31,** No. 1, 96-105.
(1984). *Probability, Random Variables, and Stochastic Processes.* Revised 2nd edition, McGraw-Hill, New York.
Parzen, E.
(1957a). On consistent estimates of the spectrum of a stationary time series. *Ann. Math. Stat.,* **28,** No. 2, 329-348. Reprinted in Parzen, 1967a, pp. 24-43.
(1957b). On choosing an estimate of the spectral density function of a stationary time series. *Ann. Math. Stat.,* **28,** No. 4, 921-932. Reprinted in Parzen, 1967a, pp. 44-55.
(1958a). Conditions that a stochastic process be ergodic. *Ann. Math. Stat.,* **29,** No. 1, 299-301. Reprinted in Parzen, 1967a, pp. 21-23.

(1958b). On asymptotically efficient consistent estimates of the spectral density function of a stationary time series. *J. Roy. Stat. Soc.*, *Ser. B*, 20, No. 2, 303-322. Reprinted in Parzen, 1967a, pp. 56-97.

(1959). Statistical inference on time series by Hilbert space methods, I. *Techn. Rep. No. 23*, Dept. Stat., Stanford Univ., Stanford, Calif. Reprinted in Parzen, 1967a, pp. 251-382.

(1960). *Modern Probability Theory and Its Applications*, Wiley, New York.

(1961). Mathematical considerations in the estimation of spectra. *Technometrics*, 3, No. 2, 167-190. Reprinted in Parzen, 1967a, pp. 132-161.

(1962a). *Stochastic Processes.* Holden-Day, San Francisco.

(1962b). Spectral analysis of asymptotically stationary time series. *Bull. Inst. Intern. Stat.*, 39, Livaison 2, 87-103. Reprinted in Parzen, 1967a, pp. 162-179.

(1963). Notes on Fourier analysis and spectral windows. *Techn. Rep. No. 48*, Dept. Stat., Stanford Univ., Stanford, Calif. Reprinted in Parzen, 1967a, pp. 190-250.

(1964). An approach to empirical time series analysis. *Radio Sci.*, 68, No. 9, 937-951. Reprinted in Parzen, 1967a, pp. 551-565.

(1967a). *Time Series Analysis Papers.* Holden-Day, San Francisco.

(1967b). The role of spectral analysis in time series analysis. *Review Int. Stat. Inst.*, 35, 125-141.

(1969). Multiple time series modeling. In *Multivariate Analysis, Vol. II* (P. R. Krishnaiah, ed.), pp. 389-409. Academic Press, New York.

(1972). Some recent advances in time series analysis. In *Statistical Models and Turbulence*, ed. by M. Rosenblatt and C. Van Atta (*Lect. Notes in Phys.*, 12), pp. 470-492. Springer, Berlin.

(1974). Some recent advances in time series modelling. *IEEE Trans. Automat. Contr.*, 19, No. 6, 723-729. Reprinted in Childers, 1978, pp. 226-233.

(1977). Multiple time series modelling: Determining the order of approximating autoregressive schemes. In *Multivariate Analysis, IV* (P. Krishnaiah, ed.), pp. 283-295. North-Holland, Amsterdam.

(1981). Modern empirical statistical spectral analysis. In *Underwater Acoustics and Signal Processing* (Proc. NATO Adv. Study Inst., Copenhagen, August 18-29, 1980, ed. by L. Bjørnø), pp. 471-497. Reidel, Dordrecht.

Peklenik, J.
(1967-68). New developments in surface characterization and measurements by means of random process analysis. *Proc. Inst. Mech. Engrs.*, 182, pt. 3K, 108-126.

Perera, W. A., and P. J. W. Rayner
(1986). Optimal design of discrete coefficient DFTs for spectral analysis. I-II. *IEE Proc., Part G*, 133, No. 1, 1-7; 8-18.

Perrin, J.
(1916). *Atoms.* Van Nostrand, Princeton, N. J.

Peschel, M.
(1961). Abschätzung des durch die endliche Mittelungszeit bedingten Fehlers der Korrelationsfunktion eines Poissonschen Sekundärprozesses. *Z. f. messen, steuern, regeln (Automatisierung)*, **4**, No. 11, 468-470.

Petersen, E. L., and S. E. Larsen
(1978). A statistical study of a composite isotropic paleotemperature series from the last 700,000 years. *Tellus*, **30**, No. 3, 193-200.

Pflug, G.
(1982). A statistically important Gaussian process. *Stochast. Process. Appl.*, **13**, No. 1, 45-57.

Phadke, M. S., and S. M. Wu
(1974). Modeling of continuous stochastic processes from discrete observations with application to sunspots data. *J. Amer. Stat. Assoc.*, **69**, No. 346, 325-329.

Phillips, O. M.
(1977). *The Dynamics of the Upper Ocean*. 2nd edition. Cambridge University Press, Cambridge.

Picinbono, B.
(1974). Properties and applications of stochastic processes with stationary nth-order increments. *Adv. Appl. Probab.*, **6**, No. 3, 512-523.

Pierson, W. J., and L. J. Tick
(1955). Stationary random processes in meteorology and oceanography. *Bull. Intern. Stat. Inst.*, **35**, 271-281.

Pinsker, M. S.
(1955). Theory of curves in a Hilbert space with stationary increments of order n. *Izv. Akad. Nauk SSSR, Ser. Mat.* (Bull. Acad. Sci. USSR. Ser. Math.), **19**, 319-344.

Pisarenko, V. F.
(1972). On the estimation of spectra by means of nonlinear functions of the covariance matrix. *Geophys. J. Roy. Astronom. Soc.*, **28**, 511-531.
(1973). The retrieval of harmonics from a covariance function. *Geophys. J. Roy. Astronom. Soc.*, **33**, 347-366.
(1977). Sampling properties of maximum entropy spectral estimates. *Vychisl. Seismol.* (Comput. Seismol., Moscow), No. 10, 118-149.

Pitt, H. R.
(1942). Some generalizations of the ergodic theorem. *Proc. Cambr. Phil. Soc.*, **38**, No. 4, 325-343.

Pólya, G.
(1949). Remarks on characteristic functions. In *Proc. (First) Berkeley Symp. Math. Stat. Probab.* (J. Neyman, ed.), pp. 115-123, University of California Press, Berkeley.

Polyak, I. I.
(1975). *Numerical Methods for Analysis of Observations*. Gidrometeoizdat, Leningrad.
(1979). *Methods for Analysis of Random Processes and Fields in the Climatology*. Gidrometeoizdat, Leningrad.

Popenko, N. V., and E. P. Tikhonov
(1972). Comparative analysis of the algorithms for computing correlation functions by the multiplicative method. In *Proc. 5th USSR Symp. on Repres. Methods and Instr. Anal. Random Process. and Fields, Sec. 3*, pp. 83-87. USSR Res. Inst. Electr. Measur. Inst., Leningrad.

Poskitt, D. S., and A. R. Tremayne
(1984). Model selection and diagnostic checking in univariate time series analysis. In *Time Series Analysis: Theory and Practice 5* (O. D. Anderson, ed.), pp. 19-38. North-Holland, Amsterdam.

Prabhu, K. M. M., and V. U. Reddy
(1980). Data windows in digital signal processing − A review. *J. Inst. Electr. Telecom. Eng.*, **26**, No. 1, 69-76.

Price, R.
(1958). A useful theorem for nonlinear devices having Gaussian inputs. *IRE Trans. Inform. Theory*, **4**, No. 2, 69-72.

Priestley, M. B.
(1962). Basic considerations in the estimation of spectra. *Technometrics*, **4**, No. 4, 551-564.
(1963). The spectrum of a continuous process derived from a discrete process. *Biometrika*, **50**, Nos. 3-4, 517-520.
(1965a). The role of bandwidth in spectral analysis. *Appl. Stat.*, **14**, No. 1, 33-47.
(1965b). Evolutionary spectra and non-stationary processes. *J. Roy. Stat. Soc., Ser. B*, **27**, No. 2, 204-237.
(1981). *Spectral Analysis and Time Series. Vols. 1 and 2.* Academic Press, London.

Priestley, M. B., and C. H. Gibson
(1965). Estimation of power spectra by a wave-analyser. *Technometrics*, **7**, No. 4, 553-559.

Privalov, I. I.
(1950). *Boundary Properties of Analytic Functions.* 2nd edition. Gos. Izdat. Tekhn.-Teoret. Liter. (GTTI), Moscow.

Privalsky, V. E.
(1973). Linear extrapolation of sea level variations using parametric model. *Izv. Akad. Nauk SSSR, Ser. Fiz. Atmosf. i Okeana* (Bull. Acad. Sci. USSR, Atmosph. and Oceanic Phys.), **9**, No. 9, 973-981.
(1976). Probability forecast of water balance elements and variations of the Caspian Sea level. *Meteorologiya i Gidrologiya* (Meteorology and Hydrology), No. 5, 70-76.
(1977). On the spectral density estimation for large-scale atmospheric processes. *Izv. Akad. Nauk SSSR, Ser. Fiz. Atmosf. i Okeana*, **12**, No. 9, 979-982.
(1985). *Climatic Variability. Stochastic Models, Predictability, Spectra.* Nauka, Moscow.

Prouza, L.
(1960). O spectru "vzorkovaného" stacionárního náhodného procesu. *Aplikace Matem.* (Prague), **5**, No. 5, 394-397.

Pugachev, V. S.
(1953). General correlation theory for random functions. *Izv. Akad. Nauk SSSR, Ser. Matem.* (Bull. Acad. Sci. USSR, Ser. Math.), 17, No. 5, 401-420.
(1965). *Theory of Random Functions and Its Application to Control Problems.* Pergamon Press, Oxford.
Pukkila, T. M., and P. R. Krishnaiah
(1986a). On the use of autoregressive order determination criteria in univariate white noise tests. *Tech. Rep. 86-15*, Center for Multivariate Analysis, Univ. of Pittsburgh, Pittsburgh, Pa.
(1986b). On the use of autoregressive order determination criteria in multivariate white noise tests. *Tech. Rep. 86-16*, Center for Multivariate Analysis, Univ. of Pittsburgh, Pittsburgh, Pa.
Rabiner, R. L., and B. Gold
(1975). *Theory and Application of Digital Signal Processing.* Prentice-Hall, Englewood Cliffs, N. J.
Rader, C. M.
(1968). Discrete Fourier transforms when the number of data samples is prime. *Proc. IEEE,* 56, No. 7, 1107-1108.
(1970). An improved algorithm for high speed autocorrelation with application to spectral estimation. *IEEE Trans. Audio and Electroacoust.,* 18, No. 4, 439-441.
Radoski, H. R., P. F. Fougere, and E. J. Zawalick
(1975). A comparison of power spectral estimates and applications of the maximum entropy method. *J. Geophys. Res.,* 80, No. 4, 619-625.
Raikov, D. A.
(1945). *Harmonic Analysis on Commutative Groups with Haar Measure.* Trudy Mat. Inst. Steklov (Works Steklov Math. Inst. Acad. Sci. USSR), No. 14, Izd. Akad. Nauk SSSR, Moscow.
Ramachandran, B.
(1967). *Advanced Theory of Characteristic Functions.* Statistical Publishing Co., Calcutta.
Rao, C. R.
(1973). *Linear Statistical Inference and Its Applications,* 2nd edition. Wiley, New York.
Rao, M. M.
(1978). Covariance analysis of nonstationary time series. In *Developments in Statistics* (P. R. Krishnaiah, ed.), *Vol. 1,* pp. 171-225. Academic Press, New York.
(1981). Representation of weakly harmonizable processes. *Proc. Nat. Acad. Sci. USA, Phys. Sci.,* 78, No. 9, 5288-5289.
(1982). Harmonizable processes: Structure theory. *L'Enseignment Math., Ser. II,* 28, Nos. 3-4, 295-351.
(1985). Harmonizable, Cramér, and Karhunen classes of processes. In *Handbook of Statistics, Vol. 5* (E. J. Hannan, P. R. Krishnaiah, and M. M. Rao, eds.), pp. 279-310. North-Holland, Amsterdam.
Rasulov, N. P.
(1976). On the asymptotically efficient estimates of regression coefficients in the case of degenerate noise spectrum. *Teor. Veroyatn. i Primen.* (Probab. Theory and Its Appl.), 21, No. 2, 324-333.

Rayner, J. N.
(1971). *An Introduction to Spectral Analysis*. Pion, London.
Reddy, N. S., and V. U. Reddy
(1980). High-speed computation of autocorrelation using rectangular transforms. *IEEE Trans. Acoust., Speech, Signal Process.*, **28**, No. 4, 481-483.
Reed, M., and B. Simon
(1972). *Methods of Modern Mathematical Physics. 1. Functional Analysis*. Academic Press, New York.
Rice, J.
(1977). On generalized shot noise. *Adv. Appl. Probab.*, **9**, No. 3, 553-565.
Rice, S. O.
(1944-45). Mathematical analysis of random noise. *Bell System Tech. J.*, **23**, No. 3, 282-332, and **24**, No. 1, 46-156. Reprinted in Wax, 1954, pp. 133-294.
Riesz, F., and B. Sz.-Nagy
(1955). *Functional Analysis*. Ungar, New York.
Ripley, B. D.
(1981). *Spatial Statistics*. Wiley, New York.
Rissanen, J.
(1978). Modeling by shortest data description. *Automatica*, **14**, No. 5, 465-471.
(1986). Order estimation by accumulated prediction errors. *J. Appl. Probab.*, **A23**, 55-61.
Robertson, H. P.
(1940). The invariant theory of isotropic turbulence. *Proc. Cambr. Phil. Soc.*, **36**, No. 2, 209-223.
Robinson, E. A.
(1960). Sums of stationary random variables. *Proc. Amer. Math. Soc.*, **11**, No. 1, 77-79.
(1967). *Multichannel Time Series Analysis with Digital Computer Programs*. Holden-Day, San Francisco.
Robinson, E. A., and M. T. Silvia
(1978). *Digital Signal Processing and Time Series Analysis*. Holden-Day, San Francisco.
Robinson, F. N. H.
(1974). *Noise and Fluctuations in Electronic Devices and Circuits*. Clarendon Press, Oxford.
Robinson, P. M.
(1977). Estimating variances and covariances of sample autocorrelations and autocovariances. *Austr. J. Stat.*, **19**, No. 3, 236-240.
Rodemich, E. R.
(1966). Spectral estimate using nonlinear functions. *Ann. Math. Stat.*, **37**, No. 5, 1237-1256.
Rokhlin, V. A.
(1949). Selected topics in metric theory of dynamic systems. *Uspekhi Mat. Nauk* (Adv. Math. Sci.), **4**, No. 2(30), 57-128.

Romanovsky, V. I.
(1932). Sur la loi sinusoidale limite, *Rend. Circ. Mat. Palermo*, **56**, 82-111.
(1933). Sur une generalization de la loi sinusoidale limite. *Rend. Circ. Mat. Palermo*, **57**, 130-136.
Rosenblatt, M.
(1961). Independence and dependence. *Proc. Fourth Berkeley Symp. Math. Stat. Probab.*, **2**, 431-443.
(1972). Probability limit theorems and some questions in fluid mechanics. In *Statistical Models and Turbulence*, ed. by M. Rosenblatt and C. Van Atta (*Lecture Notes in Physics*, 12), pp. 27-40. Springer, Berlin.
(1985). *Stationary Sequences and Random Fields*. Birkhäuser, Boston.
(1986). Parameter estimation for finite-parameter stationary random fields. *J. Appl. Prob.*, **A23**, 311-318.
Roshal, A. C.
(1976). The fast Fourier transform in the computational physics — A survey. *Izv. VUZov, Radiofizika* (Bull. Colleges Univ. USSR, ser. Radiophys.), **19**, No. 10, 1425-1454.
Roucos, S. E., and D. G. Childers
(1980). A two-dimensional maximum entropy spectral estimator. *IEEE Trans. Inform. Theory*, **26**, No. 5, 554-560.
Roy, R.
(1972). Spectral analysis for a random process on the circle. *J. Appl. Probab.*, **9**, No. 4, 745-757.
Rozanov, Yu. A.
(1959). Spectral analysis of abstract functions. *Teor. Veroyatn. i Primen.* (Probab. Theory and Its Appl.), **4**, No. 3, 291-310.
(1967). *Stationary Random Processes*. Holden-Day, San Francisco.
Rubinstein, R. Y.
(1981). *Simulation and the Monte Carlo Method*. Wiley, New York.
Rudin, W.
(1962). *Fourier Analysis on Groups*. Interscience, New York.
Rykunov, L. N.
(1967). *Microseisms. Experimental Characteristics of Natural Earth Vibrations in the Range of Periods from 0.07 to 8 Sec.* Nauka, Moscow.
Rytov, S. M.
(1976). *Introduction to Statistical Radiophysics. Part 1. Random Processes.* Nauka, Moscow. (English translation is to be published by Springer.)
Rytov, S. M., Yu. A. Kravtsov, and V. I. Tatarskii
(1978). *Introduction to Statistical Radiophysics. Part 2. Random Fields.* Nauka, Moscow. (English translation is to be published by Springer.)
Samarov, A. M.
(1977). Lower bound for risk of spectral density estimates. *Probl. Peredachi Inform.* (Probl. Inform. Transmission), **13**, No. 1, 67-72.

Sasaki, K., and T. Sato
(1977). A new definition of power spectrum of weakly non-stationary processes and its estimation. *Bull. Res. Lab. Precis. Mash. and Electron.* (Japan), No. 39, 51-59.

Schaerf, M. C.
(1964). *Estimation of the Covariance and Autoregressive Structure of a Stationary Time Series.* Techn. Rep. No. 12, Dept. Stat., Stanford University, Stanford, Calif.

Schoenberg, I. J.
(1938a). Metric spaces and completely monotone functions. *Ann. Math.*, **39**, No. 4, 811-841.
(1938b). Metric spaces and positive definite functions. *Trans. Amer. Math. Soc.*, **44**, No. 3, 522-536.
(1942). Positive definite functions on spheres. *Duke Math. J.*, **9**, No. 1, 96-108.

Schulz-DuBois, E. O., and I. Rehberg
(1981). Structure function in lieu of correlation function. *Appl. Phys.*, **24**, No. 4, 323-329.

Schuster, A.
(1898). On the investigation of hidden periodicities with application to a supposed 26-day period of meteorological phenomena. *Terr. Magn. Atmos. Elect.*, **3**, 13-41.
(1900). The periodogram of the magnetic declination as obtained from the records of the Greenwich Observatory during the years 1871-1895. *Trans. Cambr. Phil. Soc.*, **18**, 107-135.
(1906). On the periodicities of sunspots. *Phil. Trans. Roy. Soc.*, **A206**, 69-100.

Schwartz, L.
(1950-51). *Théorie des distributions. V.I-II.* Hermann, Paris (2nd edition 1957-59).
(1961). *Méthodes mathématiques pour les sciences physiques.* Hermann, Paris.

Schwarz, G.
(1978). Estimating the dimension of a model. *Ann. Stat.*, **6**, No. 2, 461-464.

Scott, P. D., and C. L. Nikias
(1981). Forward covariance least-squares algorithm: A new method in AR spectral estimation. *Electron. Lett.*, **17**, No. 3, 111-112.
(1982). Energy-weighted linear predictive spectral estimation: A new method combining robustness and high resolution. *IEEE Trans. Acoust., Speech, Signal Process.*, **30**, No. 2, 287-293.

Serebrennikov, M. G., and A. A. Pervozvansky
(1965). *Detection of the Hidden Periodicities.* Nauka, Moscow.

Shannon, C. E.
(1948). A mathematical theory of communication. *Bell System Tech. J.*, **27**, No. 3, 379-423; No. 4, 623-656.
(1949). Communication in the presence of noise. *Proc. IRE*, **37**, No. 1, 10-21.

Shapiro, H. S., and R. A. Silverman
(1959). Some spectral properties of weighted random processes. *IRE Trans. Inform. Theory*, **5**, No. 3, 123-128.
Sheppard, P. A.
(1951). Atmospheric turbulence. *Weather*, **6**, No. 2, 42-49.
Shibata, R.
(1976). Selection of the order of an autoregressive model by Akaike's information criterion. *Biometrika*, **63**, No. 1, 117-126.
Shiryaev (Shiryayev), A. N.
(1980). *Probability*. Nauka, Moscow. English translation: Shiryayev, A. N. *Probability*, Springer, New York, 1984.
Shkarofsky, I. P.
(1968). Generalized turbulence space-correlation and wave-number spectrum-function pairs. *Canad. J. Phys.*, **46**, No. 19, 2133-2153.
Silvey, S. D.
(1970). *Statistical Inference*. Chapman and Hall, London.
Skalsky, M.
(1971). A note on the estimation of the mean of a homogeneous random field. *J. Appl. Probab.*, **8**, No. 3, 626-629.
Skorohod (Skorokhod), A. V.
(1965). Constructive methods of assignment for random processes. *Uspekhi Mat. Nauk* (Adv. Math. Sci.), **20**, No. 3, 67-87.
Slepian, D., and T. T. Kadota
(1969). Four integral equations of detection theory. *SIAM J. Appl. Math.*, **17**, No. 6, 1102-1117.
Slepian, D., and H. O. Pollak
(1961). Prolate spheroidal wave functions, Fourier analysis and uncertainty — I. *Bell System Tech. J.*, **40**, No. 1, 43-64.
Slutsky, E. E.
(1927a). The summation of random causes as the source of cyclic processes. *Vopr. Konyunktury* (Problems of Economic Condit., Moscow), **3**, No. 1, 34-64. (Revised English translation is published in *Econometrica*, **5**, No. 2, 105-146, 1937.)
(1927b). Sur une théoreme limite rélatif aux séries des quantites éventuelles. *C.R. Acad. Sci.*, **185**, 169-171.
(1928). Sur les fonctions éventuelles continues, integrables et dérivables dans le sens stochastiques. *C.R. Acad. Sci.*, **187**, 878-880.
(1929a). Sur l'erreur quadratique moyenne du coefficient de correlation dans le cas des suites des épreuves non indépéndantes. *C.R. Acad. Sci.*, **189**, 612-614.
(1929b). On the mean quadratic error of the correlation coefficient in the case of homogeneous connected series. *Trudy Konyunkt. Inst.* (Proc. Inst. Economic Condit. Studies), **2**, 64-101.
(1937). Qualche proposizione relativa alla teoria delle funzioni aleatorie. *Giorn. Instituto Ital. Attuari*, **8**, No. 2, 3-19.
(1938). Sur les fonctions aléatoires presque périodiques et sur la décomposition des fonctions aléatoires stationnaires en composantes. In *Actualités scientifiques et industrielles*, No. 738, pp. 33-55. Hermann, Paris.

(1960). *Selected Works.* Izd. Akad. Nauk SSSR (Acad. Sci. USSR Press), Moscow. (This book contains, in Russian, all the above papers of this author.)

Smith, D. J.
(1981). The maximum entropy method. *Marconi Rev.,* **44**, No. 222, 137-158.

Smoluchowski, M. von
(1923). *Abhandlungen über die Brownsche Bewegung und verwandte Erscheinungen.* Ostwalds Klassiker der exakten Wissenschaften Nr. 207, Leipzig.

Solodovnikov, V. V.
(1960). *Introduction to the Statistical Dynamics of Automatic Control Systems.* Dover, New York.

Soloviev, A. P.
(1970). Estimation of the accuracy of the mathematical expectation and variance of a stationary random sequence. *Avtomat. i Telemekh.* (Automation and Remote Control), No. 8, 172-176.

Sreenivasan, K. R., A. J. Chambers, and R. A. Antonia
(1978). Accuracy of moments of velocity and scalar fluctuations in the atmospheric surface layer. *Boundary-Layer Meteor.,* **14**, No. 3, 341-352.

Stockham, T. G.
(1966). High-speed convolution and correlation. *1966 Spring Joint Computer Conf.,* *AFIPS Proc.,* **28**, 229-233, Spartan, Washington, D.C.

Stone, M. H.
(1932). *Linear Transformations in Hilbert Space.* Amer. Math. Soc., Providence, R. I.

Storch, H., von, and G. Hannoschöck
(1985). Statistical aspects of estimated principal vectors (EOFs) based on small sample sizes. *J. Climate Appl. Meteor.,* **24**, No. 7, 716-724.

Strand, O. N.
(1977). Multichannel complex maximum entropy (autoregressive) spectral analysis. *IEEE Trans. Automat. Contr.,* **22**, No. 2, 634-640.

Stratonovich, R. L.
(1967). *Topics in the Theory of Random Noise.* Gordon and Breach, New York.

Strube, H. W.
(1985). A generalization of correlation functions and the Wiener-Khinchin theorem. *Signal Process.,* **8**, No. 1, 63-74.

Stumpff, K.
(1937). *Grundlagen und Methoden der Periodenforschung.* Springer, Berlin.

Sveshnikov, A. A.
(1968). *Applied Methods of Random Function Theory,* 2nd edition. Nauka, Moscow.

Swingler, D. N.
(1979). A modified Burg algorithm for maximum entropy spectral analysis. *Proc. IEEE*, **67**, No. 9, 1368-1369.
(1984). On the optimum tapered Burg algorithm. *IEEE Trans. Acoust., Speech, Signal Process.*, **32**, No. 1, 185-186.

Taira, K., A. Takeda, and K. Ishikawa
(1971). A shipborne wave-recording system with digital data processing. *J. Oceanograph. Soc. Japan*, **27**, No. 4, 175-186.

Taqqu, M. S.
(1986a). Self-similar processes. To appear in *Encyclopedia of Statistical Sciences* (S. Kotz and N. Johnson, eds.), Wiley, New York.
(1986b). A bibliographical guide to self-similar processes and long-range dependence. In *Dependence in Probability and Statistics* (E. Eberlein and M. S. Taqqu, eds.), pp. 137-162. Birkhäuser, Boston.

Tatarskii, V. I.
(1971). *The Effects of the Turbulent Atmosphere on Wave Propagation.* Israel Program for Sci. Translations, Jerusalem, and U.S. Dept. Commerce, Nat. Tech. Inform. Service, Springfield, Va.

Taylor, F. J.
(1983). *Digital Filters Design Handbook.* Marcel Dekker, New York.

Taylor, G. I.
(1921). Diffusion by continuous movements. *Proc. London Math. Soc.*, Ser. 2, **20**, 196-211. Reprinted in Friedlander and Topper, 1961, pp. 1-16.
(1938). The spectrum of turbulence. *Proc. Roy. Soc.*, Ser. *A*, **164**, No. 919, 476-490. Reprinted in Friedlander and Topper, 1961, pp. 100-114.

Temes, G. C., and S. K. Mitra, eds.
(1973). *Modern Filter Theory and Design.* Wiley, New York.

Tempelman, A. A.
(1986). *Ergodic Theorems on Groups.* Mokslas Publishers, Vilnius.

Theodoridis, S., and D. C. Cooper
(1981). Application of the maximum entropy spectrum analysis technique to signals with spectral peaks of finite width. *Signal Process.*, **3**, No. 2, 109-122.

Thompson, P. D.
(1957). Uncertainty of initial state as a factor in the predictability of large scale atmospheric flow patterns. *Tellus*, **9**, No. 3, 275-295.

Thompson, W. E.
(1958). Measurements and power spectra of runway roughness at airports in countries of the North Atlantic Treaty Organization. *Nat. Advis. Com. Aeronaut., Tech. Note* 4303.

Thomson, D. J.
(1977). Spectrum estimation technique for characterization and development of WT4 waveguide. *Bell System Tech. J.*, **56**, No. 9, 1769-1815.
(1982). Spectrum estimation and harmonic analysis. *Proc. IEEE*, **70**, No. 9, 1055-1096.

Tikhonov, V. I.
(1982). *Statistical Radioengineering*, 2nd edition. Radio i Svyaz',
Moscow.
Titchmarsh, E. C.
(1939). *The Theory of Functions*, 2nd edition. Oxford University
Press, Oxford.
(1948). *Introduction to the Theory of Fourier Integrals*, 2nd edition.
Oxford University Press, New York.
Tjøstheim, D.
(1978). Statistical spacial modelling. *Adv. Appl. Probab.*, 10, 130-154.
(1981). Autoregressive modeling and spectral analysis of array data
in the plane. *IEEE Trans. Geosci. Remote Sens.*, 19, No. 1, 15-24.
Tong, H.
(1979). A note on a local equivalence of two recent approaches to
autoregressive order determination. *Int. J. Control*, 29, No. 3,
441-446.
Tretter, S. A., and K. Steiglitz
(1967). Power-spectrum identification in terms of rational models.
IEEE Trans. Autom. Contr., 12, No. 2, 185-188.
Troutman, B. M.
(1979). Some results in periodic autoregression. *Biometrika*, 66, No. 2,
219-228.
Tsay, R. S., and G. C. Tiao
(1985). Use of canonical analysis in time series model identification.
Biometrika, 72, No. 2, 299-315.
Tsuji, H.
(1955). The transformation equations between one- and n-
dimensional spectra in the n-dimensional isotropic vector or scalar
fluctuation field. *J. Phys. Soc. Japan*, 10, No. 4, 278-285.
Tufts, D. W., and R. Kumaresan
(1982). Estimation of frequencies of multiple sinusoids: making
linear prediction perform like maximum likelihood. *Proc. IEEE*, 70,
No. 9, 975-989.
Tukey, J. W.
(1949). The sampling theory of power spectrum estimates. *Proc.
Symp. on Applications of Autocorrelation Analysis to Phys. Problems*,
NAVEXOS-P-735, pp. 47-67. Office of Naval Research, Dept. Navy,
Washington, D.C. (Reprinted in *J. Cycle Res.*, 6, 31-52, 1957.).
(1961). Discussion of Jenkins and Parzen papers. *Technometrics*, 3,
No. 2, 191-219.
(1967). An introduction to the calculations of numerical spectrum
analysis. In *Spectral Analysis of Time Series* (B. Harris, ed.), pp.
25-46. Wiley, New York.
Uberoi, M. S., and L. S. G. Kovasznay
(1953). On mapping and measurement of random fields. *Quart. Appl.
Math.*, 10, No. 4, 375-393.
Uhlenbeck, G. E., and L. S. Ornstein
(1930). On the theory of the Brownian motion. *Phys. Rev.*, 36, 823-
841. Reprinted in Wax, 1954, pp. 93-111.

Ulrych, T. J., and T. N. Bishop
(1975). Maximum entropy spectral analysis and autoregressive decomposition. *Rev. Geophys. Space Phys.*, **13**, No. 1, 183-200. Reprinted in Childers, 1978, pp. 54-71.
Unwin, D. J., and L. W. Hepple
(1974). The statistical analysis of spatial series. *The Statistician*, **23**, Nos. 3/4, 211-227.
Van der Ziel, A.
(1959). *Fluctuation Phenomena in Semi-Conductors.* Butterworths, London.
(1970). *Noise: Sources, Characterization, Measurements.* Prentice-Hall, Englewood Cliffs, N. J.
(1976). *Noise in Measurements.* Wiley, New York.
Vanmarcke, E.
(1984). *Random Fields. Analysis and Synthesis.* MIT Press, Cambridge, Mass.
Van Trees, H. L.
(1968). *Detection, Estimation, and Modulation Theory. Part 1.* Wiley, New York.
Van Vleck, J. H.
(1943). The spectrum of clipped noise. *Rept. No. 51, Radio Res. Lab.*, Harvard University, Cambridge, Mass.
Van Vleck, J. H., and D. Middleton
(1966). The spectrum of clipped noise. *Proc. IEEE*, **54**, No. 1, 2-19.
Veltman, B. P. Th., and H. Kwakernaak
(1961). Theorie und Technik der Polaritätskorrelation für die dynamische Analyse niederfrequenter Signale und Systeme. *Regelungstechnik*, **9**, 357-364.
Veltman, B. P. Th., and A. Van den Bos
(1964). The applicability of the relay correlator and the polarity coincidence correlator in automatic control. In *Automatic and Remote Control, 1963 Proc. 2nd Congr. IFAC (Theory)*, pp. 620-627. Butterworth, London, and Oldenbourg, Münich.
Venchkovsky, L. B.
(1962). Construction of estimations for the mathematical expectation and variance by using a noncorrelated sample of a random process. *Avtomat. i Telemekh.* (Automat. and Remote Control), No. 5, 565-570.
Verbitskaya, I. N.
(1964,1966). On the conditions for applicability of the strong law of large numbers to wide sense stationary processes. *Teor. Veroyatn. i Primen.* (Probab. Theory and Its Appl.), **9**, No. 2, 365; **11**, No. 4, 715-720.
Veselova, G. P., and Yu. I. Gribanov
(1968). On the optimal sampling rate for the evaluation of the correlation functions of random processes by digital computer. *Avtomat. i Telemekh.* (Automat. and Remote Control), No. 12, 110-117.
(1969). Relay method for the determination of correlation coefficient. *Avtomat. i Telemekh.*, No. 2, 31-39.

(1970a). Statistical error in correlation function estimates for non-centered random processes. *Avtometriya* (Novosibirsk), No. 1, 116-120.
(1970b). On the errors of polarity correlators. *Avtomat. i Telemekh.*, No. 10, 54-59.
Vilenkin, N. Ya.
(1968). *Special Functions and the Theory of Group Representations.* Amer. Math. Soc., Providence, R. I.
Vilenkin, S. Ya.
(1959). On the estimation of the mean of stationary processes. *Teor. Veroyatn. i Primen.* (Probab. Theory and Its Appl.), **4**, No. 4, 451-452.
(1967). *Statistical Methods of Investigation of Automatic Control Systems.* Sovyet Radio, Moscow.
(1979). *Statistical Data Processing in Random Function Investigations.* Energiya, Moscow.
Vilenkin, S. Ya., and T. I. Dubenko
(1971). Optimal linear estimates of the mathematical expectation of a homogeneous random field. *Izv. Akad. Nauk SSSR, Ser. Tekhn. Kibernetika* (Bull. Acad. Sci. USSR, Ser. Tech. Cybernetics), No. 1, 134-141.
Vinnik, L. P.
(1968). *The Structure of Microseisms and Grouping Methods in Seismology.* Nauka, Moscow.
Vishnyakov, Yu. N., N. N. Vishnyakova, V. A. Geranin, I. R. Rudenko, and G. D. Simonova
(1974). Accuracy analysis for variance measurements in the case of non-Gaussian random processes. In *Proc. 7th USSR Symp. on Represent. Methods and Instrum. Analysis of Random Processes and Fields, Sec. 3,* pp. 182-188, USSR Res. Inst. Electr. Measuring Instr., Leningrad.
Vishnyakova, N. N., Yu. V. Gaevoi, and V. A. Geranin
(1976). Analysis of errors in correlation function measurements for non-Gaussian stationary processes. In *Proc. 9th Symp. on Represent. Methods and Instr. Analysis of Random Processes and Fields, Sec. 1,* pp. 90-97. USSR Res. Inst. Electr. Measuring Instr., Leningrad.
Vishnyakova, N. N., V. E. Gartstron, and V. A. Geranin
(1978). Errors of the correlation function measurements for non-Gaussian stationary random processes. *Prikl. Akustika* (Appl. Acoustics, Taganrog, USSR), No. 6, 101-105.
Vishnyakova, N. N., and V. A. Geranin
(1978). Present state of error theory in statistical correlation analysis of non-Gaussian random processes. In *Proc. 10th USSR Symp. on Represent. Methods and Instr. Analysis of Random Processes and Fields, Sec. 1,* pp. 20-38. USSR Res. Inst. Electr. Measuring Instr., Leningrad.
Vishnyakova, N. N., V. A. Geranin, Yu. P. Monakov, and A. N. Prodeus.
(1975). Errors of the correlation function measurements for non-Gaussian stationary random processes. In *Proc. 8th USSR Symp. on Represent. Methods and Instr. Analysis of Random Processes and Fields, Sec. 1,* pp. 90-95. USSR Res. Inst. Electr. Measuring Instr., Leningrad.

Vitale, R. A.
(1973). An asymptotically efficient estimate in time series analysis. *Quart. Appl. Math.*, **30**, No. 4, 421-440.
Volgin, V. V., and R. N. Karimov
(1967a). Selection of the discretization interval when correlation functions of random processes are calculated. *Avtomatika i Telemekh.* (Automatics and Remote Control), No. 5, 37-42.
(1967b). Selection of the realization length when correlation functions of random processes are calculated from experimental data. *Avtomatika i Telemekh.*, No. 6, 53-62.
Volkov, I. I., V. V. Motov, and S. A. Prokhorov
(1974). Correlators using approximation by parametric models. In *Proc. 7th USSR Symp. on Represent. Methods and Instr. Analysis of Random Processes and Fields, Sec. 4*, pp. 10-14. USSR Res. Inst. Electr. Measuring Instr., Leningrad.
Vovk, S. P.
(1981). On homogeneous and Lorentz-invariant generalized random fields. *Teor. Veroyatn. i Mat. Stat.* (Probab. Theory and Math. Stat., Kiev), No. 24, 10-16.
Waldmeir, M.
(1961). *The Sunspot Activity in the Years 1610-1960*. Schulthess, Zurich.
Walker, A. M.
(1965). Some asymptotic results for the periodogram of a stationary time series. *J. Austr. Math. Soc.*, **5**, No. 1, 107-128, No. 4, 512.
(1973). On the estimation of a harmonic component in a time series with stationary dependent residuals. *Adv. Appl. Probab.*, **5**, No. 2, 217-241.
Walker, G. T.
(1931). On periodicity in series of related terms. *Proc. Roy. Soc., Ser. A*, **131**, 518-532.
Wang, Ming Chen, and G. E. Uhlenbeck
(1945). On the theory of the Brownian motion II. *Rev. Mod. Phys.*, **17**, Nos. 2-3, 323-342. Reprinted in Wax, 1954, pp. 113-132.
Watson, G. N.
(1958). *A Treatise on the Theory of Bessel Functions*. Cambridge University Press, Cambridge.
Watts, D. G.
(1964). *Optimum Windows for Power Spectrum Estimation*. Tech. Summary Rep. 506. Math. Res. Center, University of Wisconsin, Madison, Wisconsin.
Wax, N., ed.
(1954). *Selected Papers on Noise and Stochastic Processes*. Dover, New York.
Webster, R. J.
(1978). A generalized hamming window. *IEEE Trans. Acoust., Speech, Signal Process.*, **26**, No. 3, 269-270.

Weil, A.
(1940). *L'Integration dans les groupes topologiques et ses applications.* Actualites Sci. et Industr. No. 869, Hermann, Paris.

Welch, P. D.
(1967). The use of fast Fourier transform for the estimation of power spectra: A method based on time averaging over short modified periodograms. *IEEE Trans. Audio Electroacoust.*, 15, No. 2, 70-73. Reprinted in Childers, 1978, pp. 17-20.

Wentzell, A. D.
(1981). *A Course in the Theory of Stochastic Processes.* McGraw-Hill, New York.

Weyl, H.
(1939). *The Classical Groups. Their Invariants and Representations.* Princeton University Press, Princeton, N. J.

Whittaker, E. T.
(1915). On the functions which are represented by the expansions of the interpolation theory. *Proc. Roy. Soc. Edinburgh, Sec. A,* 35, 181-194.

Whittle, P.
(1954a). A statistical investigation of sunspot observations with special reference to H. Alfvén's sunspot model. *Astrophys. J.,* 120, 251-260.
(1954b). On stationary processes in the plane. *Biometrika,* 41, Nos. 3-4, 434-449.
(1956). On the variation of yield variance with plot size. *Biometrika,* 43, Nos. 3-4, 337-343.
(1962). Topographic correlation, power-law covariance functions, and diffusion. *Biometrika,* 49, Nos. 3-4, 305-314.
(1963a). *Prediction and Regulation.* English Universities Press, London; also (2nd revised edition) University of Minnesota Press, Minneapolis, Minnesota, 1983.
(1963b). On the fitting of multivariate autoregressions, and the approximate canonical factorization of a spectral density matrix. *Biometrika,* 50, Nos. 1-2, 129-134.
(1963c). Stochastic processes in several dimensions. *Bull. Intern. Statist. Inst.,* 40, Book 1, 974-994.

Wiener, N.
(1930). Generalized harmonic analysis. *Acta Math.,* 55, 117-258. Reprinted in *Selected Papers of Norbert Wiener,* The MIT Press, Cambridge, Mass., 1964.
(1933). *The Fourier Integral and Certain of its Applications.* Cambridge University Press, Cambridge and Dover, New York.
(1939). The ergodic theorem. *Duke Math. J.,* 5, No. 1, 1-18.
(1949). *Extrapolation, Interpolation and Smoothing of Stationary Time Series.* MIT Press, Cambridge, Mass.

Wiener, N., and A. Wintner
(1958). Random time. *Nature,* 181, No. 4608, 561-562.

Wilks, S. S.
(1962). *Mathematical Statistics.* Wiley, New York

Williams, W. J., ed.
(1977). *Biological Signal Processing and Analysis.* Special Issue of *Proc. IEEE*, **65**, No. 5.
Wilson, G.
(1969). Factorization of the covariance generating function of a pure moving average process. *SIAM J. Numer. Anal.*, **6**, No. 1, 1-7.
Winograd, S.
(1976). On computing the discrete Fourier transform. *Proc. Nat. Acad. Sci. USA*, **73**, No. 4, 1005-1006.
(1978). On computing the discrete Fourier transform. *Math. Comput.*, **32**, No. 141, 175-199.
Wintner, A.
(1936). On a class of Fourier transforms. *Amer. J. Math.*, **58**, No. 1, 45-90; No. 2, 425.
Wold, H.
(1948). On prediction in stationary time series. *Ann. Math. Stat.*, **19**, No. 4, 558-567.
(1954). *A Study in the Analysis of Stationary Time Series.* 2nd edition. Almqvist and Wiksell, Uppsala. (1st edition was published by the same publishers in 1938.)
Wolf, D., ed.
(1978). *Noise in Physical Systems. Proc. of the 5th Intern. Conf. on Noise, Bad Nauheim, March 13-16, 1978.* Springer, Berlin.
Wolff, S. S., J. B. Thomas, and T. R. Williams
(1962). The polarity-coincidence correlator: A nonparametric detection device. *IRE Trans. Inform. Theory*, **8**, No. 1, 5-9.
Wong, E.
(1969). Homogeneous Gauss-Markov random fields. *Ann. Math. Stat.*, **40**, No. 5, 1625-1634.
(1971). *Stochastic Processes in Information and Dynamical Systems.* McGraw-Hill, New York.
Woods, J. W.
(1981). Comments on "Two-dimensional Markov estimates need not exist". *IEEE Trans. Inform. Theory*, **26**, No. 1, 129-130.
Worsdale, G. J.
(1975). Tables of cumulative distribution functions for symmetric stable distributions. *Appl. Stat.*, **24**, No. 1, 123-131.
Yadrenko, M. I.
(1983). *Spectral Theory of Random Fields.* Optimization Software Inc., Publication Division, New York.
Yaglom, A. M.
(1948). Homogeneous and isotropic turbulence in a viscous compressible fluid. *Izv. Akad. Nauk SSSR, Ser. Geograf. i Geofiz.* (Bull. Acad. Sci. USSR, Ser. Geogr. and Geophys.), **12**, No. 6, 501-522. German translation in Goering, 1958, pp. 43-69.

(1949). On problems of linear interpolation of stationary random sequences and processes. *Uspekhi Mat. Nauk* (Adv. Math. Sci.), **4**, No. 4(32), 173-178. English translation in *Selected Transl. in Math. Stat. and Probab.*, **4**, 339-344, Amer. Math. Soc., Providence, R. I., 1963, and in Kailath, 1977, pp. 135-140.

(1955a). Extrapolation, interpolation and filtering of stationary random processes with rational spectral densities. *Trudy Mosk. Mat. Obshch.* (Proc. Moscow Math. Soc.), **4**, 333-374. English translation in *Selected Transl. in Math. Stat. and Probab.*, **4**, 345-387, Amer. Math. Soc., Providence, R. I., 1963.

(1955b). Correlation theory of processes with stationary random increments of order *n*. *Mat. Sbornik* (Math. Coll.), **37**, No. 1, 141-196. English translation in *Amer. Math. Soc. Translations*, *Ser. 2*, **8**, 87-141, Amer. Math. Soc., Providence, R. I., 1958.

(1955c). *Correlation Theory for Continuous Processes and Fields with Applications to Statistical Extrapolation of Time Series and to Turbulence Theory*. Dissertation for Doctor's Degree in Phys. Math. Sci., Geophysical Institute (now Institute of Atmospheric Physics), Acad. Sci. USSR, Moscow.

(1957). Some classes of random fields in *n*-dimensional space, related to stationary random processes. *Teor. Veroyatn. i Primen.* (Probab. Theory and Its Appl.), **2**, No. 3, 292-337.

(1961). Second-order homogeneous random fields. *Proc. 4th Berkeley Symp. Math. Stat. and Probab.*, **2**, 593-622.

(1962a). *An Introduction to the Theory of Stationary Random Functions*. Prentice-Hall, Englewood Cliffs, N. J.; also Dover, New York, 1973.

(1962b). Some mathematical models generalizing the model of homogeneous and isotropic turbulence. *J. Geophys. Res.*, **67**, No. 8, 3081-3087.

(1963). Spectral representations for various classes of random functions. In *Trudy 4 Vsesoyuzn. Mat. Congr.* (Proc. 4th USSR Math. Congr.), vol. 1 (Plenary Lectures), pp. 250-273. Izd. Akad. Nauk SSSR, Leningrad.

(1965). Stationary Gaussian processes satisfying the strong mixing conditions and best predictable functionals. In *Bernoulli, Bayes, and Laplace Anniversary Volume* (L. LeCam and J. Neyman, eds.), pp. 241-252, Springer, Berlin.

(1981). *Correlation Theory of Stationary Random Functions*. Gidrometeoizdat, Leningrad.

(1985). Einstein's 1914 paper on the theory of irregularly fluctuating series of observations. *Probl. Peredachi Inform.* (Probl. of Inform. Transmission), **21**, No. 4, 101-107.

(1986a). Einstein's paper on methods for processing fluctuating series of observations and the role of such methods in meteorology. *Izv. Akad. Nauk SSSR, Ser. Fiz. Atmosf. i Okeana* (Bull. Acad. Sci. USSR, Ser. Atmosph. Oceanic Phys.), **22**, No. 1, 101-107.

(1986b). Spectra of disorderly fluctuating series of observations and the methods for their determination: Detailed comments on Einstein's note of 1914. In *Einsteinovsky Sbornik, 1982-83* (Einstein Studies, 1982-83), pp. 25-56, Nauka, Moscow.

Yaglom, A. M., and M. S. Pinsker
(1953). Random processes with stationary increments of order *n*. *Doklady Akad. Nauk SSSR* (Repts. Acad. Sci. USSR), **90**, No. 5, 731-734.

Yanovsky, B. M.
(1978). *Earth Magnetism*. Izd. LGU (Leningrad State Univ. Press), Leningrad.

Yavorskii, I. N.
(1983). Properties of estimators for the mathematical expectation and correlation function of periodically correlated random processes. *Otbor i Peredacha Inform.* (Select. and Transmiss. Inform., Kiev), No. 67, 22-28.
(1985a). Discrete sample estimates for the components of probabilistic characteristics of periodically correlated random processes. *Otbor i Peredacha Inform.*, No. 72, 17-27.
(1985b). On the statistical analysis of periodically correlated random processes. *Radiotekhn. i Electron.* (Radio Engineering and Electronics), **30**, No. 6, 1096-1104.

Yule, G. U.
(1921). On the time-correlation problem, with especial reference to the variate-difference correlation method. *J. Roy. Stat. Soc.*, **84**, 497-537.
(1927). On a method of investigating periodicities in disturbed series, with special reference to Wolfer's sunspot numbers. *Phil. Trans. Roy. Soc.*, Ser. *A*, **226**, 267-298.
(1945). On a method of studying time series based on their internal correlations. With a note by M. G. Kendall. *J. Roy. Stat. Soc.*, **108**, 208-230.

Yule, G. U., and M. G. Kendall
(1950). *An Introduction to the Theory of Statistics*. 14th edition. Griffin, London.

Yurinskii, V. V.
(1974). On strong law of large numbers for homogeneous random fields. *Mat. Zametki* (Math. Notes), **16**, No. 1, 141-149.

Zacks, S.
(1971). *The Theory of Statistical Inference*. Wiley, New York.

Zadeh, L. A.
(1951). Correlation functions and spectra of phase- and delay-modulated signals. *Proc. IRE*, **39**, No. 4, 425-427.

Zadeh, L. A., and C. A. Desoer
(1963). *Linear System Theory: The State-Space Approach*. McGraw-Hill, New York.

Zakai, M., and J. Snyders
(1970). Stationary probability measures for linear differential equations driven by white noise. *J. Differential Equations*, **8**, No. 1, 27-33.
Zernike, F.
(1932). Die Brownsche Grenze für Beobachtungsreihen. *Zs. f. Phys.*, **79**, 516-528.
Zhovinsky, V. N., and V. F. Arkhovsky
(1974). *Correlation Devices*. Energiya, Moscow.
Zhurbenko, I. G.
(1978). On local properties of a spectral density estimate. *Probl. Peredachi Inform.* (Probab. Inform. Transmission), **14**, No. 3, 85-91.
(1979). On the efficiency of estimates of a spectral density. *Scand. J. Stat.*, **6**, No. 2, 49-56.
(1980). On the efficiency of spectral density estimates for stationary processes. *Teor. Veroyatn. i Primen.* (Probab. Theory and its Appl.), **25**, No. 3, 476-489.
(1982). *Spectral Analysis of Time Series*. Izd. MGU (Moscow State University Press), Moscow. English translation is published by North-Holland, Amsterdam, 1986.
Zohar, S.
(1979). FORTRAN subroutines for the solution of Toeplitz sets of linear equations. *IEEE Trans. Acoust., Speech, Signal Process.*, **27**, 656-658.
Zolotarev, V. M.
(1954). Expression for the stable probability density with index $\alpha > 1$ in terms of the density with index $1/\alpha$. *Doklady Akad. Nauk SSSR* (Repts. Acad. Sci. USSR), **98**, No. 5, 735-738.
(1964). Representation of stable laws in terms of integrals. *Trudy Mat. Inst. Steklov* (Works of Steklov's Math. Inst., Acad. Sci. USSR), **71**, 46-50.
(1983). *One-Dimensional Stable Distributions*. Nauka, Moscow. English translation is published by Amer. Math. Soc., Providence, R. I., 1986.

NAME INDEX

This index covers the material of both Volume I and II. Page numbers of Volume I are printed in bold type while those of Volume II are printed in ordinary type.

In all cases, where an author is referred to in the text only implicitly (as et al.) the corresponding page is also indicated in the index after the author's name.

SUBJECT INDEX

This index covers the material of both Volumes I and II. Page numbers of Volume I are printed in bold type while those of Volume II are printed in ordinary type.